Acta Physica Austriaca
Supplementum X

Proceedings of the
International Symposium
"100 Years Boltzmann Equation"
in Vienna
4th—8th September 1972

1973

Springer-Verlag
Wien New York

The Boltzmann Equation
Theory and Applications

Edited by
E. G. D. Cohen, New York, and W. Thirring, Vienna

With 85 Figures and 1 Portrait

1973

Springer-Verlag
Wien New York

Prof. Dr. E. G. D. Cohen
The Rockefeller University
New York, N.Y., U.S.A.

Prof. Dr. W. Thirring
Institute for Theoretical Physics
University of Vienna, Austria

This work is subject to copyright.
All rights are reserved, whether the whole or part of the material is concerned, specifically those of translation, reprinting, re-use of illustrations, broadcasting, reproduction by photocopying machine or similar means, and storage in date banks.
© 1973 by Springer-Verlag / Wien
Library of Congress Catalog Card Number 73-76978
Printed in Austria

ISBN 3-211-81137-0 Springer-Verlag Wien-New York
ISBN 0-387-81137-0 Springer-Verlag New York-Wien

INTRODUCTION

In 1872, Boltzmann published a paper which for the first time provided a precise mathematical basis for a discussion of the approach to equilibrium. The paper dealt with the approach to equilibrium of a dilute gas and was based on an equation - the Boltzmann equation, as we call it now - for the velocity distribution function of such a gas. The Boltzmann equation still forms the basis of the kinetic theory of gases and has proved fruitful not only for the classical gases Boltzmann had in mind, but also - if properly generalized - for the electron gas in a solid and the excitation gas in a superfluid. Therefore it was felt by many of us that the Boltzmann equation was of sufficient interest, even today, to warrant a meeting, in which a review of its present status would be undertaken.

Since Boltzmann had spent a good part of his life in Vienna, this city seemed to be a natural setting for such a meeting.

The first day was devoted to historical lectures, since it was generally felt that apart from their general interest, they would furnish a good introduction to the subsequent scientific sessions. We are very much indebted to Dr. D. Flamm, who not only took the time to prepare an informative and charming lecture of the life of his grandfather, but who also kindly gave us permission to use an hitherto unknown picture of his grandfather, taken probably around 1872, as a frontispiece of this book.

Twenty-three invited lectures were presented at the Symposium and all are included in this volume. These lectures are in general reviews, which give surveys of the present status of a field, rather than reports on recent programs. Naturally a selection had to be made as to the subjects treated and a number of fields of application had to be left out. It is nevertheless hoped, that the book will provide an idea of the fruitfulness of the Boltzmann equation and thus pay a tribute to the genius of its creator.

The Symposium was sponsored by:
The International Union for Pure and Applied Physics
The Rockefeller University
The University of Trondheim
Bundesministerium für Wissenschaft und Forschung
Österreichische Akademie der Wissenschaften
Ludwig-Boltzmann-Gesellschaft
Österreichische Studiengesellschaft für Atomenergie
Österreichische Physikalische Gesellschaft.

The Organizing Committee consisted of the undersigned and Professor H. Wergeland. The Symposium was attended by seventy-five participants and sixteen observers. All the meetings, with the exception of the first one, which took place in the "Festsaal" of the Österreichische Akademie der Wissenschaften, were held in the Conference Area of the International Atomic Energy Agency in Vienna.

We would like to express our sincere appreciation to Professor Dr. E. Schmid, President of the Österreichische Akademie der Wissenschaften, and to Dr. S. Eklund, Director of the IAEA for their kind hospitality.

Finally we are very much indebted to Dr. Draxler and especially to Mrs. F. Wagner for their invaluable assistance in organizing this Symposium.

E.G.D. COHEN
W. THIRRING

CONTENTS

PARTICIPANTS IX
OBSERVERS XII
E. SCHMID, Welcome Speech 1
D. FLAMM, Life and Personality of Ludwig
          Boltzmann 3
E. BRODA, Philosophical Biography of Ludwig
          Boltzmann 17
M.J. KLEIN, The Development of Boltzmann's
          Statistical Ideas 53
G.E. UHLENBECK, The Validity and the Limitations
          of the Boltzmann Equation 107
C. CERCIGNANI, Comment to Professor Uhlenbeck's
          Paper 121
J.D. FOCH, Jr., On Higher Order Hydrodynamic
          Theories of Shock Structure 123
G.W. FORD, Sound From the Boltzmann Equation 141
E.G.D. COHEN, The Generalization of the Boltzmann
          Equation to Higher Densities 157
J.V. SENGERS, The Three-Particle Collision Term
          in the Generalized Boltzmann Equation 177
J.R. DORFMAN, Velocity Correlation Functions for
          Moderately Dense Gases 209
L. WALDMANN, On Kinetic Equations for Particles
          with Internal Degrees of Freedom 223
S. HESS, Flow-Birefringence in Gases, An Example
          of the Kinetic Theory Based on the
          Boltzmann Equation for Rotating
          Particles 247

J.J.M. BEENAKKER, Non-Equilibrium Angular Momentum
    Polarization in Rotating Molecules 267

R. KUBO, The Boltzmann Equation in Solid State
    Physics 301

K. SEEGER and H. PÖTZL, Experimental and Theoretical
    Investigations in Semiconductors Concern-
    ing the Boltzmann Equation 341

M. KAC, Some Probabilistic Aspects of the
    Boltzmann Equation 379

I. PRIGOGINE, The Statistical Interpretation of
    Non-Equilibrium Entropy 401

W.W. WOOD, A Review of Computer Studies in the
    Kinetic Theory of Fluids 451

I. KUŠČER, A Survey of Neutron Transport Theory 491

S.R. DE GROOT, Relativistic Boltzmann Theory 529

I.M. KHALATNIKOV, Kinetic Equation for Elementary
    Excitations in Quantum Systems 563

Ya.G. SINAI, Ergodic Theory 575

D. RUELLE, Ergodic Theory 609

O.E. LANFORD, Ergodic Theory and Approach to
    Equilibrium for Finite and Infinite
    Systems 619

K. PRZIBRAM, Erinnerungen an Boltzmanns
    Vorlesungen 641

PARTICIPANTS

AUERBACH, H.Th.,	Eidgenössische Technische Hochschule Zürich
BALESCU, R.,	Université Libre de Bruxelles
BAUMANN, K.,	Universität Graz
BEENAKKER, J.J.M.,	Rijksuniversiteit te Leiden
BRODA, E.,	Universität Wien
CASE, K.M.,	The Rockefeller University
CERCIGNANI, C.,	Politecnico Milano
COHEN, E.G.D.,	The Rockefeller University
COWLING, T.G.,	University of Leeds
DE BOER, J.,	Universiteit van Amsterdam
DE GROOT, S.R.,	Universiteit van Amsterdam
DE SOBRINO, L.,	University of British Columbia
DORFMAN, J.R.,	University of Maryland
EDER, G.,	Atominstitut Wien
EDER, O.J.,	Physikinstitut Seibersdorf
ERNST, M.H.,	Universiteit Nijmegen
FISCHER, D.A.V.,	International Atomic Energy Agency Vienna
FISZDON, W.,	Polish Academy of Sciences Warsaw
FLAMM, D.,	Institut für Hochenergiephysik Wien
FOCH, J.D., Jr.	University of Colorado
FORD, G.W.,	University of Michigan
GARCIA-COLIN, L.S.,	Instituto Mexicano del Petroleo
GRECOS, A.,	Université Libre de Bruxelles
GROSS, E.P.,	Brandeis University

HANSEN, J.-P.,	Université Paris - Centre d'Orsay
HARRIS, S.,	Stony Brook
HAUGE, E.H.,	Universitetet i Trondheim
HEJTMANEK, J.,	Universität Graz
HEMMER, P.C.,	Universitetet i Trondheim
HERMANN, A.,	Universität Stuttgart
HESS, S.,	Universität Erlangen-Nürnberg
HITTMAIR, O.,	Technische Hochschule Wien
ISRAEL, W.,	University of Alberta
JANCOVICI, B.,	Université Paris - Centre d'Orsay
KAC, M.,	The Rockefeller University
KANGRO, H.,	Universität Hamburg
KHALATNIKOV, I.,	Academy of Sciences Moscow
KLEIN, M.J.,	Yale University
KIEFFER, J.,	Laboratoires de Bellevue
KNAAP, H.F.P.,	Rijksuniversiteit te Leiden
KROOK, M.,	Harvard University
KUBO, R.,	University of Tokyo
KUŠČER, I.,	University of Ljubljana
LANFORD, O.E., III.	University of California
LEBOWITZ, J.L.,	Yeshiva University
LENARD, A.,	Indiana University
LIEB, E.,	M.I.T. Cambridge
MARLE, C.,	Université de Besançon
MAYER, J.E.,	University of California
PÖTZL, H.W.,	Technische Hochschule Wien
PRIGOGINE, I.,	Université Libre de Bruxelles
PUCKER, N.,	Universität Graz
PUTZ, F.,	Studiengesellschaft für Atomenergie Wien

RIBARIC, M.,	University of Ljubljana
ROSENFELD, L.,	Nordisk Institut for Teoretisk Atomfysik
ROZENTHAL, St.,	Niels Bohr Institutet
RUELLE, D.,	IHES, Bures-sur-Yvette
SEEGER, K.-H.,	Ludwig Boltzmann Institut und Universität Wien
SEITZ, F.,	The Rockefeller University
SENGERS, J.V.,	University of Maryland
SIEGERT, A.J.F.,	Northwestern University Evanston
SNIDER, R.F.,	University of British Columbia
SCHARF, G.,	Universität Zürich
THELLUNG, A.,	Universität Zürich
THIRRING, W.,	Universität Wien
TRAPPENIERS, N.,	Universiteit van Amsterdam
UHLENBECK, G.E.,	The Rockefeller University
VAN KAMPEN, N.G.,	Rijksuniversiteit Utrecht
VAN LEEUWEN, J.M.J.,	Technische Hogeschool Delft
VAN LEEUWEN, W.A.,	Universiteit van Amsterdam
VAN WEERT, C.,	Universiteit van Amsterdam
VELARDE, M.G.,	Universidad Autonoma de Madrid
WALDMANN, L.,	Universität Erlangen-Nürnberg
WOOD, W.W.,	University of California
ZWEIFEL, P.F.,	Virginia Polytechnic Institute and State University

OBSERVERS

CHOQUARD, P.,	Université de Lausanne
FESTA, R.,	Università di Genova
HOPPS, J.,	Ohio State University
KOCEVAR, P.,	Universität Graz
KONYA, A.,	University of Budapest
KOX, A.J.,	Universiteit van Amsterdam
KRIECHBAUM, M.,	Universität Graz
LEVI, A.,	Università di Genova
LOGAN, J.,	The Rockefeller University
NONNENMACHER, Th.,	Universität Ulm
POLAK, P.H.,	Universiteit van Amsterdam
QUATTROPANI, A.,	Université de Lausanne
SANCTUARY, B.C.,	Rijksuniversiteit te Leiden
SZEPFALUSY, P.,	University of Budapest
VOGL, P.,	Universität Graz
ZECH, N.,	Kernforschungszentrum Karlsruhe

WELCOME SPEECH

E. SCHMID

President of the Austrian Academy of Sciences

As President of the Austrian Academy of Sciences it is a pleasant duty for me to welcome you here on behalf of the Minister of Science and Research, Dr. Hertha Firnberg, who is at the moment abroad. As Austrians we thank you for holding this international congress, which is dedicated to one of the greatest Austrian physicists, here in Vienna. The Austrian Academy is glad to have been able to offer this hall for the opening session, in particular since Ludwig Boltzmann became member of the Academy already at the age of thirty. It was here that he announced in 1886 the statistical interpretation of the second law of thermodynamics. This symposium is going to discuss the life and the work of Ludwig Boltzmann and the further development of his ideas.

Since probably many participants are for the first time in this hall, which is one of the most beautiful here in Vienna and not open for the public, let me say a few words about it. It combines the french baroque style of Jardot, a french architect, who was brought to Vienna by Franz von Lothringen, the husband of Maria Theresia, with the italian art of the painter Guglielmi. The latter painted the ceiling fresco which represents the four faculties with in the middle a glorification of the royal couple.

Finally a few words about the musical history of this hall. Let me mention two events. On the 27th of March 1808 a famous performance of the creation of Haydn took place which led to a great triumph for the greying composer. The cream of society and art were present. Beethoven kissed the hand of Haydn, as is mentioned in a poem which appeared after this occasion; this event is also visible on a famous painting. On the 29th of October 1813 Beethoven conducted the first performance of his 7th symphony in this hall. This rich musical tradition must have made Boltzmann, a pianist himself, particularly attached to this hall.

It just remains for me to wish you a very good and successful symposium and to those of you, who came from abroad, a pleasant stay in Austria.

LIFE AND PERSONALITY OF LUDWIG BOLTZMANN

D. FLAMM

Institut für Hochenergiephysik
Österreichische Akademie der Wissenschaften, Wien

Let me start in 1872 when Boltzmann's article "Further Investigations on the Thermal Equilibrium of Gas Molecules", in which he introduces his celebrated equation was published in the proceedings of the Imperial Academy of Sciences in Vienna. He was then only 28 years old but had been for three years a full professor of mathematical physics at the University of Graz. He used short leaves to work with Bunsen and Königsberger in Heidelberg and with Kirchhoff and Helmholtz in Berlin. In January 1872 he wrote to his mother:"Yesterday I gave a talk at the Physical Society of Berlin. You can imagine that I tried to do my best in order not to bring my native country in disgrace. Hence my head was full of integrals in the last days before the talk ... By the way there was no need for much effort because most of the audience did not understand my talk anyway. However Helmholtz had come and an interesting discussion developed between myself and Helmholtz. Since you know how much I like scientific discussions you can imagine how glad I was. Especially since Helmholtz is not very accessible otherwise. Even though he had always worked in the laboratory next to me, I had not yet spoken a lot to him". Scientifically Boltzmann had a great esteem for Helmholtz. He said once:"About certain problems I can only talk to one person and this is Helmholtz, but he lives far away". Boltzmann felt like scientists of modern days that scientific progress is hampered by isolation. Thus he looked for scientific discussion whenever possible and he never lacked arguments to defend his views. Especially in later years he often had a hard stand against the opponents of the atomistic point of view. Arnold Sommerfeld wrote for instance, about the discussion at the Naturforschertagung Lübeck in 1895: "The fight between Boltzmann and Ostwald was like

the fight of a bull with a supple fencer. But this time
the bull overcame the torero in spite of all his art of
fencing. The arguments of Boltzmann won. We younger
mathematicians were all on Boltzmann's side ...." Indeed
the search for fellow physicists with whom he could discuss his ideas seems to be one of the reasons why he
moved so restlessly from one University to another in his
later years.

In 1894 Boltzmann became honorary doctor of the
University in Oxford where his ideas had a great prestige.
He suffered very much from the fact that his ideas met
little response from German speaking countries. Thus
Boltzmann welcomed Zermelo's doubts on the H-theorem as a
proof that his statistical work found some attention in
Germany. Boltzmann's desire for scientific contact is also reflected by his correspondence with many outstanding
physicists of other countries of which I only want to
mention Clark Maxwell, Lord Kelvin, Lord Rayleigh and
Bryan in Britain, H.A. Lorentz and Kamerlingh-Onnes in
the Netherlands, Svante Arrhenius and Pjerknes in Sweden.
Among his many trips the three furthest were to the
United States. In 1899 he gave four lectures on the basic
principles and equations of mechanics at the Clark University in Worchester, Massachusetts. Later he visited a
congress in St. Louis and in 1905 he delivered 30 lectures
at a summer school at the California University in Berkeley. About this last journey he wrote a fascinating description: "Reise eines deutschen Professors ins Eldorado".
Even for me as a grandson it is not easy to discover yet
unknown details about Boltzmann since 30 years separate
his life from mine.

Yet I shall try to give you an authentic description of his life and personality, by quoting what he wrote himself. First let us do a little genealogy. Great-great-grandfather Georg Friedrich Boltzmann lived in Königsberg (Kaliningrad) which nowadays is part of the USSR. Great-grandfather Samuel Ludwig moved to Berlin where he married in 1769. His son Ludwig Gottfried exchanged the capital of Prussia for the imperial capital Vienna where Boltzmann's father was born in 1802. The latter served in the imperial internal revenue service. In 1837 he married Maria Pauernfeind, the daughter of a merchant in Salzburg. Their eldest son Ludwig Eduard Boltzmann was born on the 20th of February 1844 in Vienna. This was just the night from Shrove Tuesday to Ash Wednesday and Ludwig Boltzmann used to say jokingly, that his date of birth was the reason why his temperament could change so suddenly from great joy to deep grief. His brother Albert was two years younger and died from a pulmonary disease when he was a secondary school boy. Boltzmann also had a younger sister Hedwig. The children were baptised in the catholic confession of their mother while Boltzmann's forefathers had been protestant. His primary education he received from a private tutor in the house of his parents. His father was transferred to Wels and later to Linz where the young Boltzmann entered the Gymnasium. Boltzmann was very industrious and with the exception of one term, he always was the best in his class. In Linz he received piano lessons by Anton Bruckner. One day when the maestro came to give his lesson, he put his wet raincoat on a bed, which Boltzmann's mother disliked. Bruckner was offended and never came again. When Boltzmann was fifteen his father died from tuberculosis. In 1863 Boltzmann registered at the University of Vienna where he studied mathematics and physics. The Institute of Physics had been founded only 14 years ago under Christian Doppler.

Von Ettingshausen and Josef Stefan were its directors. Among his professors was also the mathematician Petzval. Boltzmann appreciated particularly the close contact which Stefan had with his students, about which he wrote: "When I came in closer contact with Stefan the first thing was, that he gave me a copy of Maxwell's treatise and since I did not know English at that time, he also gave me an English grammar". In 1866 Boltzmann got his doctor's degree. Already one year later he qualified as a university lecturer and in 1867 he joined the Institute of Physics as assistent professor. Josef Loschmidt who also worked there became his friend. The Institute of Physics had a small laboratory in a house in the 3rd district of Vienna, Erdbergstrasse 15, but its members were full of ideas. He later said about this group: "Erdberg remained through all my life a symbol for honest and inspired experimental work. When I managed to bring some life into the physics institute in Graz, I called it jokingly Klein-Erdberg. I didn't mean that the space was small, it was probably twice as big as Stefan's insitute; but I had not yet achieved the spirit of Erdberg. Still in Munich, when the young graduate students came and did not know what to work on, I thought: how different we were in Erdberg. Today there is nice experimental equipment and people are looking for ideas how to use it. We always had enough ideas, our only worry was the experimental apparatus". Only 25 years old Boltzmann became full professor of mathematical physics at the University of Graz in 1869. Besides his theoretical ideas he persued during this time his experimental investigations on the connection between the dielectric constant and the index of refraction which he published in 1873.

In his speech in memory of Josef Loschmidt he told the following anecdote: "I planned at this time already experiments with spheres made from sulphur crystals. Since nobody could grind these spheres, Loschmidt proposed that I should do it together with him while we were waiting for tickets at the Burgtheater. He hoped that the carbon disulphide which was used for this purpose would even disperse the waiting crowd.

In 1873 Boltzmann returned to Vienna as professor of mathematics. Before his departure from Graz, Boltzmann made the aquaintance of his future wife Henriette von Aigentler, who was ten years his junior. She was very attractive, had long fair hair and blue eyes. One must not wonder that the dean of the philosophical faculty of the University of Graz found her desire to study mathematics and physics ridiculous. In his opinion the female destiny was to cook and to wipe dust which he judged as the only basis for a healthy family life. During the first term she was only allowed to attend the lectures because there were as yet no laws to ban women from the University. When the second term came the faculty had already passed a rule which excluded female students. So she made a petition to the minister of education who had been a colleague of her late father at the court of justice in Graz. The ministry waived the faculty rule, but when the next term started she had difficulties again. Finally after her engagement to Boltzmann, she followed the advice of Professor Hirzel and learned how to cook in the house of the Lord Mayor of Graz who had been a good friend of her father. Boltzmann sent a written marriage proposal to Henriette von Aigentler. He believed the written form to be the more appropriate one. From this letter we learn that hundred years ago inflation also posed a problem. He wrote: "Last year my total annual earnings were 5400 florins. This amount will be sufficient for the household, however taking into account the enormous price increase in Vienna, it is not enough to grant you many distractions and amusements".

Boltzmann was stout and not very tall. He had curly hair and blue eyes. His fiancee sometimes called him caressingly: "Mein liebes dickes Schatzerl". He had a soft heart. Each time he had to leave his fiancee when the train puffed toward the Semmering he could no longer hide his tears. But also if people asked him for a favour he could not decline. If he had to reject a student who was in bad financial situation, he had heavy remorse. In later years he never had a student fail an examination. He was very conscienctious and the administrative work, which in Vienna was much larger than in Graz, was a burden for him. In Graz A. Toepler was director of the physics institute which had just been completed. In January 1876 in a letter to Boltzmann he complains bitterly about the lack of money for his institute. Later Toepler left Graz and went to Dresden. The chair of experimental physics in Graz was quite attractive for Boltzmann for various reasons. As Toepler's successor he had his own laboratory and could lecture on physics rather than on mathematics. The burocratic burden in Graz was smaller than in Vienna. Last not least he planned to marry but had a hard time to find an appropriate appartment in Vienna. In addition his future wife was from Graz.

But even for Boltzmann it was not easy to be appointed for the chair in Graz. Among his rivals was Ernst Mach. Mach had already held a chair of mathematics in Graz from 1864 to 1867. He was now in Prag and wanted to come back. The following time was quite thrilling for the young couple. The wedding had already been fixed for the 17th of July 1876 and yet they did not know whether they would settle in Vienna or in Graz. They were also afraid that their honey-moon would be spoiled by negotiations with the ministry concerning the chair in Graz. About this unpleasant situation Boltzmann wrote: "I detest the constant secret fighting, I know much better how to integrate than how to intrigue". When five days before their wedding date they still had no affirmation from the ministry, they decided to marry anyway. They had two sons and three daughters. Boltzmann spent 14 happy years in Graz. During this time he developed his ideas on the statistical conception of nature. He was honored by the academic community and by the government. In 1878 he became dean of the faculty, in 1881 Regierungsrat, in 1885 member of the Imperial Academy of Sciences, in 1887 president of the University of Graz and in 1889 wirklicher Hofrat. Not to mention the long list of academic distinctions from other countries.

Boltzmann loved nature very much. He often made extended walks. Near Oberkroisbach he bought a farmhouse and lived with his family on the country side. He had a great knowledge of plants and maintained a herbarium and a collection of butterflies.

After Gustav Kirchhoffs death Boltzmann had the opportunity to become his successor at Berlin. His contract had already been signed by the German emperor in March of 1888, when Boltzmann asked to resolve it. Presumably the formal way of the people in Berlin made him feel uneasy since he was used to the relaxed atmosphere in Graz. For instance it is reported that after the negotiations in the ministry at a dinner with his colleagues Mrs. Helmholtz said to him: "Professor Boltzmann, I am afraid that you will not feel comfortable here in Berlin". In this year his eldest son Ludwig died at the age of eleven from appendicitis . This loss depressed him very much. In 1890 he accepted the chair of theoretical physics at the University of Munich. Here he could finally teach his main field of interest. Having taught experimental physics for fourteen years he used very demonstrative mechanical models to illustrate theoretical concepts. For instance to visualize Maxwell's theory he constructed a machine called "Bicykle".

Once a week he met a number of colleagues in the Hofbräuhaus. The mathematicians Dyk and Pringsheim, the physicists Lommel and Sohnke, the chemist Bayer, the astronomer Seeliger and the refrigeration engineer Linde belonged to this group. Boltzmann spent four happy years in Munich. During this time many students from all over the world came to study under his direction. The only drawback of his position in Munich was, that the Bavarian Government at that time did not pay a pension to university professors. His eyes became increasingly worse and so he was concerned about the future of his family.

In the introduction of his book "On Kinetic Theory of Gases" he expresses his concern: Already at the Vienna World fair in 1873 Prof. Wroblewski asked him to write this book! At this time he had refused, because he was afraid to become blind.

When Stefan died he became his successor at Vienna University in 1894. However, having been away from Vienna for 18 years he did not find such a nice group of friends and colleagues as in Munich. Especially when Ernst Mach

became professor of philosophy and history of sciences
in 1895 he had a hard stand because the former violent-
ly opposed the atomistic view of nature. This was the
reason why Boltzmann decided in 1900 to accept an appoint-
ment as professor of theoretical physics in Leipzig,
where the physico-chemist Wilhelm Ostwald had built up
a large center of research. Ostwald was his personal
friend, Ostwald's "Energetik" however, was based on Mach's
philosophy and contrary to Boltzmann's ideas. In Leipzig
Boltzmann had an even harder stand than in Vienna. The
constant struggle depressed Boltzmann since he liked a
good team-work. Furthermore he disliked the Saxon way of
life. All these circumstances caused a nervous breakdown.

In Vienna Mach retired in 1901 because of bad
health. Boltzmann's chair had remained vacant during his
stay in Leipzig. In fall 1902 he returned to Vienna. But
the establishment did not yet forgive him that he had
gone away. In a letter to his assistant Stefan Mayer he
expressed his concern about the severe cuts in the institute budget. Furthermore he had to wait until 1904 for
his re-election as a full member of the imperial academy
of sciences. In February 1903 his wife wrote to their
daughter Ida who had remained in Leipzig to finish the
Gymnasium: "Father gets worse every day. I have lost my
confidence into the future. I had imagined a better life
in Vienna". Boltzmann's health had suffered from the constant dispute with his scientific opponents. His eyes
had deteriorated to an extent that he had trouble reading. He had to employ a woman who read scientific articles
to him and his wife wrote his manuscripts.

In addition he suffered from strong asthma attacks
at night and presumably from angina pectoris. Furthermore

he was plagued by heavy headaches due to overwork. His
teaching obligations at the time were a five hours week
course on theoretical physics, the seminar on theoretical physics, and every third term a course of at least
one hour a week. Starting in 1903 Boltzmann lectured in
addition two hours a week on philosophy substituting
for Mach. You may however imagine that the enormous
teaching commitment paired with scientific work was more
than his weak health could stand in the long run. Furthermore unlike on other trips he suffered from asthma attacks
during his last stay in the United States in July 1905.
He wrote about it to his assistent Stefan Mayer:" What
I saw was very interesting and I hope to see still more
interesting things but it was also very exhausting and
tiring. Especially because the climate in California is
not so pleasant as is often reported. The change from subtropical heat to cold wheather, from dryness to fog is
hard to bear for an European. In addition the water from
the rain in the winter, which is kept in huge cisterns,
upsets my stomach. The wine one most hide like a schoolboy his cigar. This they call freedom". Boltzmann had
never given any consideration to his health but sacrificed it to his scientific work. How exhausting his desease must have been I can only guess having seen my
mother suffer from a similar illness. When even a vacation in Duino near Trieste did not bring him relief from
his painful desease he committed suicide during an attack
of depression on September 5th, 1906.

Boltzmann was an excellent teacher. Among his
students in Vienna were Paul Ehrenfest, Fritz Hasenöhrl,
Stefan Mayer and Lise Meitner. In Graz Svante Arrhenius
and Walter Nernst had been his students. Hasenöhrl reports

about him: "He never showed his superiority. Everybody was free to ask him questions or even to raise criticism. A relaxed conversation developed were he treated his student like an equal partner. Often one realized only afterwards how much one had learned from him". His colleague in Graz F. Streintz wrote:"He gave advice in every difficult situation. He even did not mind if a student disturbed him in his own apartment while he was working. The great scientist remained for hours at the disposition of his student and always kept his patience and good humor". Lise Meitner said about his lectures: "He held a course which lasted four years. It included classical mechanics, hydrodynamics, the theory of elasticity, electrodynamics and the kinetic theory of gas. The main calculations he wrote on a very big blackboard. On each side he had a small blackboard where he put auxiliary calculations. Everything was written in a clear and well organized form. I often had the impression, that one could reconstruct the whole lecture just from the blackboard. After each lecture we felt as if we would have been introduced to a completely new and wonderful world, because he was so enthusiastic about what he taught". Boltzmann's most popular lectures were the ones on philosophy. His first lecture was an enormous success. Even though the largest lecture hall had been chosen for it the people stood all the way down the stair case. Students, assistents, professors and ladies had come. The hall was ornated with twigs of silver firs and he received enthusiastic ovations. All the newspapers reported about this event. His mail was full of letters of consent. He even had an audience with emperor Franz Joseph. The emperor told Boltzmann that he was glad about his return and that he had heard how crowded his lectures were. Also Boltzmann's seminars were very stimulating. Ehrenfest, for instance, writes in the introduction to his thesis that his work resulted from a suggestion of Boltzmann during a seminar.

Not only scientific problems but also promising new technical inventions fascinated Boltzmann. His enthusiasm was not always shared by his colleagues. When Nernst had invented his electric lamp he sent a few samples to Boltzmann. On one lamp Boltzmann wrote a verse:
"Da Du den sprödesten Stoff Dir gewählt
Und ihn zwangst, den elektrischen Strom zu leiten,
Schufst Du das glänzendste Licht".
He invited the members of the Physical Society in Vienna to his house for a demonstration of the lamp. 55 invitations were sent out. Sandwiches and 50ℓ of beer waited for the guests, but only seven came.

Boltzmann also supported Kress who in 1880 had invented an aeroplane. In 1894 he even gave a talk about it in the "Gesellschaft Deutscher Naturforscher und Ärzte". Unfortunately Kress nevertheless shared the fate of most Austrian inventors. He ran out of money and was forgotten. Boltzmann also constructed an electric sewing machine for his wife, who made dresses for the whole family. Boltzmann was fond of his children. Once his youngest daughter, my mother, desired to have a small monkey. But his wife disliked animals in the house. As a compromise Boltzmann bought rabbits for his daughter and installed a cage for them in his own library.

The room of his youngest daughter was next to his study and in the evening he used to tap at her door as a sign of friendship.

Boltzmann's hobbies included scating in wintertime and swimming in summertime. He used to rise early in the morning. In his later years he often worked already at five o'clock in the morning. Nevertheless he enjoyed parties until late at night.

There were regular parties in Boltzmann's house to which he also invited his graduate students. He liked social life. At parties he was a favorite guest since he used to entertain the people with his splendid humor. At such occasions he wrote funny poems. One of these is called "Beethoven in heaven".

Boltzmann was a great connoisseur of classical literature which he liked to quote. His book "Populäre Schriften" he dedicated to Friedrich Schiller who was his preferred poet. His favorite composer was Ludwig van Beethoven. Boltzmann liked to play Beethoven's symphonies on the piano in the form adapted by Liszt. Together with friends and his son Arthur he used to play chamber music. He liked also philharmonic concerts and had a subscription for the Vienna Opera.

Boltzmann loved nature and arts but the aim of his life remained always his scientific work. His endeavor can best be characterized by the words he has put at the beginning of his book "Principles of Mechanics":

>Bring' vor, was wahr ist;
>Schreib' so, daß es klar ist
>Und verficht's, bis es mit dir gar ist.

PHILOSOPHICAL BIOGRAPHY OF L. BOLTZMANN

E. BRODA

Institute of Physical Chemistry, University of Vienna

ABSTRACT

During the second half of his working life, Boltzmann's centre of interest shifted to philosophy. In his philosophy, he generalized his experience as a physicist and a protagonist of atomistics. He dealt with the essence of physical theories and the mechanism of their evolution. Much attention was paid to the main problem of epistemology, the relationship of existence and consciousness. Boltzmann followed Mach in teaching natural philosophy in Vienna University. But in contrast to him Boltzmann insisted on the necessity of accepting the reality of the external world, unless one wanted to embrace solipsism. Boltzmann called his philosophy realism, and, later, materialism. The evolution of theories and ideas was considered from the point of view of Darwinism, whose fiery supporter Boltzmann was. The laws of thinking are, though largely inborn, by no means infallible. They developed by proving their worth in practice, and need readjustment when they do not fit experience any longer. Our ideas about Euclidian space were given as an example. The Darwinist approach was also applied by Boltzmann to the evolution of ethical and aesthetical feelings and concepts. Boltzmann's philosophical views, put forward with great temperament, led to his increasing intellectual isolation. An attempt is now made by the author to assess Boltzmann's position within contemporary Austrian society. In the author's view, the feeling of isolation greatly contributed to Boltzmann's suicide.

I.

The pronounced and distinctive philosophical views of Ludwig Boltzmann merit particular attention as they are based on his experience as a pathfinder and leader in theoretical physics. Before outlining them, one difficulty must be pointed out. I must translate many passages into English, and the translation must be faithful. However, German style tends to be less precise than English or French style. Moreover, Boltzmann was, it must regretfully be admitted, at times loose and inconsistent in his terminology. This will inevitably appear from the translation. Yet I hope that Boltzmann's meaning will always be clear.

In particular, the term "mechanics" was used broadly by Boltzmann. It often appears as a synonym for "natural science". Certainly the term was not meant to exclude electromagnetism or chemistry. As will be seen later, even Darwin's theory was, in 1886 and again in 1900, called a "mechanical theory", and kings rule, according to Boltzmann, by grace of mechanical laws (1900). A more subtle, but not wholly consistent, use of the term is found in Boltzmann's critique of Ostwald's lecture on Materialism in 1895, and again in Boltzmann's lecture on the Indispensability of Atomism in 1897.

Though there are many indications of Boltzmann's early interest in philosophical problems, he devoted the main part of his efforts to philosophy only after he had more or less completed his contribution to physical research. The last of the major papers in his principal field, statistical theory, appeared in 1898, while his philosophical views were expressed in papers and lectures from 1886 onwards, first in a lecture on the Second Law of Thermodynamics. The change in emphasis came half-way in Boltzmann's working life -- nel mezzo del cammin di sua vita, as Dante said. The philosophical work is conveniently collected in the volume "Popular Writings" [1,4]; incidentally, not all these writings are popular in the sense of easy.

II.

Boltzmann's principal achievements in physics lay in statistical atomistics, and culminated in the interpretation of the Second Law of Thermodynamics. But the idea of the atom was at that time still strongly opposed by influential scientists. They included the colleague of Boltzmann in Vienna, the physicist and philosopher Ernst Mach, and the co-founder of physical chemistry, Wilhelm Ostwald of Leipzig - incidentally, a personal friend of Boltzmann. Boltzmann spoke of the hostile attitude in Germany against the gas theory, i.e., against atomistics [5]. "The attacks against the gas theory multiplied", and "it had gone out of fashion". The struggle was not easy. As Planck [6] put it in 1946, looking back to his youth: "One just could not prevail against the authority of men like Ostwald, Helm and Mach". Boltzmann felt it right, in 1897, to use the words borrowed from Galileo: "Yet I hope to be able assuredly to say of the molecules: Still they are moving" [7].

In his lecture on the Second Law in 1886 Boltzmann said: "Perhaps the atom hypothesis will be replaced by a different one one day - perhaps, but not probably. This is not the place to list all reasons for so saying. I need not recall the conclusions of the genius (William) Thomson who calculated in the most diverse ways, always in a rather satisfactory manner, of how many of those individuals (atoms. E.B.) one cubic millimetre of water consists. I need not mention that ..... by means of the atom hypothesis the prediction of the dependence of the viscosity of gases on temperature was possible, the prediction of the absolute and relative value of the diffusion constant and of thermal conductivity - predictions certainly on a level with that of the existence of the planet Neptune by <u>Leverrier</u> and that of conic refraction by <u>Hamilton</u>".

In the paper "On the Indispensability of Atomistics in Science" (1897) we read: "Contemporary atomistics gives a fully adequate picture of all mechanical phenomena. Considering the compactness of the field we shall hardly expect to find there phenomena that will not fit into the frame of the picture. The picture further includes the phenomena of heat. The fact that this cannot be demonstrated quite so clearly is due only to the difficulties in computing molecular motion. In any case, all essential facts are found in the features of our picture. It has proved enormously useful also in the interpretation of the facts of crystallography, of the constant proportions of the masses in chemical compounds, of chemical isomerism and the relationship between optical rotation and chemical constitution, etc. Atomism is, moreover, capable of further large-scale development". In his polemic against <u>Ostwald</u> in 1904 Boltzmann wrote: "We have here successes, to which all philosophical (meaning: metaphysical, E.B.) views of nature from <u>Hegel</u> to <u>Ostwald</u> can oppose nothing".

Of course, atoms may be attributed only qualities justified by experience, as pointed out in 1897: "Simple consideration as well as experience show that it is hopelessly difficult to find the right pictures of the world by mere guessing into the blue. Rather, the pictures always form slowly from individual lucky ideas by fitting. Rightly epistemology turns against the activities of the many lighthearted producers of hypotheses (Hypothesenschmiede) who hope to find a hypothesis explaining the whole of nature with little effort, as well as against the dogmatic and metaphysical derivation of atomistics".

An example for a dogmatic view: "The reproach that the observed immutability of atoms, lasting only limited time, has been generalized without reason would certainly be justified if one tried to prove, as used to be done, the immutability of atoms a priori. We include it (immutability, E.B.) in our picture merely to represent as many phenomena as possible.... We are ready to drop immutability in cases where another assumption would represent the phenomena better". Nor need atoms be considered as indivisible, as pointed out in the lecture "On Statistical Mechanics", St. Louis, 1904.

While theories based on models may at first be arbitrary and imprecise, they may also be more fertile than phenomenological theories. "The ideas of Faraday were much less clear than the previous, highly precise, hypotheses. Many a mathematician of the old school thought little of the theories of Faraday, without, however, arriving through the clarity of his views at equally great discoveries"(On the Methods of Theoretical Physics, 1892). "In any case, phenomenology cannot represent nature without transgressing experience... Experience, says Goethe, is always only half experience" (On the Evolution of the Methods of Theoretical Physics in Recent Times, 1899). The merits of the "deductive" and the "inductive" methods in mechanics (perhaps not a fortunate terminology) were compared in much detail in Boltzmann's Clark University lectures in 1899.

III.

The kinetic and statistical theory of assemblies of atoms served Boltzmann as an excellent starting point to discuss the essence of physical theory. While Boltzmann was a fine experimentalist when opportunity arose, he considered himself, of course, primarily as a theoretician. In the lecture "On the Importance of Theories" (1890), Boltzmann spoke of the "idea that fills my thought and my action, the elaboration of theory" and continued: "No sacrifice for it is too much for me, for theory, the content of my whole life".

He further explained: "I should not be a true theoretician, if I did not ask first: What is theory?". He answered: "The layman notices first that it is hard to understand, surrounded by a heap of equations that have no meaning for the uninitiated. But these are not the essential thing, and the true theoretician does without them as far as he can. Whatever may be said in words, he expresses in words, while it is just in the books of the practicians that equations figure for mere decoration .... I hold the view that the task of theory is the construction of a picture of the external world that exists merely within ourselves, and which has to serve as a guiding star to all our thoughts and experiments. The task is, in a way, the perfection of the processes of thought, the execution on a large scale of what happens on a small scale within us in the production of any idea".

"It is a peculiar tendency of the human mind to create such a picture and to fit it more and more to the external world.... The continuous perfection of the picture is the main task of the theory. Imagination is its cradle, observing intellect its mentor..... How childlike were the first theories about the Universe..... No wonder that these theories were derided by the empiricists and practical men, and yet they contained the germs of the later powerful theories.... Today it may be said that theory has conquered the world".

After the role of theory in science had been praised where "the stars bend to the laws that were not given to them but learned from them by the human mind", the overwhelming influence of theory on human practice was emphasized. The equations of theory guide the ships, allow the construction of Brooklyn bridge and of the Eiffel tower, theoretical chemists have become rich men through the practical application of their syntheses, and electrotechnics is applied theory. "Perhaps the time is not distant when every household bill will glorify those great electricians (Ohm, Ampere), and in the next century perhaps every cook will know with how many volt-amperes the meat is to be roasted, and how many ohms her lamp has. It is precisely the practical technologist who, as a rule, treats the complicated equations of electricity with more certainty than some beginner in science, as he must pay for any error not only through the reproach of his teacher, but with cash".

"One is almost tempted to maintain that theory, in addition to its intellectual mission, is the most practical thing; in a way the quintessence of practice, as the precision of its conclusions cannot be equalled through any routine in guessing or trying - true, because of the hiddenness of its (theory's. E.B.) paths, only for him who walks them securely. A single error in sign may raise the results a thousandfold while the empiricist never errs to that extent".

These enthusiastic words are followed by a warning: "I called theory a purely intellectual internal picture, and we have seen how capable it is of high perfection. How then could it not happen that on continuing immersion into theory one comes to think of the picture as of the really existing thing?.... Thus it may happen to the mathematician that he, always occupied with his equations and dazzled by their internal perfection, takes their mutual relationships for what truly exists, and that he turns away from the real world. Then the lament of the poet applies to him as well: that his works are written with his heart blood and that highest wisdom borders on highest folly".

The criterion of practice was also emphasized in 1902 in his last inaugural lecture in Vienna: "What does it mean to understand a mechanism wholly correctly? Everyone knows that the practical criterion is that one knows how to treat it correctly. But I go even further and maintain that this is also the only tenable definition for the understanding of a mechanism".

How does one theory supplant another? In his obituary to his beloved teacher Josef Stefan (1895) Boltzmann draws a dramatic picture of the struggle of theories: "The layman perhaps imagines that new fundamental causes of phenomena and fundamental ideas are added to those already found, and nature is, in continuous development, understood more and more. Here he is, however, mistaken. The evolution of theoretical physics has always proceeded in jumps. Often a theory has been developed through many years, even through more than a century, so that it gave a fairly lucid picture of a certain class of phenomena. Then new phenomena became known that contradicted the theory. Attempts at fitting the theory to these phenomena had no success. A struggle between the supporters of the old and of a quite new view arose, until finally the latter generally prevailed. It used to be said that the old view was recognized as wrong. This sounds as if the new view had to be absolutely correct, and, on the other hand, as if the old view, because wrong, had been wholly useless. To avoid the appearance of these two claims, it is rather stated today: The new view is a better, a more perfect, picture, a more useful description of the facts. Thereby it is clearly expressed that the old theory as well was of use by also giving, in part, a picture of the facts. Moreover, that the possibility is not excluded that the new theory, too, may be supplanted by an even more useful one".

This idea of a progress of theory through contradictions might be called after <u>Hegel</u>, a philosopher not exactly appreciated by Boltzmann, dialectical. In his lecture on the Development of the Methods of Theoretical Physics, in 1899, Boltzmann said: "When we consider more closely the evolution of theory, it strikes us first that it is by no means as continuous as one might expect, that it is rather beset with discontinuities, and that, at least seemingly, it does not occur along the simplest path logically given. Often certain methods up to now yielded the most beautiful results, and it may have been believed by some that the evolution of science to infinity would consist in nothing else but in their (of the methods. E.B.) continuous application. In contrast, they (the methods. E.B.) suddenly prove exhausted, and one strives to look for quite new and disparate ones. Probably a struggle between the supporters of the old methods and the innovators then develops. The point of view of the former is called by their opponents an obsolete and defeated one, while they themselves denigrate the innovators as spoilers of genuine

classical science.... This is, by the way, a process that
is by no means limited to theoretical physics, but which
appears to take place in the evolutionary history of all
branches of intellectual activity...Therefore we shall not
be surprised any more that theoretical physics is no exception to this general law of evolution.... It follows
that it cannot be our task to find a theory that is absolutely correct, but rather a picture that is as simple as
possible and represents the phenomenon as well as possible.
It is even conceivable that two different theories are
equally simple and fit the phenomena equally well, and
therefore are, albeit totally different, equally correct.
The assertion that a theory is the only correct one can
merely be the expression of our subjective conviction that
no other picture can exist that is equally simple and
equally well fitting".

IV.

Boltzmann had been led to a consideration of the
essence of theory from his work in physics. His concern
with theory led him further on into the field of philosophy.
While retaining the chair of theoretical physics, in 1903
Boltzmann undertook to teach, in the University, natural
philosophy as successor to Mach ("Lehrauftrag"). He described
his way as a philosopher in his inaugural lecture: "I have
often pondered the enormous field of philosophy. Infinite
it appears to me, and my power feeble. A man's life would
be little to achieve some successes there, the untired
activity of a teacher from youth to old age insufficient
to hand them over to posterity, and for me this is to be

a part time task in addition to another teaching subject that by itself requires one's whole power?.... Perhaps I have been selected, because, though I did not write about logic, I belong to a science where one has the opportunity for daily practice in the most precise logic".

Boltzmann did not think highly of German idealism. He continued: "While only hesitantly I have followed the call to interfere with philosophy, philosophers have all the more often interfered with natural science. A long time ago already they entered my realm. I did not even understand what they meant, and so I wanted better information about the fundamental teachings of all philosophy. To draw from the most profound depths, I took Hegel into my hands: but what unclear thoughtless mass of words was I to find there! My unlucky star guided me from Hegel to Schopenhauer".

(We may interpolate that in 1905 Boltzmann gave a lecture to the Philosophical Society of Vienna whose original title, borrowed from Schopenhauer himself who used it in respect of Hegel, was "Proof that Schopenhauer is a mindless, ignorant philosophaster who scribbles nonsense and debases the heads through empty word trash fundamentally and for ever". In the same lecture, the thoughts of all (!) philosophers, including Kant, for whom Boltzmann had much respect, are called basically untenable, and it is stated that the liberation of mankind from the mental headache, called metaphysics, is the goal.)

Yet for Boltzmann there was more in philosophy than Hegel and Schopenhauer. He continued: "The bent towards philosophy appears to be inborn, ineradicably. Not only Robert Mayer, who was a philosopher through and through, but also Maxwell, Helmholtz, Kirchhoff and Ostwald willingly made sacrifices to philosophy and recognized her problems as the highest so that today she stands again as the queen of the sciences".

Schiller had told the scientists and philosophers of his time: "Enmity be between you, it is still too early for an alliance". But, Boltzmann said in his lecture "On Statistical Mechanics" in 1904, "the time for alliance has come now".

V.

For Boltzmann the central problem of philosophy was that of epistomology, the problem of the relationship of existence and consciousness. To anticipate his point of view, we may quote Boltzmann's view of bishop Berkeley. After his arrival in Berkeley, California, in 1905, he remarked: "The name of Berkeley is that of a highly esteemed English philosopher, who is even credited with being the inventor of the greatest folly ever hatched by a human brain, of philosophical idealism that denies the existence of the material world".

In a more restrained way Boltzmann had written in 1896, in connection with Ostwald's lecture against materialism: "Altogether the distrust of ideas derived from direct sensual perception has led to an extreme, opposed to the previous naive belief. Only sensual perception is given to us, and therefore - so they say - we must not go a step beyond. But if one were logical, one would have to ask further: Are yesterday's sensual perceptions given to us, too? Immediately given is, after all, only the one sensual perception or the one thought that we think in this moment. If one were logical, one would have to deny not only all other beings except one's own "I", but even all ideas that one had in all previous times. How do I know them? From memory. Yet how do I know that not memory alone exists, but the perception remembered never existed - as happens again and again with the mentally deranged, and from time to time with normal persons as well. If one does not want to conclude that only the idea exists that I have at the moment, and nothing else - a conclusion refuted by the utility of the knowledge for our actions - one must at last, in spite of all necessary caution, admit our capacity to draw conclusions from perceptions in respect to something that we do not perceive, conclusions, it is true, that we must correct whenever they contradict perceptions".

In Boltzmann's paper "On the Problem of the Objective Existence of the Processes in Inanimate Nature" (1897) it is emphasized that fellow humans react just like ourselves, for instance to fire. Therefore our views are simplified rather than complicated by ascribing other humans sensations like our own, and by breaking the monopoly granted to the thinking subject by subjective idealism. But having admitted the existence of fellow humans, one step only, a necessary step, remains to the acceptance of the objective existence of non-living matter.

"We therefore name these sensations of others by signs and words analogous to our own (we imagine them), because in this way a good picture of the course of many complexes of sensations is obtained, our view of the world is simplified. To express that these are imagined sensations, we say that these are not our own, but those of other people.... By analogy to the sensations of others, the processes of non-living nature also exist only as imagined by us, i.e., we mark them by certain thoughts and word signs, because in this way the construction of a world view is facilitated that is capable of predicting our future sensations. Thus the processes in non-living nature are on the same level, in this respect, with the sensations of other people, and the non-living things with other people - only the signs and the rules of their combinations differ far more from those that we use for the representation of our own sensations. "A non-living thing exists or does not exist" has the same meaning as "a man exists or does not exist". Therefore it would be a complete error to believe it proven in this way that matter is more of a thought thing (a product of thought. E.B.) than another human being is".

After having pointed out the importance of communication through language whereby the equality of the perception of the world by different humans is established, Boltzmann concludes that one must "take an objective point of view", as the saying goes. "It turns out that the ideas connected with "existence" and "non-existence", in their majority, remain applicable without change. Those people or non-living things that I only fancy, i.e. conceive without being required to do so by the regularity of complexes of sensations, do not exist for other people either, they have no objective existence".

Man has no monopoly in perception: "First, the analogy between the sensations of man with those of the highest animals is so perfect that we absolutely must ascribe objective existence to the latter sensations. But where is the limit? True, occasionally one hears doubts whether insects, or divisible animals, certain worms, have sensations. But a definite limit where sensation stops cannot be given. At last we come to organisms so simple that their world views and thoughts are zero. Unless we suddenly deny the predicate of existence to the sensations of animals beneath a certain level, which would be most inappropriate, we must ascribe existence even to the sensations of this unthinking organized matter where sensations are hard to demonstrate.... But then it would appear to me as an unjustified and inappropriate jump to deny this predicate to non-organized matter".

More evidence for the objective existence of the external world follows: "We see, moreover, that the series of sensations and acts of will that we call individual human beings break off again and again, that the individuals die, while the matter to which those expressions of the mind were linked remains. The subjective world view, which considers matter as mere expression of equations between the complexes of human sensations, seeks first to define the fluid and complicated by terms, and only later to use these pictures to represent the simple and the constant, i.e., matter. It takes the Egyptian pyramids, the Acropolis of Athens as mere equations that have existed between the sensations of generations through thousands of years".

"But beside it a simpler (objective) world view must be possible that begins with what is more constant and that represents the transitory through the laws applying to the constant. When we are following our thoughts logically, i.e., in accordance with the rules that always have led to confirmation by experience, we arrive at the results that the planet Mars resembles Earth in size.... and it does not appear impossible that on planets of other suns the most magnificant scenery exists without ever having produced sensations in a living being".

In his obituary to his great friend Josef Loschmidt (1895) Boltzmann emphasized that the older expression "The molecules exist" and the newer phrase "Our relevant ideas are a simple and useful picture of the phenomena observed" essentially mean the same thing.

Boltzmann usually termed his philosophy "realism". At the end of his 1897 lecture , his words were: "The idealist compares the assertion that matter exists just like our sensations to the view of the child that a beaten stone suffers pain. The realist compares the assertion that one can never conceive how mental phenomena can be represented through matter or even through a play of atoms with the opinion of an uneducated person who maintains that the Sun cannot be 20 million miles (German miles! E.B.) from the Earth, as he cannot imagine it. As ideology (idealism. E.B.) is only a view for one individual, but not for mankind, to me the terminology of realism appears more useful than that of idealism if we want to include the animals, nay the Universe". In his Schopenhauer lecture in 1905 Boltzmann directly called his view materialism: "Idealism maintains only the existence of the "I", the existence of various ideas, and seeks to explain matter therefrom. Materialism starts from the existence of matter, and seeks to explain the sensations therefrom".

These opinions of Boltzmann, strongly differed from those of his Vienna colleague as physicist and philosopher, Ernst Mach. To him, atoms and molecules "were mere concepts to represent economically, and to illustrate, the regular relationships of our sensual perceptions and ideas", as Boltzmann put it in his polemic against Ostwald (1904). As we know well, the idea of thought economy plays a central part in <u>Mach</u>'s philosophy. According to <u>Mach</u>, an external world independent of man cannot be an object of science. Hence science must limit itself to order economically our thoughts about our sensations. While Boltzmann readily appreciated the importance of economy in classifying and ordering the results of science, he did not see in economy an object in itself. He said in his paper "On the Methods of Theoretical Physics" (1892):

"In mathematics and geometry it was no doubt at first the need for economy of labour that led from purely analytical to constructive methods and to illustration by models. While this need appears purely practical and self-evident, we find ourselves here already in a field where a whole family of modern methodological speculations grew up, expressed in the most precise and spirited way by <u>Mach</u>. He maintains directly that the aim of science is nothing but economy of labour".

"Almost with equal justification one could, noticing that in business greatest economy is desirable, consider this economy as the purpose of shops and money - which would, in fact, in a certain sense be correct. Yet one will not easily term as mere economy the investigation of distances, movements, sizes, physical and chemical properties of fixed stars, the invention of microscopes, and the discovery therewith of the causes of our diseases".

Here we may refer to Boltzmann's views on God. While I do not propose to discuss the problem now, the confession of pantheism in the lecture on Objective Existence (1897) may be mentioned: "It is true that only a madman denies the existence of God, but it is equally true that all our pictures of God are only insufficient anthropomorphisms, i.e. that what we imagine as God does not exist in the way in which it is imagined. Hence when one person says that he is convinced of the existence of God, and the other, that he does not believe in God, perhaps both of them think the same, without suspecting it. "A further, somewhat dark, passage on religion is found in Boltzmann's inaugural lecture in Leipzig (1900).

VI.

Boltzmann was a younger contemporary of <u>Darwin</u>, whom he greatly admired. The century will be called <u>Darwin</u>'s century, he said in his lecture on the Second Law in 1886. "In my view all salvation (alles Heil) for philosophy is to be expected from <u>Darwin</u>'s teachings" he maintained in 1905 in the Schopenhauer lecture.

Boltzmann reasoned that the power of theoretical thinking can ultimately be explained on an evolutionist basis. Only mental processes that proved their worth during the long march of living matter from its beginning to the emergence of man could maintain themselves. More exactly, only the capacity for processes that made possible the correct understanding of nature could be transmitted to progeny. In his 1886 lecture Boltzmann said:

"We make the hypothesis that complexes of atoms evolved that could multiply by formation of similar complexes around themselves. Among the larger masses so formed those were most vital that could proliferate by division, further those that tended to move to places with favourable conditions for life. This was much promoted by receptivity for external impressions, chemical quality and movement of the surrounding medium, light and shadow, etc. Sensitivity led to the development of sensory nerves, motility to motor nerves. Sensations with which, through inheritance, the always strongly forceful report to the centre to escape them is linked, are called pain. Quite crude signs for external objects remained within the individual, they developed into complicated signs for complicated conditions .... When the individual has a developed memory capacity of this kind, we define it as consciousness. Here is a continuous bridge between the closely connected, clearly conscious ideas to the unconcious reflex movements. Does not our feeling tell us again that consciousness is also something quite different? But I have withdrawn the word from feeling. If the hypothesis explains all relevant phenomena, feeling will have to submit, as it did in the question of the axial rotation of the Earth (i.e. after Copernicus.E.B.)

Eleven years later, in his article on Objective Existence, Boltzmann proposed: "Finally, the close connection of the mental with the physical is given to us by experience. Through experience it is highly probable that a material process in the brain corresponds to any mental process, i.e. the latter is unambiguously coordinated to the former, and that the mental processes always are genuine material processes.... Then it must be possible to predict all mental processes from the pictures serving the representation of brain processes. The brain is considered by us as the instrument, the organ for the production of world pictures, which because of their great utility for the preservation of the species according to Darwin's theory developed to particular perfection in man, as in the giraffe the neck, and in the stork the beak developed to unusual length.... As soon as we follow this view, we have to assume that the pictures and laws that serve to represent processes in non-living nature are sufficient to represent unambiguously all mental processes as well. We say briefly: The mental processes are identical with certain material processes in the brain (realism)".

As emphasized again in the Schopenhauer lecture, man owes the laws of his mind to evolution: "What would be the position of what are called laws of thinking in logic? Now, in the sense of Darwin these laws of thinking are nothing else but inherited habits of thinking". After pointing out that the laws serve us to influence nature, Boltzmann continues: "This influence has been much improved through conservation and suitable ordering of the memory pictures and the acquisition and practice of speech. This improvement is the criterion of truth".

"The method to combine the memory pictures and the words, spoken audibly or not, has been perfected more and more and has been inherited so that solid laws of thinking have developed. It is quite true that any knowledge would stop and perception would have no context if we did not bring with us these laws of thinking.... It is possible to call these laws of thinking aprioristic because they are inborn to each individual through the experience of the species during many thousands of years. But it appears to us as a logical blunder of Kant to conclude that they (the laws. E.B.) are infallible in all cases".

"This blunder is fully understandable on the basis of Darwin's theory. Only what was certain has been inherited. So these laws of thinking have obtained such an appearance of infallibility that it was believed justified to subject to their judgment even experience.... In a like manner it used to be assumed that our ear and eye are absolutely perfect because indeed they have been developed to amazing perfection. Today it is known that this is erroneous, and that they (eye and ear. E.B.) are not perfect. In analogy I should deny that our laws of thinking are perfect. They do not differ from all the other inherited habits". The laws of Euclidian space and the law of conservation of energy are mentioned as examples of laws of thinking that require experimental confirmation.

In a famous book on the principles of mechanics, Heinrich Hertz had proposed that the pictures constructed by us must conform with the laws of thought. Boltzmann wrote in 1899 ("On the Basic Principles and Equations of Mechanics"): "I want to raise objections against this proposition. Certainly we must bring with us a rich treasure of laws of thinking. Without them, experience would be quite useless; we could not even fix it through internal pictures. Almost without exception, these laws of thought are inborn, but nevertheless they suffer modifications through education, teaching and one's own experience. They are not quite the same for a child, for a plain uneducated person and for a scientist. We shall appreciate this when we compare the direction of thought of a naive people like the Greeks with that of the medieval scholastics and this again with the present direction. Surely there are laws of thinking that proved their worth in such a way without exception that we have absolute confidence in them, that we consider them aprioristic unchangeable principles of thought. But nevertheless I think that they developed slowly. Their first source consisted of primitive experience of mankind in the original condition, gradually they became stronger and clearer through complicated experience, until finally they assumed their present sharp formulation; but I should not recognize the laws of thinking as absolutely supreme judges. We cannot know whether they will not suffer one or the other modification. One may recall how definitely children or uneducated persons are convinced that it must be possible to distinguish up and down in all places of space through mere feeling, and how they believe to be able to deduce therefrom the impossibility of antipodes. If such people wrote logic, they would certainly consider it a law of thought that is evident a priori. Likewise in the beginning aprioristic objections were raised against the theory of <u>Copernicus</u>. In the history of science there are many cases where theorems were either proved or refuted through evidence which was thought to correspond the laws of thinking, while now we are convinced of its futility. Therefore I should like to modify the request of <u>Hertz</u>.... The sole and ultimate verdict about the fitness of the pictures is based on their representation, as simply as possible and completely correctly, of experience.... Precisely here is the test for the correctness of the laws of thinking".

Also in his St. Louis lecture on Statistical Mechanics in 1904 Boltzmann referred to "excessive confidence in the so-called laws of thinking". "True, it is certain that we could not have experience if we had not certain congenital forms of connecting perceptions, i.e. of thinking. If we like to call them laws of thinking they are really aprioristic in so far as they are present before any experience in our soul (mind. E.B.), or, if we prefer, in our brain. But nothing appears to me less well motivated than a conclusion from aprioristic thought in this sense to absolute certainty, to infallibility. These laws of thinking have formed according to the same laws of evolution as the optical apparatus of the eye, the acoustical apparatus of the ear, the pumping device of the heart. In the course of the evolution of mankind anything unsuited has been left behind, and in this way that conformity and perfection has arisen which can easily be taken for infallibility. .... But as soon as contradictions apparently cannot be removed we must immediately try to examine, to enlarge and to modify what we call our laws of thinking, but what is nothing but inherited concepts, to which we became used and which proved useful through ages to signify practical requirements. Just as numberless artificial inventions, created quite consciously, have been added to the inherited inventions of the roller, the cart, the plough, we must likewise deliberately and consciously bring better order into the inherited ideas. It cannot be our task to cite what is given (experimentally. E.B.) before the judge's bench of the laws of thinking, but we must rather adapt our thoughts, ideas and concepts to what is given"....

"True, our inborn laws of thinking are the precondition of our complicated experiences, but this was not so in the most primitive beings. With them, they (the laws. E.B.) arose slowly through their simple experiences, and they were inherited by more highly organized beings. So it is to be explained that synthetic propositions exist there that are inherited from our ancestors and are inborn to us, and hence aprioristic. Their overpowering force, but not their infallibility is a consequence".

Clearly Boltzmann was an evolutionist not only because <u>Darwin</u> had put forward a powerful theory for a most important area of science that was not Boltzmann's. Rather, Boltzmann considered <u>Darwin</u>'s methods as the key for the understanding of truth or falsehood of scientific theories. It must be critically pointed out that Boltzmann did not distinguish between biological and cultural evolution, i.e., between what is physiologically fixed, and what is handed down to descendants in human society through example and precept.

VII.

Problems of ethics are likewise treated from an evolutionist point of view. Boltzmann held that only those actions qualified as good that furthered the survival of the species. Another morality could not have maintained itself either among animals or among men. Thus Boltzmann did not seek the roots of ethical behaviour in conscious thought, as do positivists, and even less outside, as does religious tradition. During evolution in morality as well as in cognition conscious and directed mental processes must have replaced instinctive behaviour.

In his Schopenhauer lecture, Boltzmann stated: "Ethics must ask when the individual may maintain his own will, and when he must subordinate it to that of others, so that the existence of the family, the tribe, of mankind - and therefore of all together - be furthered as much as possible. But this inborn trend to enquire overreaches itself when the question is asked whether life itself ought to be furthered or inhibited.... If morality of any kind had the effect that the tribe following it decays, it would be refuted thereby. In the last analysis not logic, not philosophy, not metaphysics decides whether something is true or untrue, but action (practice. E.B.). In the beginning, there was action. Whatever directs us to correct action, is true".

Ostwald in 1904 put forward a mathematical expression for "happiness", namely, $E^2 - W^2$. E is the energy spent intentionally and successfully, W the energy spent with repulsion. Boltzmann did not think highly of Ostwald's attempt. In a lecture devoted to its refutation (1905), he took the opportunity of dealing, in more detail, with the evolution of morality: "As regards the concept of happiness, I derive it from Darwin's theory. Here it does not matter to us whether during the millions of years in the enormous mass of water on earth the first protoplasm evolved "by accident" in moist mud, whether egg cells, spores or other germs, as dust or embedded in meteorites, some time arrived on earth from space. More highly developed individuals will hardly have fallen from the skies. Thus at first there were only quite simple individuals, simple cells or lumps of protoplasm. Constant movement, so-called Brownian movement, is shown, as is well known, by all small lumps; growth through absorption of similar components and subsequent proliferation by division is fully imaginable mechanically. Similarly it can be understood that the rapid movements were influenced and modified by the environment. Lumps where modifications took place in the sense that on an average (preferentially) they moved towards substances better suitable for absorption (better food) succeeded better in growth, and more often in propagation. Hence they soon overran the others".

"In this simple process, easily understood mechanically, we have all in essence - inheritance, selection, sensual perception, will, joy and pain. Only quantitative increase, with application of the same principle, is needed to come, through the whole empire of the plants and animals, to mankind with its thinking and feeling, will and action, joy and pain, artistic and scientific enquiry, generosity and vice".

"Cells that had associated in larger societies where division of labour took place, and which again segregated cells with similar tendencies, had greater chances in the struggle for existence, especially if certain cells in the case of harmful influences did not rest until the working cells had removed them (these influences. E.B.) as far as possible (pain). The activity of such cells was particularly effective, if, in an instance where the complete removal of the harmful influences had not been successful, it continued and left tension that relaxed very slowly only, and the tension burdened the memory cells and incited the movement cells to even more energetic and purposeful collaboration in the case of a return of similar (harmful. E.B.) circumstances. This condition is called discomfort, feeling of unhappiness. The contrary, full freedom from such gnawing after-effects, the exhortation to the memory cells that the movement cells act likewise in similar future cases, is called lasting joy, feeling of happiness".

"True, we have by no means exhausted all gradations of happiness in highly organized beings. Not even a beginning has been made for a physiology of happiness; yet the point of view has been fixed from which one must consider the relevant phenomena if one does not merely want to produce beautifully sounding, exalting, poetical, inspiring phrases, but explain them (the phenomena. E.B.) scientifically".

Thus values are judged according to the role which objects or relationships play in respect to evolving life. Any other measure is rejected. There is no reason why this natural explanation of ethical values should be considered as an act of debasement or as a source of deception.

"We believe therefore" Boltzmann said in the paper on Objective Existence (1897), clearly without approving of such belief, "that we must be as dead and listless as these machines (clocks or dynamos. E.B.) if our mental processes could be represented exhaustively through pictures of material processes in the brain. Obviously this is why this view appears dreary and desolate to some. But without any reason, I think. Precisely the origin of violent feelings of pain and joy is explained by Darwin's theory, as they are needed to obtain the reaction energy required for the survival of the species. The whole intensity, diversity and richness of intellectual and emotional life cannot be due to greater nobility and loftiness of these processes than of those in dead machines, but only to their greater richness and diversity and to the fact that our own "I" belongs to the same species. As one will not doubt that mental processes follow quite definite laws, I could not consider it in any way discouraging if these were identical with laws responsible for equally complicated material processes. Noble and lofty is for our subjective feeling whatever furthers and exalts our species. These ideas (nobility and loftiness. E.B.) have no objective existence. Hence if material processes can be equally diverse and complicated as our mental processes - there is no reason to doubt it - I do not see how the noble and lofty character of our mental processes or our passionate interest in them could be touched in any way by the assertion that our mental processes could be exhaustively represented by the picture of material processes in the brain. We know that a clock is incapable of sensations, i.e., that a mechanism so simple cannot represent anything somehow similar to sensations. But what does one want to express by saying that it follows from the qualitative difference between our sensations and material processes that the course of the former can never at all represented by a combination, however complicated, of those pictures that also represent to us the processes in non-living nature?"

The origin not only of the moral sense but also of the aesthetic sense was treated in Boltzmann's inaugural lecture in Leipzig (1900): "We still have to remember the most wonderful mechanical theory in the field of the biological sciences, namely, the theory of <u>Darwin</u>. It undertakes to explain the whole diversity of the world of plants and animals through the purely mechanical principle of heredity - which admittedly in itself is dark like all fundamental principles of mechanics".

"The explanation of the wonderful beauty of flowers, of the richness of forms in the insect world, of the fitness of the structure of the organs in the animal and human body, all this herewith becomes the domain of mechanics. We understand that it has been useful and important to our species that certain sensations flattered us and were sought by us, others repelled us. We see what an advantage it has been to construct, as precisely as possible, pictures of the environment in our mind and to distinguish strongly what of these pictures agreed with experience, as true, from what did not agree, as wrong. Thus we can explain mechanically the origin of the concept of beauty as well as of that of truth."

"We understand, too, why only individuals could continue to exist who abhorred with all intensity of their nerve power and tried to prevent certain highly pernicious influences, but strove with equal nerve power towards others that were needed for their own preservation or that of the species. Thus we understand how the whole intensity and power of our emotional life evolved joy and pain, hatred and love, desire and fear, delight and despair. We can get rid of the whole scale of our passions just as little as of the diseases of our body, but we learn to understand and to bear them".

Boltzmann was conscious of the obstacles in the realization of the moral ideas - also by himself. We hear the self-irony when Boltzmann in the same lecture stated: "The enthusiastic love for liberty of <u>Cato</u>, <u>Brutus</u> and <u>Verrina</u> sprang from sentiments that germinated in their hearts through purely mechanical causes. Again it is to be explained through mechanics that we live at our ease in a well-ordered monarchy, and yet like to see that our sons read <u>Schiller</u> and <u>Plutarch</u> and draw inspiration from the words and deeds of enthusiastic republicans. This again we cannot change; but we learn to understand and bear it. The god, by whose grace the kings rule, is the fundamental law of mechanics".

Though Boltzmann did not intend to play <u>Brutus</u> against <u>Franz Joseph</u>, his militant mind accepted no compromise in science. He concluded his oration on <u>Kirchhoff</u> to Graz University (1887): "Nay, the most distant posterity will not grudge admiration to the great men that our century begot. If anything were to equal this admiration, it is the surprise that the same century could not get rid of so much ridiculous pedantry, so much inherited nonsense and foolish superstition.... Does it not sound louder than ever, the bleating of all obscurantists, the enemies of free speech and enquiry against the new theorem of <u>Pythagoras</u>, the teachings of <u>Darwin</u>.... But hail! It is the storm that heralds the coming of spring. But before, light-hearted jests are premature, arm for the bitter bloody struggle."

VIII.

We have seen that in philosophy Boltzmann was essentially an autodidact. Probably he discussed his philosophical views with fellow physicists, but no important school of philosophy existed in Austria during most of his time. Boltzmann was in contact with the remarkable psychologist Brentano who had been appointed to a chair in philosophy in Vienna University in 1874, but had to resign because he married in spite of having been a Catholic priest before. Results of the interaction with Brentano are not obvious, however.

In later years, Boltzmann was more often seen in the Philosophical Society than in lectures on physics [9]. But nothing is known about his influence on philosophers, and even among the physicists Boltzmann's views on philosophy were forgotten. Among his students we may name Arrhenius, Nernst, Hasenöhrl, Smoluchowski, Lise Meitner, Ehrenfest, Przibram, among his students' students Schrödinger and Hans Thirring. Though the students were often imprinted by Boltzmann as a man, and not only as an expert, they tended to forget his deep attachment to philosophy. In the celebrated Vienna Circle of the nineteen twenties, where progressive scientists and philosophers combined, Mach's influence certainly was more evident than Boltzmann's - in spite of Mach's opposition not only against atomistics, but also against the theory of relativity.

Not that no interest was paid to Boltzmann's views

in his time. On the contrary, crowds wanted to listen to his views on natural philosophy [10]. Some may even have looked at his lectures as a show. But I think that official Vienna, under the Habsburgs, did not favour the passionate philosophy of a fighting mind. This philosophy had, as we have seen, political aspects. On the basis of his general views, Boltzmann may be called a radical democrat and a - however resigned - republican. That was why he admired the America of his day, as can be seen from the "Journey of a German Professor to the Eldorado" (1905). "Freedom breeds colossuses", he quoted his hero Schiller.

True, Vienna was, within limits, a liberal place - though the example of Brentano does show the limits. Somehow in old Vienna Slavs, Hungarians, Italians and Jews coexisted with the original population. There was also a lot of contact between the landed aristocracy, the immensely rich bourgeoisie, often Jewish, whose children were numerous among the intellectuals, and the austromarxist Socialists. The same applied, incidentally, to Prague.

But power was, and had to remain, in the hands of the aristocracy, presided over by the Habsburgs and interconnected with the then rigid Church, the military, the high bureaucracy and centralized finance. This block of conservative forces had no use for philosophical militancy. The established leaders of the arts and the sciences mostly were chosen accordingly, or they adapted themselves. They were conservatives or liberals, but certainly no radicals. This situation has been discussed by me elsewhere [8].

In Habsburg Austria, sceptical and antiorthodox ideas had to arise outside the establishment, typically in modest flats or even in coffee houses. Psychoanalysis, modernistic music, art nouveau (Jugendstil) and advanced literary criticism of the Karl <u>Kraus</u> kind may be mentioned. But as far as we know, Boltzmann, a University professor if ever there was one, had no personal contact with this milieu. Among his colleagues, <u>Helmholtz</u> was the one with whom he could talk best [11], although the ways of the Prussian Geheimrat (Secret Councillor) chilled him. But <u>Helmholtz</u> lived in Berlin and died as early as 1894.

Boltzmann might have found a response among Socialists. Boltzmann's realist-materialist philosophy of practice, his views on the role of theory and on evolution strikingly recall those of <u>Engels</u>, as expressed in the "Anti-Dühring", though Boltzmann probably was not aware of it and may never have heard of <u>Engels</u>. But the concept of classes and the possibility of an interpretation of history on their basis is not mentioned in the work of Boltzmann, and altogether Socialist thought or action was miles away from his world. This lack of communication is also expressed in Boltzmann's high-handed unconditional dismissal, in 1903, of <u>Hegel</u>.

For Boltzmann, his science and his philosophy were a unity, as is implied in his grand, all-embracing use of the term "mechanics". In all aspects of this unity, from atomism to evolutionism, to philosophical realism-materialism and to republicanism, he felt increasingly isolated. Other men might have reacted differently to hostility, outspoken and also silent, to their world view. But Boltzmann was passionate and vulnerable at the same time. Moreover, his health was failing in various ways in later years. This is a factor probably overemphasized by some authors, but it may have accentuated the feeling of isolation.

For long, Boltzmann had suffered from rapid changes

between cheerfulness and sadness - he joked about it himself when cheerful. Physiologically conditioned gloom and the cold feeling of frustration may have conspired to produce a mood that on a dark day led one of the greatest thinkers of all times to lay hand at himself.

References

[1] L. Boltzmann, Populäre Schriften, Leipzig 1905. Whenever in the present paper Boltzmann is quoted by paper title and year only, the source is found in that volume and in [2]. Many of the passages from Boltzmann's papers have been selected and quoted before by me [3].

[2] L. Boltzmann, Wissenschaftliche Abhandlungen, ed. F. Hasenöhrl, Vols. 1-3, Leipzig 1909

[3] E. Broda, L. Boltzmann, Mensch-Physiker-Philosoph, Vienna 1955 and Berlin 1957.

[4] Concerning the social setting, the following point may be made. Already in his life time, Boltzmann was greatly respected as a physicist, nationally and internationally. He was, for instance, a member of many Academies. Also the Austrian civil servants, among whom were enlightened men, appreciated Boltzmann as a physicist. He was even given the title Hofrat (Court Councillor). Twice the Austrian patriot was recalled to his beloved country from Germany - the second time only on his word of honour never to leave Austria again. But characteristically, in spite of his magnificent title, he never appears to have been consulted by Government or Emperor even in professional matters, although he held that great men ought to take part in public life (see Obituary to Loschmidt).

Kaiser Wilhelm the Second was certainly an unpleasant and dangerous person - but he did want to know the scientific and technical views of Fritz Haber, Richard Willstätter and Otto Hahn. For Franz Joseph and his entourage, science was not interesting. The same was true of the heir presumptive, the Archduke Franz Ferdinand, later assassinated in Saraievo, however much he otherwise was opposed to his uncle [8].

[5] L. Boltzmann, Vorlesungen über Gastheorie, Preface to Vol.II, Leipzig 1898.
[6] M. Planck, Vorträge und Erinnerungen, Stuttgart 1949.
[7] No. 123, in [2].
[8] These problems have been treated in a separate essay: E. Broda, Warum ist es in Österreich um die Naturwissenschaft so schlecht bestellt?, in H. Fischer, ed., Versäumnisse und Chancen, Vienna 1967. I am referring to this essay because of the discussion of the past. The passages about the present are largely obsolete now.
[9] A. Lampa, quoted in [3].
[10] H. Marek, née Broda, quoted in [3].
[11] W. Kienzl, quoted in [3].

THE DEVELOPMENT OF BOLTZMANN'S STATISTICAL IDEAS

M.J. KLEIN

Yale University, New Haven

ABSTRACT

In 1866 Boltzmann began his scientific career with an attempt to give a purely mechanical explanation of the second law of thermodynamics. He gradually recognized the need to introduce statistical concepts in order to understand irreversibility and the second law. This paper follows the development of Boltzmann's views, tracing the increasingly important role that he assigned to statistical concepts. Boltzmann's critics played an important part in forcing him to clarify his own thinking about statistical mechanics, and the successive criticisms made over a period of a quarter of a century are described and analyzed. There is also an attempt to estimate the degree to which Boltzmann's ideas were generally understood and accepted by his contemporaries.

1.   Ludwig Boltzmann's first significant paper [1] was published in 1866; his last [2] appeared, posthumously, in 1907. Both were concerned with the kinetic theory of gases, the subject to which Boltzmann devoted his greatest efforts during the forty years of his career [3]. His work in this field alone covers an extremely wide range of problems, extending from the most general mechanical theorems to forbiddingly detailed and complicated calculations of the viscosity coefficient of a gas. Despite the variety and the sheer quantity of this work, there is no difficulty in identifying Boltzmann's most important contribution to physics: it is surely his explanation of the irreversibility of natural processes, the irreversibility expressed in the second law of thermodynamics. It was Boltzmann who showed how one could understand macroscopic irreversibility by applying statistics to a gas composed of a vast number of molecules, even though the individual molecular motions are described by the reversible laws of mechanics. This demonstration - that thermodynamic irreversibility is, in Erwin Schrödinger's words, "the pure embodiment of the statistical law itself," and that "events move in the direction in which they are most likely to move" - is one of the great achievements of nineteenth century science [4].

Boltzmann's explanation of the second law has become a cornerstone of statistical mechanics. After the creation of such a major physical theory, and especially after it has been clarified, developed, applied, and taught to generations of aspiring physicists, it tends to acquire a certain air of inevitability. It becomes hard to imagine that there was a time when the theory did not exist, that its concepts did not always seem to be the obvious ones to use in describing the world, that it presented genuine and legitimate difficulties to those contemporary scientists who did not immediately accept it. This process has gone very far in the case, for example, of classical mechanics. Few physicists will really believe that some of the most acute minds of the seventeenth century found Newton's theory of the solar system unacceptable [5], or that it took Newton himself two decades to clarify the concept of force [6]. While statistical mechanics has never acquired that appearance of a completed and unalterable structure which is so characteristic of mechanics or thermodynamics [7], the fact that it had a history, that it did not come into being all at once, is often forgotten.

In this paper I want to discuss one aspect of the history of statistical mechanics - namely, the development of Boltzmann's own views on the meaning of the second law. Although he did not make any attempt to describe this development in his later writings, his interpretation of the second law changed profoundly through the years. That in itself is hardly surprising, but the circumstances in which his views altered and deepened, step by step, are worth exploring. For it was often the pressure of external criticism that forced Boltzmann to re-examine his position and refine his understanding. The vigor of his polemical style sometimes conceals the extent to which he learned from his critics, but those critics - from Rudolf Clausius to Ernst Zermelo - played an essential part in Boltzmann's development.

There is another, and, in some ways, even more important, aspect to the story of Boltzmann's evolving ideas on how to interpret the second law of thermodynamics.

The great unsettled question of late nineteenth century physics was the status of the mechanical world view. For two hundred years physicists had been striving to construct mechanical explanations. In 1690 Christiaan Huygens had referred to "the true Philosophy, in which one conceives the causes of all natural effects in terms of mechanical motions," [8] and two centuries later Heinrich Hertz wrote: "All physicists agree that the problem of physics consists in tracing the phenomena of nature back to the simple laws of mechanics" [9]. But by the time Hertz's words appeared in print in 1894, physicists did not all agree that their goal was the explanation of the natural world in mechanical terms. Mechanism had become suspect for a variety of reasons. Even many of its supporters looked on it with newly critical eyes. Alternatives to mechanism were proposed and hotly debated during the 1890's. No one took a more active and a more central part in these debates than Ludwig Boltzmann, the great defender of the twin flags of mechanism and atomism against the onslaughts of all who would replace them by the standards of energetics, phenomenology, or any other barbaric creed. [10].

These battles were not fought over purely philosophical issues. Boltzmann felt that his most prized scientific accomplishments were as much at stake as his general ideas on physics. He was forced to state and defend his fundamental belief in the mechanical world view, precisely in connection with his explanation of the second law. Boltzmann's most significant and most cogent defense of mechanism came, not in his polemics against Wilhelm Ostwald and the other energeticists, but in his detailed arguments reconciling thermodynamic irreversibility and mechanical reversibility. With the help of statistics he was able to hold on to the mechanical world view, or at least a modified form of it, and could continue his efforts to understand the particulars of how the world worked.

2.   Boltzmann began his paper on "The Mechanical Meaning of the Second Law of Thermodynamics" [11] by referring to the "peculiarly exceptional position" of the second law and to the "roundabout and uncertain methods" by which physicists had tried to establish it. He contrasted it with the first law, whose identity with the law of energy conservation had been known "for a long time already". This "long time" was less than twenty years, but then Boltzmann was only twenty-two himself. He announced his aim in no uncertain terms: "It is the purpose of this article to give a purely analytical, completely general proof of the second law of thermodynamics, as well as to discover the theorem in mechanics that corresponds to it".

The problem Boltzmann had seized upon was a live issue in physics in 1866. Only the year before, in a paper which might well have been the inspiration for Boltzmann's work, Rudolf Clausius had introduced the concept of entropy, restated the second law in its now familiar form, and developed much of the analytical machinery of thermodynamics [12]. Fifteen years had gone by since Clausius first recognized that two basic laws were necessary as foundations for the theory of heat, and after much effort he had now found an extremely compact way of stating the second law [13]. (One feels the impact of those fifteen years in Clausius's remark that "the second law is much harder for the mind to grasp than the first"). In his new formulation the second law consisted of two statements. First, for every thermodynamic system there exists a function S, its entropy, which is a function of the variables characterizing the equilibrium state of the system. The entropy function, S, is defined by the differential relationship,

$$TdS = đQ , \qquad (1)$$

where đQ is the heat supplied to the system in a reversible process and T is the absolute temperature. The notation đQ emphasizes the inexactness of this differential [14]. Second, for irreversible processes one has only the inequality,

$$TdS > đQ . \qquad (2)$$

Boltzmann wanted to derive these results as purely mechanical theorems. This would have meant constructing functions of the coordinates and momenta of the particles composing a suitably general mechanical system, functions that could represent the thermodynamic quantities, temperature and entropy, and the two modes of energy transfer, heat and work. He would also have had to distinguish mechanically between reversible and irreversible processes - that is, to give a purely mechanical characterization of thermodynamic equilibrium. The final stage would have been the proof, from mechanical principles, of a theorem relating these quantities as their thermodynamic equivalents are related by the second law. All this would have been closely parallel to Hermann von Helmholtz's original discussion of what had become the first law of thermodynamics [15].

Although Boltzmann claimed to have done all this, his actual accomplishment was much more modest. He limited his discussion to gases, not surprisingly, but he also made the highly restrictive assumption that the system was strictly periodic. After a time τ the entire molecular configuration would be exactly repeated, so that all molecular orbits had to be closed. Under these circumstances Boltzmann proved a theorem, a generalized form of the principle of least action, on the basis of which he obtained a mechanical counterpart for the entropy, S. His expression had the form,

$$S = \Sigma \ln(T\tau)^2 + \text{constant} \tag{3}$$

where the sum is over all molecules. The absolute temperature, T, is the kinetic energy of one molecule averaged over the period,

$$T = \frac{1}{\tau} \int_0^\tau (\tfrac{1}{2}mv^2) \, dt . \tag{4}$$

Since the period appears explicitly in the entropy equation it seems reasonable to guess that the periodicity assumption could not be dispensed with. Boltzmann tried, nevertheless, to extend his argument to nonperiodic systems, but could only conclude lamely that "if the orbits are not closed in a finite time, one may still regard them as closed in an infinite time" [16]. When it came to the irreversible aspect of the second law, Boltzmann had little to say. His brief discussion of one particular case gave no insight into the molecular basis of irreversibility.

Boltzmann's paper appeared in 1866. Five years later Rudolf Clausius published an article entitled,"On the Reduction of the Second Law of Thermodynamics to General Mechanical Principles" [17]. His title sounds much like Boltzmann's, and for good reason: Clausius had independently discovered the same theorem, although his approach was somewhat different. As one can readily imagine Boltzmann did not hesitate to point out his obvious priority in establishing this mechanical interpretation of the second law [18]. He demonstrated the identity of Clausius's equations and his own, reproducing verbatim some ten pages of his original article to reinforce his claim. His concluding remark did nothing to endear him to Clausius. "I think I have established my priority", he wrote, and then added, "I can, finally, only express my pleasure that an authority with Mr. Clausius's reputation is helping to spread the knowledge of my work on thermodynamics" [19].

Clausius had, of course, overlooked Boltzmann's paper. He had moved from Zürich to Würzburg in 1867 and then to Bonn in 1869, and the "extraordinary demands" on his time and energy prevented him from keeping up with the current literature [20]. While Clausius granted Boltzmann's evident priority for all results common to both papers, he was not convinced that Boltzmann's arguments were as general or as sound as his own, particularly when it came to the distiction between heat and work at the molecular level. Boltzmann did not respond to these criticisms, probably because his work had taken a new direction.

3.   When Boltzmann made his original attempt to understand the second law in mechanical terms, he had apparently not yet read James Clerk Maxwell's early papers on the theory of gases [21], but his ignorance of Maxwell's work did not last long. Boltzmann has described how, during his student years and before he had learned any English, his teacher, Josef Stefan, handed him Maxwell's papers on electromagnetism and an English grammar book [22]. And so by the time Maxwell's articles "On the Dynamical Theory of Gases" appeared [23], Boltzmann was prepared to appreciate them. His study of Maxwell led him directly to a new approach to the study of gases, one in which the molecular distribution function rather than the complete set of molecular variables played a central part.

None of Maxwell's predecessors, not even Clausius, had ever raised the question of how molecular velocities are distributed in a gas. Maxwell not only raised it but immediately showed "that the velocities are distributed among the particles according to the same law as the errors are distributed among the observations in the theory of the 'method of least squares' [24]. Maxwell's two derivations of the velocity distribution and his repeated use of the distribution function in analyzing the transport properties of gases opened a whole range of further studies to his enthusiastic young reader in Vienna.

In 1868 Boltzmann published the first of a series of lengthy memoirs in which he greatly extended Maxwell's results [25]. These are the papers in which Maxwell's distribution law for point molecules was elaborated into the general Maxwell-Boltzmann law for complex molecules in the presence of an external field of force, and in which the theorem of the equipartition of energy also received a corresponding generalization [26]. Boltzmann's stated purpose in these papers was to analyze the nature of thermodynamic equilibrium. The meaning attached to the function giving the law according to which molecular velocities are distributed at equilibrium is obviously critical, but Boltzmann used two such meanings. He spoke of the distribution function as determining the fraction of any suitably long time interval during which the velocity of any particular molecule had values within prescribed limits. He also described it as determining the fraction of the total number of molecules in the gas which had velocities within the prescribed limits at any given moment [27]. Apparently Boltzmann saw no need, at first, for an analysis of the assumed equivalence of these two quite different meanings assigned to the distribution function. He soon realized, however, that a definite hypothesis was called for, a hypothesis that he thought "not improbable" for real bodies composed of atoms moving with "the motion we call heat" [28]. His hypothesis was that, in the course of time, the atomic coordinates and velocities take on all possible values consistent with the fixed total energy of the gas. Years later, in 1911, Paul and Tatyana Ehrenfest gave this its now familiar name of the ergodic hypothesis [29].

By 1871 Boltzmann was ready for a new attack on the problem of understanding the second law [30]. This time his starting point was the very general distribution law he had derived for any system maintained at a given temperature T. If E is the total energy of the system as a function of the coordinates $x_1, y_1, z_1, \ldots, x_r, y_r, z_r$ and velocity components $u_1, v_1, w_1, \ldots, u_r, v_r, w_r$ of the r atoms of the system, one can write E in the form

$$E = \sum_{i=1}^{r} \frac{1}{2} m_i (u_i^2 + v_i^2 + w_i^2) + \chi(x_1,\ldots,z_r) \qquad (5)$$

where χ is the potential energy of the system. The distribution law is given by

$$dp = \frac{\exp(-hE)\, d\omega}{\int \exp(-hE)\, d\omega} \qquad (6)$$

where dp is the probability (in either of the senses equated by the ergodic hypothesis) of finding the atomic coordinates and velocities with the values $x_1,\ldots, w_r$ in the range $d\omega$, which is just the product $dx_1\ldots dw_r$. The integral runs over all allowed values of the variables. Boltzmann had previously argued that the temperature could be set equal to the average kinetic energy of one atom, where the average is over the distribution of equation (6),

$$T = \int \frac{1}{2} m_i (u_i^2 + v_i^2 + w_i^2)\, dp \equiv \langle \frac{1}{2} m_i (u_i^2 + v_i^2 + w_i^2) \rangle \qquad (7)$$

It is worth noting that Boltzmann's T can be read equally well as temperature and as (average) kinetic energy, which is equivalent to saying that he had no need for what we call Boltzmann's constant [31]. The distribution parameter h can be directly evaluated as (3/2T).

The essential point in establishing the second law is to make a clear distinction between heat and work. Although this was not really possible when the system was described directly by its atomic variables, it could be done with the help of the distribution function. Boltzmann examined the average energy $\langle E \rangle$, where

$$\langle E \rangle \equiv \int E\, dp = \frac{3r}{2h} + \langle \chi \rangle. \qquad (8)$$

The average energy can be changed by changing the distribution without otherwise affecting the nature of the system, that is to say, by changing the temperature or the parameter h. One can also change <E> by changing the potential energy χ, which depends on the strength of the external force and which, as Boltzmann showed, can also be considered as depending on the volume. The variation in <E> is given by the expression

$$\delta <E> = -\frac{3r}{2h^2}\delta h + \delta <\chi> . \qquad (9)$$

The first term clearly comes from a temperature change only, but the second includes changes in both the distribution function and in the potential energy function itself. These latter changes, however, are what one means by external work. The heat δQ added to the system can therefore be written as the difference between the energy change $\delta <E>$ and the work done on the system $<\delta\chi>$,

$$\delta Q = \delta <E> - <\delta\chi> = -\frac{3r}{2h^2}\delta h + \delta<\chi> - <\delta\chi> . \qquad (10)$$

This equation can be written more explicitly in the form

$$\delta Q = -\frac{3r}{2h^2}\delta h + \delta[\frac{\int \chi \exp(-h\chi)d\sigma}{\int \exp(-h\chi)d\sigma}] - \frac{\int \delta\chi \exp(-h\chi)d\sigma}{\int \exp(-h\chi)d\sigma} \qquad (11)$$

where $d\sigma$ is just $dx_1...dz_r$. This expression for δQ is not a complete differential but if one divides by the temperature, that is by (3/2h), one can show that an exact differential results, namely,

$$\frac{\delta Q}{T} = \delta[-r\ln h + \frac{2h}{3}<\chi> + \frac{2}{3}\ln(\int \exp(-h\chi)d\sigma) + \text{constant}]. \qquad (12)$$

The function inside the square brackets is the entropy S of the system, and it can be written more compactly in the form,

$$S = \frac{<E>}{T} + \frac{2}{3}\ln \left[\int \exp(-hE) d\omega\right] + \text{constant} . \qquad (13)$$

Boltzmann calculated the entropy of an ideal monatomic gas by this method to show that it did indeed give the familiar thermodynamic result. As a second example he discussed a simple model of a solid, in which each atom is bound to its equilibrium position by a force proportional to its displacement, showing how the Dulong-Petit rule for the specific heat could readily be deduced [32].

4. In 1872, the year after he had published the "analytical proof of the second law of thermodynamics" which I have just described, Boltzmann wrote a long memoir with the uninformative title, "Further Studies on the Thermal Equilibrium of Gas Molecules" [33]. This memoir contains the Boltzmann Equation whose centennial we are celebrating here, but that is only one of its remarkable features. For it was in this memoir that Boltzmann also gave the first derivation of the irreversible increase of entropy on the joint basis of the laws of mechanics and the laws of probability. Even these two major results do not exhaust the riches of this work, as we shall see.

The first problem Boltzmann posed was the proof of the uniqueness of Maxwell's velocity distribution law as a description of the equilibrium state. Maxwell had shown that his distribution was stationary, that is, that it would not change as a result of collisions between the molecules, but Boltzmann was not satisfied with Maxwell's proof that it was the only such distribution. He set out to prove that "whatever the state of the gas may have been initially, it must always approach the limiting distribution found by Maxwell" [34]. Let us look at the structure of his argument for a monatomic gas, considering, as he did, distributions which are spatially uniform and which depend only on the magnitude of the molecular velocity.

On the basis of what he called an "exact consideration of the collision process" [35], Boltzmann derived the basic partial differential equation for the distribution function $f(x,t)$, where $f(x,t)dx$ represents the number of molecules per unit volume whose kinetic energies at time t lie in the interval x to x + dx. This equation, Boltzmann's first version of the Boltzmann Equation, took the form,

$$\frac{\partial f(x,t)}{\partial t} = \int_0^\infty \int_0^{x+x'} [\frac{f(\xi,t)}{\sqrt{\xi}} \frac{f(x+x'-\xi,t)}{\sqrt{x+x'-\xi}} - \frac{f(x,t)}{\sqrt{x}} \frac{f(x',t)}{\sqrt{x'}}]$$
$$\cdot \sqrt{xx'}\ \psi(x,x',\xi)\ dx'd\xi\ . \qquad (14)$$

The function $\psi(x,x',\xi)$ depends on the nature of the intermolecular force, and is defined by the equation giving the number of collisions dn in time τ in which a molecule A having energy between x and x+dx collides with a molecule B having energy between x' and x'+dx' to produce a situation where A has energy between ξ and $\xi+d\xi$,

$$dn = \tau\ f(x,t)dx\ f(x',t)dx'\ d\xi\ \psi(x,x',\xi)\ . \qquad (15)$$

The unfamiliar square roots in the equation arise from Boltzmann's choice of energy rather than velocity as the basic variable.

Boltzmann pointed out that the Maxwell distribution,

$$f(x,t) = f_0(x) = C \sqrt{x} \exp(-hx) \qquad (16)$$

is indeed stationary in the sense that it makes $\partial f/\partial t$ vanish, according to equation (14). To demonstrate the uniqueness of the Maxwell distribution he introduced an auxiliary quantity E (which he later called H), defined by the equation

$$E \equiv \int_0^\infty f(x,t) \{ \ln[\frac{f(x,t)}{\sqrt{x}}] - 1 \} \, dx . \qquad (17)$$

By considering the symmetrical character of the collision process and the possibility of inverse collisions, Boltzmann proved that E could only decrease in time,

$$\frac{dE}{dt} \leq 0 . \qquad (18)$$

Since E cannot go on decreasing to infinity, the distribution function $f(x,t)$ must approach a form for which E has its minimum value and the time derivative of E vanishes. This, he showed, can occur only for the Maxwell distribution.

What he had done up to this point was, as Boltzmann remarked, "only a mathematical trick for giving a rigorous proof of a theorem which had not previously been properly proved" [36]. This "trick" acquired considerable significance, nevertheless, when one realized that the same methods could be applied to a gas whose molecules were arbitrarily complex structures [37]. Here too one could define a quantity E whose evolution in time was strictly monotonic. And, what was even more impressive, this quantity E was proportional (with a negative proportionality constant) to the entropy of the gas in the form given by Boltzmann the previous year. This implied, as Boltzmann wrote, an entirely new approach to proving the second law, an approach that could deal with the entropy increase in irreversible processes as well as with the existence of the entropy as an equilibrium state function.

Boltzmann made these remarks in the middle of a paragraph, a third of the way through his hundred page memoir. He returned to this aspect of his study at the end of the work, where he emphasized the generality of the result [38]. The "complex molecule" in the gas could really be taken to be a macroscopic body interacting with its surroundings. To eliminate any effects of the particular initial phases in the motions of the body and its surroundings, Boltzmann proposed that one consider a large number of identically constructed systems, distributed over their possible phases. The proof already given for polyatomic gases could then be directly carried over to this very general case, and so Boltzmann thought he could at last justifiably claim to have given a complete proof of the second law of thermodynamics.

Before going on to see what became of this claim I must at least point to some of the other subjects that Boltzmann considered in the course of his "Further Studies". He devoted a whole section of his memoir to an alternate derivation of the theory sketched above for a monatomic gas, a derivation he considered to be "much clearer and much more intuitive" [39]. The basic idea was to treat the energy as a discrete variable rather than a continuous one, so that the Boltzmann Equation, equation (14) above, is replaced by a set of ordinary differential equations in time, nonlinear coupled stochastic equations as we might call them. Boltzmann always preferred to think in discrete terms when he could, and argued that this way of clarifying a problem had historical precedents that included Lagrange and Riemann. His general analysis of the relationship between the discrete and the continuous became an important aspect of his defense of atomism [40], but that discussion has no place here. We shall, however, have to look at Boltzmann's later use of his favorite illustrative example, a system whose possible energy values are restricted to the set $0, \varepsilon, 2\varepsilon, \ldots, p\varepsilon$, where p is an integer and ε is a small energy unit.

Only a few pages of Boltzmann's memoir deal with the calculation of the transport properties of gases [41]. It was here, however, that he put the Boltzmann Equation into its more familiar form, where the distribution function f is considered to be a function of molecular velocity and is allowed to vary with position. In other words $f(\vec{r},\vec{v},t)d\vec{v}$ is the number of molecules, per unit volume located at $\vec{r}$, whose velocities lie between $\vec{v}$ and $\vec{v} + d\vec{v}$, at time t. The vector notation is, of course, anachronistic. The basic equation then reads

$$\frac{\partial f}{\partial t} + \vec{v} \cdot \overrightarrow{\text{grad } f} + \frac{\vec{F}}{m} \cdot \overrightarrow{\text{grad}_v f} = \text{collision terms} \quad (19)$$

where $\vec{F}$ is the external force on the molecule of mass m and the operation $\overrightarrow{\text{grad}_v}$ is the gradient taken with respect to the velocity dependence of f. His calculations showed how the coefficients of viscosity, heat conduction and diffusion could be found, using the basic equation as starting point, and that the results agreed with Maxwell's for an inverse fifth power repulsive force between the molecules. Boltzmann also made some cautionary remarks about the greater complexity of the situation for other force laws.

Despite the importance of this aspect of Boltzmann's work I want to return to my main theme - his struggle with the second law of thermodynamics.

5. Boltzmann seems to have been satisfied with the derivation and discussion of the second law in his papers of 1871 and 1872. In any case he turned his attention to other matters, even becoming an experimentalist for a time in order to confirm Maxwell's predicted relationship between the index of refraction and the dielectric constant of any insulator [42]. It was not until 1875 that Boltzmann returned to the theory of gases, showing how his results could be generalized to include the kinetic effects of external forces [43]. This work prompted a critical response from his colleague and former teacher Josef Loschmidt [44]. Loschmidt was troubled by several aspects of Boltzmann's work, even though he was himself a staunch atomist, best remembered for his determination of the number of molecules in a gas. His deepest concern was over the "heat death" of the universe which was often

discussed in connection with the second law, but his specific criticisms dealt with Boltzmann's treatment of the thermal equilibrium of a gas in a constant gravitational field. Boltzmann and Loschmidt debated this subject for several years in the pages of the <u>Wiener Berichte</u>, with Boltzmann repeatedly trying to explain Loschmidt's errors to him [45]. One of Loschmidt's arguments had to be taken very seriously, however, and it forced Boltzmann to rethink the very basis of his proof of the second law.

When he responded to this argument Boltzmann began by paying a handsome tribute to his critic [46]. "In his article on the state of thermal equilibrium of a system of bodies in a gravitational field", Boltzmann wrote, "Loschmidt has stated a proposition which involves a doubt about the possibility of a purely mechanical proof of the second law. Because it is most ingeniously devised and because it seems to me to be of great importance for the proper understanding of the second law, I want to try first to restate it in other words, since Loschmidt's original statement may be difficult for physicists to understand due to its rather philosophical wording".

In Boltzmann's revised formulation Loschmidt's objection seems clear and compelling. His essential point was that one could never derive the irreversible approach to equilibrium and the associated monotonic increase of entropy from the reversible laws of mechanics. These observed characteristics of the natural world must, therefore, depend on the initial conditions that actually prevail. For, suppose one has an isolated mechanical system that proceeds from some initial state (i) at $t = 0$ to another, more mixed or more uniform, state (f) at $t = \tau$. This is an entropy increasing process. Consider, however an initial state (-f) at t = 0, in which all particles have the positions they would have attained at t = τ in the former motion, and also the velocities they would have attained except that the directions of these velocities are reversed. The motion that will ensue in this latter case is just the original motion occurring in the opposite sequence in time, since the laws of mechanics are unaffected by reversing the sign of the time variable. After a time τ the particles will have reached the former initial state except that they will be moving with reversed velocities, (-i). This second process is one that proceeds away from equilibrium to less uniform states and is associated with a decrease in entropy. Because one could always construct such an entropy decreasing process from each entropy increasing process, Loschmidt argued that the second law must depend on the special initial conditions in the world and not on the laws governing molecular motions.

Boltzmann granted that this was a "very seductive" argument, but he thought it was, after all, only an "interesting sophism". Loschmidt was quite correct in asserting that entropy decreasing processes existed, and he had given a clever prescription for finding one class of them. One could not, in fact, prove that entropy increased "with absolute necessity". "The theory of probability itself teaches us that", wrote Boltzmann, "since every molecular distribution, no matter how nonuniform it may be, even if it is exceedingly improbable, is still not absolutely impossible". That one could find improbable situations in which the entropy would decrease did not contradict the fact that for the overwhelming majority of initial states the entropy could be counted on to increase. Loschmidt's objection had only made it clearer than ever "how intimately the second law is related to the theory of probability". But the improbabilities involved in entropy decreasing processes were tantamount to impossibility; one should not expect to find a mixture of nitrogen and oxygen gases separating into the pure gases again a month after they were mixed just because there is a nonzero probability of its taking place.

"One could even calculate the probabilities of the various states," Boltzmann wrote, "from the ratios of the number of ways in which their distributions could be achieved, which would perhaps lead to an interesting method of calculating thermal equilibrium". He developed this remark into a major study a few months later, but before we turn to that some comments are in order.

It seems to me that Boltzmann's reply to Loschmidt's reversibility objection is somewhat disingenuous, or at least misleading. I find no indication in his 1872 memoir that Boltzmann conceived of possible exceptions to the H-theorem, as he later called it. His argument made essential use of the distribution function, to be sure, but his conclusion was presented in absolute form: "It is accordingly rigorously proved that, whatever the initial distribution of kinetic energy may have been, it must always necessarily approach the Maxwellian form after a very long time has elapsed" [47]. Yet Boltzmann gave no indication in his answer to Loschmidt five years later that there had been a change in his views, a deepening in his understanding, as a result of reflecting on Loschmidt's objection. Boltzmann wrote in 1877 as if he were merely elaborating on what was at least implicitly present in the 1872 memoir. It is true that he had been emphasizing the importance of "probability" since 1871, but his own insight into this concept certainly went a lot further after he had come to terms with Loschmidt's criticism [48].

6.   Although Boltzmann did not know it, his new insight into the probabilistic nature of the second law had already been formulated a decade earlier, and it was even available in a popular textbook on the <u>Theory of Heat</u>. It was Maxwell who had first seen the peculiarly nonmechanical character of the second law and expressed it in several striking images. Boltzmann who learned so much from Maxwell, might have learned this too. In a letter dated 11 December 1867 Maxwell wrote to his friend Peter Guthrie Tait that it might be possible to pick a hole in the second law of thermodynamics [49]. One could imagine - or at least Maxwell could imagine - " a finite being who knows the paths and velocities of all the molecules by simple inspection but who can do no work except open and close a hole". Such a being could take advantage of the fact that a gas at any temperature has molecules of all velocities, although the fraction having large velocities increases with increasing temperature. By exercising precisely those powers Maxwell assigned him this being could allow only sufficiently fast molecules to pass from a cold gas to a hot one, and only sufficiently slow ones to go the other way. The net effect would be that "the hot system has got hotter and the cold colder and yet no work has been done, only the intelligence of a very observant and neat-fingered being has been employed".

The point of Maxwell's strange creation, dubbed a "demon" by William Thomson, was to show that exceptions to the second law are possible. Tait showed Maxwell's letter to Thomson - all three were friends - and Thomson added a comment in pencil: "Very good. Another way is to reverse the motion of every particle of the Universe and to preside over the unstable motion thus produced" [50]. Maxwell himself remarked that the "chief end" of the demon was "to show that the 2nd Law of Thermodynamics has only a statistical certainty" [51].

Several years later Maxwell repeated both arguments, his and Thomson's, at greater length in a letter to John William Strutt, the future Lord Rayleigh [52]. "If this world is a purely dynamical system, and if you accurately reverse the motion of every particle of it at the same instant, then all things will happen backwards to the beginning of things, the raindrops will collect themselves from the ground and fly up to the clouds, etc. etc. and men will see their friends passing from the grave to the cradle till we ourselves become the reverse of born, whatever that is. ... The possibility of executing this experiment is doubtful, but I do not think it requires such a feat to upset the 2nd law of thermodynamics".

He went on to repeat the demonic argument, referring to the "doorkeeper, very intelligent and exceedingly quick, with microscopic eyes," and finally pointing to the "Moral": "The 2nd law of thermodynamics has the same degree of truth as the statement that if you throw a tumblerful of water into the sea, you cannot get the same tumblerful of water out again".

When Maxwell wrote his textbook <u>Theory of Heat</u> in 1871, he included a discussion of the demon under the heading "Limitation of the Second Law of Thermodynamics" [53]. The validity of the second law was statistical, "drawn from our experience of bodies consisting of an immense number of molecules". There was no reason to expect it to be "applicable to the more delicate observations and experiments which we may suppose made by one who can perceive and handle the individual molecules which we deal with only in large masses". Maxwell was certainly ready to admit the existence of entropy decreasing processes, if one looked at systems composed of only a small number of molecules. He summed matters up in a review in 1878 [54]: "Hence the second law of thermodynamics is continually being violated, and that to a considerable extent, in any sufficiently small group of molecules belonging to a real body. As the number of molecules in the group is increased, the deviation from the mean of the whole becomes smaller and less frequent; and when the number is increased till the group includes a sensible portion of the body, the probability of a measurable variation from the mean occurring in a finite number of years becomes so small that it may be regarded as practically an impossibility". Small wonder that Maxwell flatly rejected attempts "to deduce the second law from purely dynamical principles ... without the introduction of any element of probability ... no deduction of this kind, however apparently satisfactory can be a sufficient explanation of the second law."

Still another of Boltzmann's contemporaries saw this point very clearly. Josiah Willard Gibbs's memoir "On the Equilibrium of Heterogeneous Substances" is usually considered to be the ultimate formulation of classical thermodynamics in its "naked and pure" form, free of any "hankering or yearning after mechanics" and atomism, as Georg Helm, a major spokesman of the energeticists described it [55]. But in the midst of his thermodynamics Gibbs turned up the paradox on diffusion which still bears his name. His discussion of this apparent paradox contains the same analysis of the possibilities inherent in the reversible laws of mechanics applied to molecular motions that Thomson, Maxwell and Loschmidt had seen. He concluded by remarking: "In other words, the impossibility of an uncompensated decrease of entropy seems to be reduced to improbability" [56]. Years later Boltzmann used this sentence as the motto for the second volume of his Lectures on the Theory of Gases, happy to have such a succinct statement of the view that he had made so much his own. For it was Boltzmann, and not Maxwell or Gibbs, who worked out precisely how the second law was related to probability, creating the subject of statistical mechanics.

7.   In his answer to Loschmidt, Boltzmann had suggested that one might develop "an interesting method of calculating thermal equilibrium" by directly determining the probabilities of the states of a thermodynamic system. The long memoir in which this fruitful suggestion was worked out proved to be the culmination of his studies of probability and the second law [57]. Boltzmann had already admitted that the second law had only a statistical certainty, just as Maxwell had argued earlier. But he was now ready to claim that the second law is a direct expres-

sion of the laws of probability: that the entropy of a state measures its probability and that entropy increases because systems evolve from less to more probable states. Let us examine the structure of Boltzmann's argument justifying these assertions.

He began with his favorite discrete model, "an unrealizable fiction" which could, nevertheless, be used to bring out the essential ideas. This was a collection of N particles whose individual energies were restricted to the set $0, \varepsilon, 2\varepsilon, \ldots, p\varepsilon$, in other words to a finite set of integral multiples of an energy unit ε. The total energy of this "gas" was fixed and equal to $\lambda\varepsilon$, where λ is also an integer. The molecular distribution would be given by the set of numbers $w_0, w_1, w_2, \ldots, w_p$, where w_r is the number of particles that have energy $r\varepsilon$.

All previous studies of the molecular distribution had been based on an analysis of how it changed with time, that is to say, with how it was modified as a result of molecular collisions. In this paper, however, Boltzmann abandoned his kinetic approach. He set out to determine the probability of a distribution in a way "completely independent of whether or how that distribution has come about" [58]. The new method was to be a direct counting of the number of distinct ways in which the distribution could be realized, a combinatorial approach that completely bypassed all questions of kinetics or collision mechanisms.

For the discrete example the method was straightforward. A complete specification of the molecular state, or a complexion as Boltzmann called it, would require a listing of the energy of each individual molecule. One would have to be able to say, for example, that molecule number one has energy 3ϵ, molecule number two has energy 9ϵ, and so on. Evidently a molecular distribution, where one specifies only the numbers w_o, $w_1,\ldots,$ w_p, is generally compatible with a number of different complecions. This number, P, the permutability measure, or number of complexions for the given distribution, is given by the equation

$$P = \frac{N!}{w_o!\, w_1!\ldots w_p!} \qquad (20)$$

Boltzmann took the quantity P as proportional to the probability of the distribution $\{w_o, w_1,\ldots, w_p\}$. More exactly, he set the probability W equal to the ratio of P for the given distribution to the sum of the P values for all allowed distributions, that is all distributions compatible with the fixed total number of particles

$$\sum_{r=0}^{p} w_r = N \qquad (21)$$

and the fixed total energy

$$\sum_{r=0}^{p} r\epsilon w_r = \lambda\epsilon . \qquad (22)$$

When he took P, the number of complexions compatible with the given distribution $\{w_o, w_1,\ldots w_p\}$ as a measure of the probability of that distribution, Boltzmann emphasized that he was assuming that any particular complexion was as likely to occur as any other [59]. Whatever the collision mechanism it would not introduce any special preferences; it would be essentially random. (To quote Boltzmann's example, the drawing of the successive integers 1, 2, 3, 4, 5 in a game of Lotto is no more and no less likely than the drawing of any other specific, ordered set of five integers).

The most probable distribution is the one for which P is largest. It is determined by finding the set $\{w_0, w_1, \ldots w_p\}$ which makes P a maximum subject to the restrictions expressed in equations (21) and (22). When the numbers involved are large enough to justify using Stirling's approximation for the factorials the resulting most probable distribution can be adequately represented by an exponential; w_r is proportional to $\exp(-\beta r\varepsilon)$ in the most probable distribution, where β is a parameter determined by the energy constraint.

Boltzmann went on to treat the case of the gas, using the same method that he had introduced for the special model. To calculate the permutability measure he had to introduce a discreteness artificially, and then remove it by a limiting process, going over from sums to integrals [60]. In finding the most probable distribution for the discrete model Boltzmann had actually maximized lnP rather than P itself, using the equation

$$\ln P = -\sum_r w_r \ln w_r + \text{constant} \qquad (23)$$

which follows from equation (20), when Stirling's approximation is valid; that is, $\ln(w_r!) \simeq w_r \ln w_r - w_r$. The equivalent expression for a simple gas whose molecules can have any velocities became

$$\ln P = -\int \ldots \int f \ln f \, dx\,dy\,dz\,du\,dv\,dw + \text{constant} \qquad (24)$$

where f is the molecular distribution function. That is,

$$f(x,y,z,u,v,w) \, dxdydzdudvdw$$

is the number of molecules whose coordinates lie between x and x + dx, y and y + dy, z and z + dz, and whose velocity components lie in the intervals u to u + du, v to v + dv, w to w + dw. He now found the most probable distribution by maximizing the lnP of equation (24) subject to the constraints on the number of particles and the total energy. His basic result was that the most probable distribution is the Maxwell-Boltzmann distribution, the already familiar description of the equilibrium state, now derived from a completely new starting point. The equilibrium distribution was not only the unique stationary distribution as Boltzmann had shown in 1872, it was also the distribution most likely to occur, in the sense that it could be achieved in the largest number of ways.

In deriving this result Boltzmann pointed out that one has to be careful in assigning the underlying a priori probabilities or statistical weights [61]. Each complexion of the discrete model had been assigned equal probability, but the analogous step for the gas was less obvious. The expression for lnP in equation (24), which Boltzmann actually used, is based on the assumption that equal weights (or a priori probabilities) are assigned to equal volumes in the molecular phase space, whose axes are x,y,z,u,v, and w. One could, however, imagine instead assigning equal weights to equal energy intervals, but this would lead to a most probable distribution different from the Maxwell-Boltzmann law. (Boltzmann demonstrated this false result before deriving the correct one, perhaps to produce the kind of dramatic effect he admired so much in Maxwell's writing). The guiding principle for selecting the correct weight function was Liouville's theorem, as Boltzmann showed.

The logarithm of the permutability measure P rather than P itself had been used as the function to be maximized simply for technical reasons, but lnP turned out to have a real significance of its own. Boltzmann evaluated lnP explicitly for a simple gas in equilibrium, where f is the Maxwell distribution, and he showed that it was essentially the same as the entropy of the gas as calculated directly with the help of thermodynamics [62]. The two functions differed only by a definite scale factor, and an undetermined additive constant of no thermodynamic significance. The function lnP was also the same, apart from sign, as Boltzmann's function E, whose monotonically increasing behavior as a result of collisions he had demonstrated in 1872 [63].

Boltzmann emphasized that lnP was well defined whether or not the system is in equilibrium, so that it could serve as a suitable generalization of the entropy. He then asserted the following general theorem: "Suppose we are given an arbitrary system of bodies which undergo an arbitrary change of state; neither the initial nor the final state is necessarily an equilibrium state. Then the total permutability measure of all bodies will continually increase in the course of the change of state and can at best remain constant so long as all the bodies are infinitely close to equilibrium throughout the change of state (reversible change of state)[64]". This was Boltzmann's reformulation of the second law as the statement that systems go from less to more probable states, a statement he discussed at length and applied to diffusion processes the following year.

8. Boltzmann once again felt he had settled the problem of the foundations of the second law. He wrote nothing on the subject for a number of years, returning to it only occasionally to elaborate particular points, until he was challenged by a barrage of criticism in the 1890's. That criticism, to which I shall return shortly, showed that Boltzmann had not made his fundamental ideas clear to the physicists who read his work. Since Boltzmann was considered to be a splendid teacher [65] and obviously devoted much effort to his writing, one wants to know why there was this lack of comprehension.

One reason was the sheer quantity of his scientific writing. His memoirs appeared at the rate of four or five a year and some of the most important ones ran to fifty, eighty, or a hundred pages. Maxwell, who should have been Boltzmann's most interested and most perceptive reader, seems to have lost patience early. "By the study of Boltzmann I have been unable to understand him", he wrote to Tait in 1873 [66]. "He could not understand me on account of my shortness, and his length was and is an equal stumbling-block to me. Hence I am very much inclined to join the glorious company of supplanters and to put the whole business in about six lines". Maxwell apparently never read any of the papers Boltzmann wrote after about 1870.

A more profound reason why Boltzmann's ideas were not widely understood was the way in which he changed his point of view without informing his readers. We have already seen this in his response to Loschmidt. The remark also applies to the paper just discussed, in which Boltzmann made a major change in his use of the concept, probability. In his earlier studies he had always been concerned with the statistics of the molecule. The question had always been, what is the probability that a molecule has such and such a property, with the answer determined by the molecular distribution function f. In his new analysis Boltzmann was concerned with the statistics of the gas as a whole. The question was now, what is the probability that the gas is in a state characterized by a certain distribution, and the answer was determined by the permutability measure P.

This fundamental change in approach must have puzzled Boltzmann's contemporaries. Even the kinetic theorists in England who usually followed Boltzmann's work ignored the 1877 memoir in their books on the theory of gases [67]. No real clarification came until the Ehrenfests' <u>Encyklopädie</u> article in 1911 [68]. They described Boltzmann's shift from a theory using the 6 dimensional phase space (μ-space) of a single molecule, to a theory using the 6N dimensional phase space (Γ-space) of the whole gas of N particles; it was a shift from a theory emphasizing kinetics and based on the special assumptions about collisions underlying the Boltzmann Equation [69], to a theory emphasizing combinatorial statistics and independent of collision analysis. But the Ehrenfests also had to deal with the fact that Boltzmann had elaborated his views again in the 1890's in response to his critics, so that his works were indeed "very far from presenting a systematic treatment" [70].

9. There was another difficulty that interfered with general acceptance of the whole kinetic theory of gases, and this one had nothing to do with the problems of read-

ing and understanding Boltzmann. I refer to the uncertain status of the theorem of the equipartition of energy, a result anticipated by Clausius and developed by both Maxwell and Boltzmann [71]. One consequence of this theorem is that c_v, the molar heat capacity of a gas measured at constant volume, should have the value

$$c_v = \frac{1}{2} nR \qquad (25)$$

where R is the gas constant and n is an integer equal to the number of quadratic terms in the expression for the energy of a single molecule. This means that the ratio γ of the heat capacity at constant pressure to that at constant volume would also have to be determined by n,

$$\gamma = \frac{n + 2}{n} . \qquad (26)$$

For a molecule which is a simple particle γ should be 5/3, whereas a rigid molecule with six degrees of freedom should have γ equal to 4/3. All the common elementary gases known to Clausius and Maxwell had a γ very close to 1.4, however, which led Maxwell to remark at the end of his first paper on the kinetic theory that he had "proved that a system of such [rigid] particles could not possibly satisfy the known relation between the two specific heats of all gases" [72].

If one also remembered that molecules must have a complex internal structure in order to produce the complicated assortment of spectral frequencies that were observed, matters became even worse. It was indeed "the greatest difficulty which the molecular theory has yet encountered", as Maxwell wrote in 1875 [73]. It is true that the heat capacity of mercury vapor, a monatomic gas, was measured the following year and that its γ was 5/3, just as the theory predicted for a point molecule [74]. Boltzmann immediately proposed that the diatomic molecules of common gases might be rigid linear structures constructed out of two point atoms a fixed distance apart, or perhaps such molecules might be smooth ellipsoids of revolution, whose axial rotation would be unaffected by collisions [75]. But Boltzmann's suggestions did not satisfy other physicists and the equipartition difficulties continued to be one of the two "Nineteenth Century Clouds over the Dynamical Theory of Heat and Light" [76].

There was enough interest in these difficulties in England so that the British Association for the Advancement of Science appointed a committee to report on "The Present State of Our Knowledge of Thermodynamics". The committee consisted of Joseph Larmor and George Bryan; Bryan's lengthy report was submitted in two parts, appearing in 1891 and 1894. The first report [77] covered the various attempts to relate the second law to the principles of dynamics, which included the early work of Boltzmann and Clausius as well as the later statistical theories. Bryan also analyzed the monocycle analogies introduced by Helmholtz in 1884 and discussed and generalized at length by Boltzmann [78]. Bryan's second report, on "The Laws of Distribution of Energy and Their Limitations" [79] was presented to the British Association at its annual meeting which, in 1894, was held at Oxford. Boltzmann himself was there, presenting two short papers and taking part in the discussion which followed Bryan's report. He must have enjoyed himself immensely. He had an audience that was really interested in the kinetic theory of gases and generally sympathetic to his point of view. Many of its members were well prepared to argue technical details as well as general principles, and generally to carry on a vigorous debate. No wonder Boltzmann referred to it as "the unforgettable meeting of the British Association at Oxford" [80].

There was much more to be said than would fit into the scheduled hours of the meeting, and so the discussion was carried on in the columns of Nature during much of the following year, with half a dozen correspondents contributing twenty or so Letters to the Editor [81]. Two main points came up over and over. The first was the difficulty in reconciling the specific heats of gases with the obvious existence of the many internal molecular motions that gave rise to spectra, if the equipartition theorem were to be maintained. Despite a variety of suggestions involving the aether, this difficulty remained completely unresolved. The second issue that was debated at length was the apparent paradox involved in explaining the irreversible second law on the basis of a reversible mechanics, the paradox that Loschmidt had raised and Boltzmann had answered almost twenty years earlier.

By 1894, however, Boltzmann was prepared to deal with this criticism even more effectively, and his discussion [82] brought out one aspect of the equilibrium distribution never previously emphasized. His analysis of 1877 had shown that the Maxwell-Boltzmann distribution was the most probable distribution. It was actually by far the most probable distribution in the sense that any distribution whose form was appreciably different from Maxwellian had a probability negligibly small compared to that of the Maxwell distribution. The probabilities of the distributions had been determined by their respective permutability measures, but Boltzmann now reinterpreted probability once again. He took the probability of a distribution to be the fraction of any sufficiently long time interval during which one could expect to find the gas described by this distribution. This meant that if one examined H (the negative of the entropy, formerly called E by Boltzmann), it would have a value at or very near its minimum value almost all the time, since the minimum occurs for the Maxwell distribution and that distribution is by far the most probable. If H were found with a value appreciably larger than its minimum, one could be almost certain that it would decrease, whether one followed it forward or backward in time. In other words any serious departure of H from its minimum value was most likely to show up as a spike or local maximum in the curve for H as a function of time. Thus in a time-reversed motion of the kind proposed by Loschmidt, H would indeed increase for a while, back to its initial value in the original, unreversed motion, but it would then almost surely decrease to its minimum value. Following up a remark by Edward P. Culverwell [83], Boltzmann compared the H-curve to a succession of inverted trees, mostly very low and with horizontal or nearly horizontal branches, occasionally higher with inclined branches. "The improbability of such a tree increases enormously with its height," Boltzmann concluded [84].

10. Only a year after the Oxford meeting, Boltzmann was involved again in public debate. This time he travelled to Lübeck for the September 1895 meeting of the Ger-

man Society of Scientists and Physicians [85]. The subject for debate was the view of physical theory held by the school of energeticists, presented at Lübeck by Ostwald and Helm. Boltzmann was there to defend his own position. This time, however, it was not the latest ideas in the kinetic theory that needed explaining to a sympathetic audience, but the mechanistic world view itself that needed protecting against a basically hostile attitude. Ostwald and his colleagues insisted that science should abandon the attempt to understand the world in mechanical terms, and that it should give up atomism as well, in favor of a new world view in which energy would be the only fundamental concept. Boltzmann's discussion seems to have persuaded most of the audience: "The arguments of Boltzmann broke through", as Arnold Sommerfeld described it. The details of the debate and of Boltzmann's subsequent writings against the position of the energeticists would be out of place here. What is relevant is Boltzmann's embattled position in defense of the approach to physics that had been the basis of his whole scientific life.

Early in 1896 the <u>Annalen der Physik</u> printed yet another attack on Boltzmann's position [86]. This one came from Ernst Zermelo, a young mathematician in Berlin, who was serving as Max Planck's assistant in the Institute for Theoretical Physics. Planck had already expressed negative views about the reduction of the second law to a theorem in statistical mechanics, and had stated his preference for an approach based only on the two laws themselves. Although he, too, criticized the blunders and inadequacies in the work of the energeticists, Planck did not ally himself with Boltzmann [87]. And now Planck's assistant, Zermelo, had launched a new criticism of Boltzmann's views.

Zermelo based his discussion on a theorem proved a few years earlier by Henri Poincaré [88]. This theorem asserted that, in any system of particles acting on each other by arbitrary forces which depend only on their positions in space, any configuration (specified by the coordinates and velocities of the particles) will recur within arbitrarily specified limits, and will recur infinitely often. The theorem requires only that the coordinates and velocities are bounded, and it holds for almost all initial configurations. Zermelo argued that Poincaré's theorem automatically excluded irreversibility for mechanical systems. The entropy of a system would have to return to its initial value infinitely often, and no state of thermodynamic equilibrium could ever be attained. The only way out, according to Zermelo, would be to make the unappealing assumption that natural systems are always found in those singular configurations to which Poincaré's theorem does not apply. Since these singular configurations constitute a set of measure zero this was hardly an acceptable solution. Zermelo's conclusion was harsh: physics would have to choose between the second law of thermodynamics and the mechanical interpretation of nature.

Boltzmann replied immediately and rather sharply [89]. He repeated yet another time that what was at stake was not a purely mechanical explanation of the second law, but an explanation making essential use of the laws of probability. "Now Mr. Zermelo's article shows that my papers on this subject have not been understood", Boltzmann went on bitterly, "but I am pleased with his article in spite of that because it is the first evidence that these papers have received any attention at all in Germany".

He immediately conceded that Poincaré's theorem was "obviously correct", but denied the conclusions Zermelo had drawn from it. Once again, Boltzmann described schematically how the H-function must vary with time. (In his second reply to Zermelo, Boltzmann included a sketch showing how H was either in the immediate neighborhood of its minimum or else it was near a local maximum [90]. He warned his readers, however, that the sketch must be taken "sehr cum grano"). The point Boltzmann stressed most heavily was the typical or representative character of the Maxwell distribution. Zermelo thought that only a very few singular initial conditions would allow the system to approach the equilibrium distribution. Quite the contrary, replied Boltzmann. The system will almost always have an essentially Maxwellian distribution, although certain peculiar initial conditions (such as the molecules all moving in parallel paths perpendicular to the walls), will not allow the system to approach equilibrium.

As for the recurrence itself, it did not bother Boltzmann at all. He simply pointed out that the time necessary for a return to the initial state would be inconceivably long. As usual he had a nice illustrative example. The time in seconds for a gas of 10^{18} particles to return to its initial state within specified intervals, would be a number having some 10^{18} digits, according to Boltzmann's calculation. To allow one to appreciate the size of this number, he remarked that if every star visible through the best telescopes had as many planets as does the sun, and if every planet had as many inhabitants as the earth, each inhabitant living for 10^{18} years, then the total number of seconds in _all_ these lives would be a number with less than 50 digits. In other words the recurrence times for macroscopic systems were so long that one would not expect ever to experience such a recurrence. The apparent irreversibility of the world is therefore not at all inconsistent with the existence of Poincaré recurrences, because their time scale is so inordinately long. The recurrence times for sufficiently small systems might, however, well be observable, if such systems could ever be studied experimentally.

Boltzmann saw no reason to give up statistical mechanics in particular or the mechanical world view in general on the strength of Poincaré's demonstration of a general quasi-periodicity. He compared Zermelo's willingness to do so to the attitude of a dice player who claims a die is false just because he has never thrown one thousand successive ones, even though he had a nonzero probability of doing so. And he concluded with another sharp rebuke to Zermelo: "All the objections raised against the mechanical world view are therefore aimless and are based on erroneous ideas. But anyone who is incapable of overcoming the difficulties involved in obtaining a clear understanding of the theorems of gas theory ought indeed to take Mr. Zermelo's advice and decide to give it up completely" [91].

11. Boltzmann never wrote a systematic exposition of the statistical mechanics he created. His <u>Lectures on the Theory of Gases</u> appeared in two parts, in 1896 and 1898, but this book was written with other goals in mind. The first volume emphasizes kinetic theory and transport phenomena, in accordance with Boltzmann's attempt "above all to make clearly comprehensible the path-breaking works

of Clausius and Maxwell" [92]. There is, to be sure, a brief discussion of the H-theorem which includes a few pages on Boltzmann's combinatorial method and its consequences. This appears under the misleading heading, "Mathematical meaning of the quantity H" [93].

The second volume of the <u>Gas Theory</u> was delayed for a while. Boltzmann discovered that the parenthetic remarks on mechanics that he wanted to include kept expanding. They took up "first a whole paragraph, then a section", and he finally decided to publish a separate book on mechanics. Boltzmann made use of lectures he had prepared but then put aside, and so described his book as "Lectures on Mechanics Not Delivered at the University of Vienna" [94].

His plans for the second volume on gases had originally not included the most subtle and difficult aspects of the subject, but events made him change his mind. In the Foreword he wrote in August 1898 for the second volume [95] Boltzmann commented that: " It was just at this time [1896] that the attacks on the theory of gases began to increase". Which of the many attacks he had in mind is not quite clear. The energeticists were still writing against mechanical physics and molecular theories despite their defeat at Lübeck. Boltzmann's two exchanges with Zermelo took place in 1896, with two supplementary papers written in 1897 and 1898 [96]. In 1897 Planck published his <u>Treatise on Thermodynamics</u>, noting in his Preface that "obstacles, at present insurmountable" stood in the way of further progress in the kinetic theory, and referring to the "essential difficulties ... in the mechanical interpretation of thermodynamics", presumably the difficulties found by

Zermelo [97]. Planck had made similar negative remarks about Boltzmann's statistical theory of irreversibility earlier that year in a paper proposing a new electromagnetic foundation for the second law [98].

In any event Boltzmann sounded shaken by the attacks. " I am conscious of being only an individual struggling weakly against the stream of time", he wrote. "But it still remains in my power to contribute in such a way that, when the theory of gases is again revived, not too much will have to be rediscovered" [99]. In this mood Boltzmann decided to add to his already promised discussion of the van der Waals theory, polyatomic gases, and dissociation, "the parts that are most difficult and most subject to misunderstanding". This material is, presumably, what he included in his final supplementary chapter, which deals once again with the objections that had been raised by Zermelo and others, but only in a fragmentary way.

Boltzmann's last work, the article on "Kinetic Theory of Matter" for Volume V of the Encyklopädie der mathematischen Wissenschaften, referred only briefly to the fundamentals of the statistical theory [100]. We know however, that Boltzmann had promised to write a separate article in Volume IV of the Encyklopädie, the volume on Mechanics edited by Felix Klein. Klein had extracted Boltzmann's promise to write on the foundations of statistical mechanics by threatening to give the job to Zermelo if he refused [101]. This article remained unwritten at Boltzmann's death and Klein passed the task along to the Ehrenfests [102]. One wonders what Boltzmann would have written, whether he would have presented the structure of the subject in the same way that his former student did, emphasizing the ambiguities, shifts of meaning, unstated assumptions, and unproven assertions in his many papers. Boltzmann would surely have approved the Ehrenfests' statement that there were no basic inconsistencies in statistical mechanics - that all objections were essentially misguided - and he would have enjoyed the urn model with which they clarified some of his own ideas [103]. But most of all Boltzmann would have rejoiced in the twentieth century's justification of his unswerving faith in the truth of his statistical mechanics.

NOTES

[1] Ludwig Boltzmann, "Über die mechanische Bedeutung des zweiten Hauptsatzes der Wärmetheorie", Wiener Berichte 53 (1866), 195. Reprinted in L. Boltzmann, Wissenschaftliche Abhandlungen, ed. F. Hasenöhrl (Leipzig, 1909), 1, 9-33. This collection will be referred to as Wiss. Abh. and page references to Boltzmann's work will be made to this edition.

[2] L. Boltzmann and J. Nabl, "Kinetische Theorie der Materie", Encyklopädie der mathematischen Wissenschaften, Vol.V, Part 8 (Leipzig, 1907).

[3] See Engelbert Broda, Ludwig Boltzmann: Mensch, Physiker, Philosoph (Berlin, 1955) and René Dugas, La théorie physique au sens de Boltzmann et ses prolongements modernes (Neuchâtel, 1959).

[4] Erwin Schrödinger, "The Statistical Law in Nature", Nature 153 (1944), 704.

[5] See, for example, E.J. Dijksterhuis, The Mechanization of the World Picture, transl. C. Dikshoorn (Oxford, 1961), p. 479.

[6] John Herivel, The Background to Newton's "Principia" (Oxford, 1965).

[7] George E. Uhlenbeck, "Structure of Statistical Mechanics", Proceedings of the 1966 Midwest Conference on Theoretical Physics (Bloomington, 1966), p. 1.

[8] Christiaan Huygens, Treatise on Light, transl. S.P. Thompson (London, 1912), P. 3.

[9] Heinrich Hertz, *The Principles of Mechanics Presented in a New Form*, transl. D.E. Jones and J.T. Walley (London, 1899), Author's Preface.

[10] L. Boltzmann, *Populäre Schriften* (Leipzig, 1905). Also see Broda and Dugas, *op. cit.*, note 3.

[11] L. Boltzmann, *op. cit.*, note 1.

[12] Rudolf Clausius, "Über verschiedene für die Anwendung bequeme Formen der Hauptgleichungen der mechanischen Wärmetheorie", *Pogg. Ann.* 125 (1865), 353.

[13] See Martin J. Klein, "Gibbs on Clausius", *Historical Studies in the Physical Sciences*, ed R. McCormmach (Philadelphia, 1969) 1, 127.

[14] This notation was introduced by Carl Neumann in his *Vorlesungen über die mechanische Theorie der Wärme* (Leipzig, 1875), p. ix.

[15] Hermann von Helmholtz, *Über die Erhaltung der Kraft* (Berlin, 1847).

[16] L. Boltzmann, *op. cit.*, note 1, p. 30. Boltzmann later found other and stronger arguments against the applicability of this purely mechanical interpretation to nonperiodic systems. See his "Bemerkungen über einige Probleme der mechanischen Wärmetheorie", *Wiener Berichte* 75 (1877), 62, especially Section III, "Bemerkungen zur mechanischen Bedeutung des zweiten Hauptsatzes der Wärmetheorie", *Wiss. Abh.* 2, 122-148.

[17]   R. Clausius, "Über die Zurückführung des zweiten Hauptsatzes der mechanischen Wärmetheorie auf allgemeine mechanische Principien", Pogg. Ann. 142 (1871), 433.

[18]   L. Boltzmann, "Zur Priorität der Auffindung der Beziehung zwischen dem zweiten Hauptsatze der mechanischen Wärmetheorie und dem Prinzip der kleinsten Wirkung", Pogg. Ann. 143 (1871), 211; Wiss. Abh. 1, 228-236.

[19]   Ibid., p. 236.

[20]   R. Clausius, "Bemerkungen zu der Prioritätsreclamation des Hrn. Boltzmann", Pogg. Ann. 144 (1871), 265.

[21]   See L. Boltzmann, op. cit., note 1, especially pp. 20-23.

[22]   L. Boltzmann, Populäre Schriften, p. 96.

[23]   James Clerk Maxwell, "On the Dynamical Theory of Gases", Phil. Trans. Roy. Soc. 157 (1867), 49. Reprinted in The Scientific Papers of James Clerk Maxwell, ed. W.D. Niven (Cambridge, 1890), 2, 26-78.

[24]   J.C. Maxwell, "Illustrations of the Dynamical Theory of Gases", Phil. Mag. 19 (1860), 19; 20 (1860), 21,33. See Scientific Papers 1, 382.

[25]   L. Boltzmann, "Studien über das Gleichgewicht der lebendigen Kraft zwischen bewegten materiellen Punkten", Wiener Berichte 58 (1868), 517; Wiss. Abh. 1, 49-96.

[26] (a) L. Boltzmann, "Über das Wärmegleichgewicht zwischen mehratomigen Gasmolekülen", Wiener Berichte 63 (1871), 397; Wiss. Abh. 1, 237-258.
(b) L. Boltzmann, "Einige allgemeine Sätze über Wärmegleichgewicht", Wiener Berichte 63 (1871), 679; Wiss. Abh. 1, 259-287.

[27] L. Boltzmann, op. cit., note 25, p. 50.

[28] L. Boltzmann, op. cit., note 26(b), p. 284.

[29] (a) Paul and Tatyana Ehrenfest, "Begriffliche Grundlagen der statistischen Auffassung in der Mechanik", Encyklopädie der mathematischen Wissenschaften, Vol. IV, Part 32 (Leipzig, 1911). Reprinted in Paul Ehrenfest, Collected Scientific Papers (Amsterdam, 1959), pp. 213-300. English translation by M.J. Moravcsik, The Conceptual Foundations of Statistical Mechanics (Ithaca, 1959).
(b) Stephen G. Brush, "Foundations of Statistical Mechanics 1845-1915", Archive for History of Exact Sciences 4 (1967), 145-183.
(c) Hannelore Bernhardt, "Über die Entwicklung und Bedeutung der Ergodenhypothese in den Anfängen der statistischen Mechanik", N.T.M. 8 (1971), 13-25.

[30] L. Boltzmann, "Analytischer Beweis des zweiten Hauptsatzes der mechanischen Wärmetheorie aus den Sätzen über das Gleichgewicht der lebendigen Kraft", Wiener Berichte 63 (1871), 712; Wiss. Abh. 1, 288-308.

[31] Max Planck properly claimed credit for introducing this constant. See his Wissenschaftliche Selbstbiographie (Leipzig, 1948).

[32] Edward E. Daub, "Probability and Thermodynamics: The Reduction of the Second Law", *Isis* 60 (1969), 318-330.

[33] L. Boltzmann, "Weitere Studien über das Wärmegleichgewicht unter Gasmolekülen", *Wiener Berichte* 66 (1872), 275; *Wiss. Abh.* 1, 316-402.

[34] *Ibid.*, p. 320.

[35] *Ibid.*, p. 324.

[36] *Ibid.*, p. 345.

[37] Boltzmann's proof required correction as H.A. Lorentz pointed out.
(a) See H.A. Lorentz, "Über das Gleichgewicht der lebendigen Kraft unter Gasmolekülen", *Wiener Berichte* 95 (1887), 115. L. Boltzmann, "Neuer Beweis zweier Sätze über das Wärmegleichgewicht unter mehratomigen Gasmolekülen", *Wiener Berichte* 95 (1887), 153; *Wiss. Abh.* 3, 272-282.
(b) See also L. Boltzmann, *Lectures on Gas Theory*, transl. S.G. Brush (Berkeley, 1964), pp. 49-55, 412-443. Richard C. Tolman, *The Principles of Statistical Mechanics* (Oxford, 1938), pp. 99-165.

[38] L. Boltzmann, *op. cit.*, note 33, p. 401.

[39] *Ibid.*, p. 346.

[40] See L. Boltzmann, *Populäre Schriften*, pp. 141-161. Also see R. Dugas, *op. cit.*, note 3, pp. 25-29, 101-109; E. Broda, *op. cit.*, note 3, pp. 41-49.

[41] L. Boltzmann, *op. cit.*, note 33, p. 361-369.

[42] L. Boltzmann, <u>Wiss. Abh.</u> <u>1</u>, 403-615.

[43] L. Boltzmann, "Über das Wärmegleichgewicht von Gasen, auf welche äußere Kräfte wirken", <u>Wiener Berichte</u> <u>72</u> (1875), 427; <u>Wiss. Abh.</u> <u>2</u>, 1-30.

[44] Josef Loschmidt, "Über den Zustand des Wärmegleichgewichtes eines Systems von Körpern mit Rücksicht auf die Schwerkraft", <u>Wiener Berichte</u> <u>73</u> (1876), 139.

[45] See the extensive discussion in R. Dugas, <u>op. cit.</u>, note 3, pp. 158-184. Also L. Boltzmann, "Über die Aufstellung und Integration der Gleichungen, welche die Molekularbewegung in Gasen bestimmen", <u>Wiener Berichte</u> <u>74</u> (1876), 503; <u>Wiss. Abh.</u> <u>2</u>, 55-102. "Weitere Bemerkungen über einige Probleme der mechanischen Wärmetheorie", <u>Wiener Berichte</u> <u>78</u> (1878), 7, especially Section II, "Über das Wärmegleichgewicht in einem schweren Gas", <u>Wiss. Abh.</u> <u>2</u>, 264-288.

[46] L. Boltzmann, "Bemerkungen über einige Probleme der mechanischen Wärmetheorie", <u>Wiener Berichte</u> <u>75</u> (1877), 62, especially Section II, "Über die Beziehung eines allgemeinen mechanischen Satzes zum zweiten Hauptsatz der Wärmetheorie", <u>Wiss. Abh.</u> <u>2</u>,116-122.

[47] L. Boltzmann, <u>op. cit.</u>, note 33, p. 345.

[48] See R. Dugas, <u>op. cit.</u>, note 3, pp. 185-191. Also see H. Bernhardt, "Der Umkehreinwand gegen das H-Theorem und Boltzmanns statistische Deutung der Entropie", <u>N.T.M.</u> <u>4.</u> (1967), No.10, 35-44.

[49] J.C. Maxwell to P.G. Tait, 11 December 1867. Quoted in Cargill Gilston Knott, *Life and Scientific Work of Peter Guthrie Tait* (Cambridge, 1911), p. 213. See also M.J. Klein, "Maxwell, His Demon, and the Second Law of Thermodynamics", *American Scientist* **58** (1970), 84-97.

[50] C.G. Knott, *op. cit.*, p. 214.

[51] *Ibid.*, p. 215.

[52] J.C. Maxwell to J.W. Strutt, 6 December 1870. Quoted in Robert John Strutt, *Life of John William Strutt, Third Baron Rayleigh* (Reprinted Madison, Wisconsin, 1968), p.47.

[53] J.C. Maxwell, *Theory of Heat* (Reprinted New York, 1872), p. 308.

[54] J.C. Maxwell, "Tait's *Thermodynamics*", *Nature* **17** (1878), 257. *Scientific Papers* **2**, 660-671.

[55] Georg Helm, *Die Energetik* (Leipzig, 1898), p. 146.

[56] Josiah Willard Gibbs, *The Scientific Papers* (Reprinted New York, 1961), **1**, 167.

[57] L. Boltzmann, "Über die Beziehung zwischen dem zweiten Hauptsatze der mechanischen Wärmetheorie und der Wahrscheinlichkeitsrechnung respektive den Sätzen über das Wärmegleichgewicht", *Wiener Berichte* **76** (1877), 373; *Wiss. Abh.* **2**, 164-223.

[58] *Ibid.*, p. 168.

[59] *Ibid.*, p. 165.

[60] Ibid., pp. 186-190.

[61] Ibid., pp. 190-195.

[62] Ibid., pp. 215-217.

[63] Boltzmann had already shown that the functional E, defined by equation (17), could also be written in terms of velocity variables in essentially the form of equation (24). See L. Boltzmann, op. cit., note 33, p. 362.

[64] L. Boltzmann, op. cit., note 57, p. 218.

[65] See E. Broda, op. cit., note 3, pp. 9-15.

[66] J.C. Maxwell to P.G. Tait, August 1873. Quoted in Knott, op. cit., note 49, p. 114.

[67] See H.W. Watson, A Treatise on the Kinetic Theory of Gases (Oxford, 2nd ed. 1893) and S.H. Burbury, A Treatise on the Kinetic Theory of Gases (Cambridge, 1899).

[68] P. and T. Ehrenfest, op. cit., note 29(a). Translation pp. 17-39.

[69] Ibid., pp. 9-16, 40-42, for discussions of the Stosszahlansatz.

[70] Ibid., p.3.

[71] See, for example, S.G. Brush, op. cit., note 29(b).

[72] J.C. Maxwell, Scientific Papers, 1, 409.

[73] J.C. Maxwell, "On the Dynamical Evidence of the Molecular Constitution of Bodies", Nature 11 (1875), 357. Scientific Papers 2, 433.

[74] A. Kundt and E. Warburg, "Über die spezifische Wärme des Quecksilbergases", Pogg. Ann. 157 (1876), 353.

[75] L. Boltzmann, "Über die Natur der Gasmoleküle", Wiener Berichte 74 (1876), 553; Wiss. Abh. 2, 103-110.

[76] William Thomson, Lord Kelvin, Baltimore Lectures on Molecular Dynamics and the Wave Theory of Light (London, 1904), pp. 486-527.

[77] G.H. Bryan, "Researches Relating to the Connection of the Second Law with Dynamical Principles", B.A.A.S. Report 61 (1891), 85.

[78] For a discussion of the work on monocycles and references see M.J. Klein, "Mechanical Explanation at the End of the Nineteenth Century", Centaurus 17 (1972).

[79] G.H. Bryan, "The Laws of Distribution of Energy and Their Limitations", B.A.A.S. Report 64 (1894), 64.

[80] L. Boltzmann, Lectures on Gas Theory, p. 22.

[81] See Nature 50 (1894), 51 (1894-95). Full references are given in M.J. Klein, Paul Ehrenfest. The Making of a Theoretical Physicist (Amsterdam, 1970), pp. 111-112.

[82] L. Boltzmann, "On Certain Questions of the Theory of Gases," Nature 51 (1895), 413; Wiss.Abh. 3, 535-544.

[83] E.P. Culverwell, "Professor Boltzmann's Letter on the Kinetic Theory of Gases", Nature 51 (1895), 581.

[84] L. Boltzmann, "Reply to Culverwell", Nature 51 (1895), 581; Wiss. Abh. 3, 545.

[85] For a recent discussion and references see Erwin N. Hiebert, "The Energetics Controversy and the New Thermodynamics" in Duane H.D. Roller, ed. Perspectives in the History of Science and Technology (Norman, Oklahoma, 1971), pp. 67-86.

[86] Ernst Zermelo, "Über einen Satz der Dynamik und die mechanische Wärmetheorie", Wied. Ann. 57 (1896), 485-494.

[87] See M. Planck, op. cit., note 31.

[88] Henri Poincaré, "Sur le problème des trois corps et les équations de la dynamique", Acta Math. 13 (1890) 67.

[89] L. Boltzmann, "Entgegnung auf die wärmetheoretischen Betrachtungen des Hrn. E. Zermelo", Wied. Ann. 57 (1896), 773; Wiss. Abh. 3, 567-578. See also H. Bernhardt, "Der Wiederkehreinwand gegen Boltzmanns H-Theorem und der Begriff der Irreversibilität", N.T.M. 6 (1969), No. 2, 27-36.

[90] L. Boltzmann, "Zu Hrn. Zermelos Abhandlung 'Über die mechanische Erklärung irreversibler Vorgänge'", Wied. Ann. 60 (1897), 392; Wiss. Abh. 3, 579-586.

[91] L. Boltzmann, op. cit., note 89, p. 576.

[92] L. Boltzmann, Lectures on Gas Theory, p. 22.

[93] Ibid., pp. 55-62.

[94] L. Boltzmann, Vorlesungen über die Prinzipe der Mechanik I. Teil (Leipzig, 1897), p. iii.

[95] L. Boltzmann, Lectures on Gas Theory, p. 215.

[96] L. Boltzmann, "Über einen mechanischen Satz Poincaré's", Wiener Berichte 106 (1897), 12; Wiss. Abh. 3, 587-595. "Über die sogenannte H-Kurve", Math. Ann. 50 (1898), 325; Wiss. Abh. 3, 629-637.

[97] M. Planck, Treatise on Thermodynamics, transl. Alexander Ogg (Reprinted New York, n.d.), p. viii.

[98] See M.J. Klein, "Max Planck and the Beginnings of the Quantum Theory", Archive for History of Exact Sciences 1 (1962), 459-479; "Thermodynamics and Quanta in Planck's Work", Physics Today 19 (1966), No. 11, 23.

[99] L. Boltzmann, Lectures on Gas Theory, p. 216.

[100] L. Boltzmann, op. cit., note 2, pp. 512-522.

[101] L. Boltzmann, Populäre Schriften, pp. 405-407.

[102] See M.J. Klein, Paul Ehrenfest, pp. 81-83, 119-140.

[103] P. and T. Ehrenfest, "Über zwei bekannte Einwände gegen das Boltzmannsche H-Theorem", Phys. Z. 8(1907), 311. Scientific Papers, pp.146-149. See also M.J. Klein, Paul Ehrenfest, pp. 114-119.

[104] I gratefully acknowledge the support for this work provided by a grant from the National Science Foundation.

THE VALIDITY AND THE LIMITATIONS OF THE BOLTZMANN-EQUATION

G.E. UHLENBECK

The Rockefeller University, New York

ABSTRACT

A series of questions arising from the attempts to generalize the Boltzmann equation for monatomic molecules. A discussion of the Chapman-Enskog development. The question of the generalization of the Boltzmann equation to the quantum theory and the theory of relativity.

1. Introduction.

The Boltzmann equation has become such a generally accepted and central part of statistical mechanics, that it almost seems blasphemy to question its validity and to seek out its limitations. It is also almost a miracle how the equation has withstood all criticisms and how it has overcome or could be adapted to the quantum-theoretical revolution. Furthermore, it turned out to be very difficult to make significant generalizations, to go beyond the limits set by the old master! The many attempts to do so are, I think, at least in part responsible for the remarkable revival of interest in the kinetic theory of matter, say in the last quarter century. In the coming days you will hear more about these recent developments. I think it is fair to say that many of these developments originated from questions about the limitations and even about the validity of the Boltzmann equation. It seemed to me appropriate therefore and perhaps useful to present in this first lecture a series of such questions. It may serve as a kind of table of contents for this conference!

2. The Classical Boltzmann Equation for Monatomic Gases.

Let me remind you of the classical Boltzmann argument to show the approach to equilibrium. [1] For a monatomic gas the distribution function $f(\vec{r},\vec{v},t)$ of the molecules in the phase space (µ-space) of a molecule changes in time according to the Boltzmann equation:

$$\frac{\partial f}{\partial t} + \vec{v} \cdot \frac{\partial f}{\partial \vec{r}} + \vec{a} \cdot \frac{\partial f}{\partial \vec{v}} = \int d\vec{v}_1 \int d\Omega g I(g,\theta) [f'f'_1 - f f_1] \quad (1)$$

The notation is by now almost standard. The indices refer to the velocity variables in the binary collision $(\vec{v},\vec{v}_1) \rightleftarrows (\vec{v}',\vec{v}_1')$; $I(g,\theta)$ is the differential cross-section for a collision in which the relative velocity $g = |\vec{v} - \vec{v}_1| = |\vec{v}' - \vec{v}_1'|$ turns over the angle θ. Finally $\vec{a}$ is the acceleration due to an outside potential $U(\vec{r})$, so that $m\vec{a} = -\vec{\nabla}U$. Now Boltzmann proves that any initial distribution $f(\vec{r},\vec{v},0)$ for sufficiently general $U(\vec{r})$ will become the Maxwell-Boltzmann distribution:

$$f_o = A\, e^{-\beta(\frac{m}{2}v^2 + U(\vec{r}))} \qquad (2)$$

where the constants A and β must be determined by the given total number of molecules and total energy.

The proof goes by means of the famous H-theorem. Define:

$$H(t) = \int d\vec{r} \int d\vec{v}\, f\, \ell n\, f \qquad (3)$$

Then Boltzmann proves from (1) that $dH/dt \leq 0$ and that H is only constant if for <u>all</u> collisions

$$ff_1 = f'f_1' \qquad (4)$$

This characterizes therefore the equilibrium state, which in addition must fulfill (as seen from (1)) the condition:

$$\frac{\partial f}{\partial t} + \vec{v} \cdot \frac{\partial f}{\partial \vec{r}} + \vec{a} \cdot \frac{\partial f}{\partial \vec{v}} = 0 \qquad (5)$$

The equations (4) and (5) determine uniquely the M-B distribution (2) if $U(\vec{r})$ is sufficiently general. For special forms of $U(\vec{r})$ there are a host of other solutions, which Boltzmann discussed in great detail [2] but which have found so far no applications and which are therefore only of limited interest.

You know that with this explanation of the zeroth law of thermodynamics Boltzmann was able to interpret the notion of temperature and the second law of thermodynamics. One finds that in equilibrium $\beta = 1/k_B T$ and except for a (conventional) constant $S = - k_B H$ where S is the entropy. These were the successes which aroused the admiration <u>and</u> the suspicion of Boltzmann's contemporaries!

Let me now start my list of questions.

I. <u>Can the Restriction to Monatomic Gases Be Removed</u>?

That with non-central intermolecular forces there are difficulties Lorentz pointed out in 1887 [3], and Boltzmann then tried hard to generalize his H-theorem to the case of arbitrary polyatomic molecules [4]. Even if one accepts these generalizations (and I have some doubts about them), it is, I think, fair to say that these purely classical considerations could not be satisfactory, since we now know that at least for the description of the internal degrees of freedom of a molecule the quantum theory is essential. You will hear more about it from Dr. Waldman and Dr. Beenakker.

II. <u>Can the Restriction to Binary Collisions Be Removed</u>?

Clearly the Boltzmann equation (1) only holds for gases of normal density. More precisely if d is the range of the intermolecular force (or the size of a molecule) and λ a mean free path, then one must assume that $d \ll \lambda$. Or one can say that the time $\tau = d/<v>$ <u>of</u> a collision must be very short compared to the time $t_o = \lambda/<v>$ <u>between</u> collisions, so that the molecules are free most of the

time and only binary collisions need to be considered.
Many attempts have been made to derive the Boltzmann
equation from the general Liouville equation and then
by expansion in the small parameter d/λ and/or by
making assumptions about the two particle and higher
distribution functions to find the higher density corrections. You will hear more about it from Dr. Cohen.
The main surprise is that the Boltzmann equation can
not be considered as the first term in a virial expansion similar to the density expansion of the equation
of state of a non-ideal gas in equilibrium. In fact
there are even doubts whether at higher density there
exists a closed kinetic equation for the first distribution function $f(\vec{r},\vec{v},t)$. This would make the Boltzmann equation even more miraculous!

III. Mathematical Questions.

Boltzmann's argument is of course not mathematically rigorous since it assumes the existence and
the unicity of the solution of the initial value problem
for the non-linear integrodifferential equation (1),
and the continued uniform convergence of the integral
defining $H(t)$. Carleman was able to make the proof
mathematically respectable at least for the spatially
uniform case and for purely repulsive intermolecular
forces [5]. I don't know whether Carleman's work has
been continued. In this context, I think one should also consider the question of the boundary conditions
when solid bodies are immersed in the gas. The simplest
description is by the so-called Maxwell boundary condition. Maxwell assumed that a fraction γ of the molecules
hitting the solid surface is adsorbed and then re-emitted with a Maxwell distribution determined by the motion

and the temperature of the solid, while the remaining fraction $(1 - \gamma)$ is specularly reflected. Although this is surely an over-simplified picture, it should be a good example of a possible boundary condition, and one wonders whether Boltzmann's theorem would remain valid. I don't know whether such questions have been considered for the general case, but for the linearized Boltzmann equation in which $f = f_o[1 + h(\vec{r},\vec{v},t)]$ with $h \ll 1$, all mathematical qualms can and have been resolved. We may hear about that later at this conference.

IV. Can the Boltzmann Equation Describe the Fluctuations of the Properties of the Gas?

The answer is clearly no! The Boltzmann equation is a causal equation and not a stochastic equation. Boltzmann knew of course very well that his equation described in some sense the most probable or average change of the distribution function and that $f(\vec{r},\vec{v},t)$ was really a fluctuating or stochastic variable. He always insisted that the fluctuations would resolve the irreversibility paradoxes [6], but when in 1877 he developed a truly statistical method [7] he did not so far as I know, return to his equation in order to give it a proper probabilistic interpretation. The Ehrenfest's showed later by their famous simple model [8], that indeed the paradoxes disappeared if the H theorem was interpreted stochastically, and in their Encyclopedia article they suggest how Boltzmann's H(t) curve might be related to the family of fluctuating H-curves [9]. They also emphasized that the Stoszzahl Ansatz is a _statistical_ assumption, but a truly probabilistic interpretation of the Stoszzahl Ansatz is of rather recent origin. It came through the so-called master equation approach which will be discussed by Dr. Kac. Let me only still mention that for the linearized

Boltzmann equation the extension to a stochastic equation has recently been achieved by appealing to the general theory of stationary, Gaussian Markov processes.

3. The Chapman-Enskog Development.

The most significant development of the kinetic theory in the post-Boltzmann era is, in my opinion, the work of Chapman and Enskog [10] in which for the first time a systematic derivation of the hydrodynamical equations from the Boltzmann equation was achieved. Let me remind you of the main ideas. The essential physical insight was that the approach to equilibrium goes in <u>two temporal stages</u>. As already suggested by Boltzmann's proof, one expects that because of the sharp repulsive forces between the molecules the distribution in <u>velocities</u> will become close to the Maxwell distribution after a small number of collisions. Or, in other words after a time of order $\lambda/<v>$ $f(\vec{r},\vec{v},t)$ will be close to the local Maxwell distribution:

$$f^{(o)} = n \left(\frac{m}{2\pi k_B T}\right)^{3/2} \exp\left[-\frac{m}{2k_B T}(\vec{v}-\vec{u})^2\right] \qquad (6)$$

where the local particle density n, average velocity $\vec{u}$, and temperature T are still slowly varying functions of $\vec{r}$ and t. Their slow relaxation to the constant equlibrium values is what the Chapman-Enskog theory tries to describe by a successive approximation method in which the distribution function $f(\vec{r},\vec{v},t)$ is expanded in the form:

$$f = f^{(o)}[1 + a_1(\vec{v})(\lambda \nabla) + a_2(\vec{v})(\lambda \nabla)^2 + \ldots] \qquad (7)$$

where λ is again a mean free path, and where the gradient operator ∇ acts on the macroscopic variables n, $\vec{u}$, and T. The expansion parameter in (7) is therefore the relative change of the macroscopic variables over a mean free path. The successive approximations can be found explicitly from the Boltzmann equation and then by introducing the expansion (7) into the general conservation laws for the number density, momentum, and energy (which also follow from the Boltzmann equation), one obtains the successive order hydrodynamical equations. In zeroth order ($f = f^{(o)}$) one gets the Euler or ideal fluid equations and in first order the Navier-Stokes equations with explicit expressions for the transport coefficients in terms of the collision cross section $I(g,\theta)$ or in terms of the intermolecular potential. I think there is no doubt that these expressions agree with experiment. They are in a way the first really "hard" experimental predictions from the Boltzmann equation. This triumph of the Chapman-Enskog theory leads immediately to the question:

V. What Is the Physical Significance of the Higher Order Hydrodynamical Equations?

Note that since the expansion (7) is in inverse powers of the density, these equations should describe the flow properties of the gas better if the gas is more dilute or if the flow varies very quickly in space and time. You will hear from Dr. Ford and from Dr. Foch about the experimental confirmation of these equations, expecially for high frequency sound propagation and for the structure of shock waves. Let me only say that in this way one can test not only the linearized Boltzmann equation beyond the Navier-Stokes approximation, but also aspects of the non-linear Boltzmann equation.

The Chapman-Enskog expansion leads to a number of mathematical problems just as every successive approximation method does. They can perhaps be summarized in the question:

VI. <u>In Which Sense Does the Chapman-Enskog Expansion Approximate the Solution of the Initial Value Problem for the Boltzmann Equation</u>?

Note that the higher order hydrodynamical equations are of the first order in the time derivatives of the five hydrodynamical variables n, $\vec{u}$, and T, so that presumably the initial values of these variables will determine the time evolution. This is much less "information" than the initial value of the distribution function $f(\vec{r},\vec{v},t)$, which is required for the Boltzmann equation. This is the so-called <u>Hilbert causality paradox</u>, [11] which, I think, shows the subtlety of the question.

4. Some Remarks About the Generalization to the Quantum and the Relativity Theory.

I mentioned already the need of the quantum theory to describe the internal degrees of freedom of polyatomic molecules. For monatomic molecules there is an obvious quantum mechanical effect since the collision cross-section $I(g,\theta)$ must of course be calculated from the quantum scattering theory. If the mean De Broglie wavelength $\Lambda = h/(mkT)^{1/2}$ becomes comparable to the size d of the molecule, then there will be deviations from the classical theory, which will affect the values and the temperature dependence of the transport coefficients. For a light gas like Helium at low temperatures, these effects have been analyzed and checked with experiment. I think there is no doubt that with this interpretation of $I(g,\theta)$

the Boltzmann equation remains valid.

There is, however, a second and more subtle quantum mechanical effect, namely the effect of the Bose or Fermi statistics. It is well known that for bosons and fermions one must modify the Stoszzahl Ansatz to insure that one gets in equilibrium the Bose or Fermi distribution instead of the Maxwell-Boltzmann distribution. Pauli [12] has shown how to generalize the H-theorem and also the extension of the Chapman-Enskog theory has been worked out. However, I think, there is <u>no empirical evidence</u>, that this gives a satisfactory description say of an interacting Bose gas. The reason is that usually $\Lambda \lesssim d$, so that the restriction to binary interactions ($nd^3 \ll 1$) implies $n\Lambda^3 \ll 1$, which makes the effect of the statistics negligible. Only for the academic but very interesting model of a Bose gas of very small hard spheres could one ask the question:

VII. <u>Does the Boltzmann Equation for a Hard Sphere Bose Gas Describe the Approach to Equilibrium Even in the Condensed Region</u>?

Because of the existence of persistent currents, the answer seems to me doubtful.

Finally, let me conclude with another academic but quite interesting question:

VIII. <u>Can the Boltzmann Equation Be Generalized to Relativistic Particles of Given Rest Mass</u>?

It is clear that for such a generalization the interaction between the particles must be strictly local. It turns out that with such an assumption the Boltzmann equation can be written in covariant form and that Boltzmann's argument and also the Chapman-Enskog development

can be generalized. You will hear more about it from
Dr. de Groot.

This is only a sample of the problems, inspired
by the Boltzmann equation, which will be discussed at
this conference. I hope that after hearing these discussions, you will agree with me that to doubt the value
of the Boltzmann equation is indeed blasphemy, that in
fact a phrase used by Boltzmann to characterize the Maxwell equations may as well be used to characterize his
equation:

Ist es ein Gott, der dieses Zeichen schuf?

NOTES

[1] L. Boltzmann, Wien.Ber. $\underline{66}$, 275 (1872)[= Wiss.Abh. Vol.1, p.316].
Extension to Outside Forces in Wiss.Abh. Vol.2, p.1 (1875).

[2] L. Boltzmann, Wiss.Abh. Vol.2, p.83 (1876).

[3] I refer to the non-existence of restituting collisions (often also called inverse collisions) for non-spherical molecules. See H.A. Lorentz, Wien.Ber. $\underline{95}$, 115 (1887)[= Coll. Papers, Vol.6, p.74].

[4] This is the so-called cycle proof of the H-theorem. See L. Boltzmann, Wien.Ber. $\underline{95}$, 153 (1887)[= Wiss. Abh. Vol.3, p.272]; also Vorlesungen über Gastheorie, Vol.2, Ch. VII.

[5] T. Carleman, Problèmes Mathematiques dans la Théorie Cinetique des Gaz (Publications Scientifiques de l'Institut Mittag-Leffler, Vol.2, Uppsala, 1957).

[6] I refer of course to the famous Umkehr Einwand of Loschmidt (Wien.Ber. $\underline{73}$, 139 (1876)) and Wiederkehr Einwand of Zermelo (Wied.Ann. $\underline{57}$, 485 (1896)).

[7] L. Boltzmann, Wien.Ber. $\underline{76}$, 373 (1877)[= Wiss.Abh. Vol.2, p.112].

[8] I refer of course to the two urn or "dog-flea" model. See P. and T. Ehrenfest, Phys. Zeitschr. $\underline{8}$, 311 (1907) [= Coll. Papers p. 146].

[9] P. and T. Ehrenfest, Enz. der Math. Wiss. Vol. IV, Art. 32, 1912. [= Coll. Papers, p.213] English translation by M.J. Moravcsik, Cornell Univ. Press, 1959. Compare esp. sections 14c and 14d.

[10] S. Chapman, Phil. Trans. Roy. Soc. A 216, 279 (1916); A 217, 115 (1917).
D. Enskog, Dissertation Univ. of Uppsala, 1917.

[11] D. Hilbert, Grundzüge einer allgemeinen Theorie der linearen Integralgleichungen (Teubner, Leipzig, 1912) Ch. 22.

[12] W. Pauli, Probleme der modernen Physik (Hirzel Verlag, Leipzig, 1928) p.30 [= Coll. Papers, Vol.I, p.549].

COMMENT TO PROFESSOR UHLENBECK'S PAPER

C. CERCIGNANI

Politecnico di Milano, Milano

In his talk, Professor Uhlenbeck pointed out that an extension of the H-theorem to the case in which solid boundaries are present would be desirable. I am glad to point out that such an extension is now available. For the case of Maxwell's boundary conditions, the proof is contained in my book [1], but much more general results are available [2-4]. The most general result is the following [4]: if the boundary conditions are linear in the distribution function and compatible with the Maxwellian distribution at equilibrium, no particles are absorbed at the wall, the time of interaction between a gas molecule and the wall is negligible, then, for any convex function C, at any point of a solid boundary:

$$\int \xi_n f_w C(f/f_w) d\underline{\xi} \leq 0$$

where f_w is the wall Maxwellian, ξ_n the component of the molecular velocity along the normal (pointing into the gas). By taking $C(f/f_w) = (f/f_w)\log(f/f_w)$ we obtain [4]

$$\int \xi_n f \log f \, d\underline{\xi} \leq -\frac{1}{RT_w}[q_n]_{solid}$$

where T_w is the wall Maxwellian and q_n is the normal flux from the solid wall into the gas. From this result the H-theorem follows in the form:

$$\frac{dH}{dt} \leq -\frac{1}{R} \int \frac{d^*Q}{T_w}$$

where d^*Q is the heat received by the gas from a point of the boundary where the temperature is T_w and integration extends to all the boundaries of the region filled with gas. This result is the Clausius inequality, which one would expect from the second principle of thermodynamics.

References

[1] C. Cercignani, "Mathematical Methods in Kinetic Theory". Plenum Press, New York (1969).
[2] J. Darrozès and J.P. Guiraud, Compt. Rend. <u>A262</u>, 1368 (1966).
[3] C. Cercignani and M. Lampis, "Transport Theory and Statistical Physics, <u>1</u>, 101 (1971).
[4] C. Cercignani, "Transport Theory and Statistical Physics", <u>2</u>, 27 (1972).

ON HIGHER ORDER HYDRODYNAMIC THEORIES OF SHOCK STRUCTURE

J.D. FOCH, Jr.

Department of Aerospace Engineering Sciences
University of Colorado
Boulder, Colorado 80302

ABSTRACT

The Burnett equations for shock structure in a Maxwell gas (inverse fifth power repulsion) are not amenable to numerical integration by the usual methods (fourth order Runge-Kutta) above a Mach number of 1.9. The super-Burnett equations are not amenable to numerical integration by the usual methods for any Mach number. It is not yet clear whether these negative results indicate a fundamental difficulty in the Chapman-Enskog solution of the Boltzmann equation.

Introduction [1]

The problem of working out a satisfactory theory of shock structure is of interest primarily because such a theory will lead to tests of the nonlinear features of the Boltzmann collision operator. I say "will" because the experimentalists are ahead of the theoreticians, as usual [2,3]. What I want to tell you about today are some efforts with Mr. Charles Simon to base a theory of shock structure on the higher order hydrodynamic equations of the Chapman-Enskog development. We have restricted ourselves to a Maxwell gas (inverse fifth power repulsion) in order to take advantage of the well known simplifications of the collision operator, but even so we are a long way from solving the problem.

Outline of the Hydrodynamic Theories

Consider stationary one dimensional flow through a normal shock wave. If we choose the z-axis as the direction of the flow, the differential equations expressing conservation of mass, momentum and (total) energy are

$$\frac{d}{dz}[\rho u] = 0 , \tag{1}$$

$$\frac{d}{dz}[\rho u^2 + p - \sigma] = 0 , \tag{2}$$

$$\frac{d}{dz}[\rho(\frac{u^2}{2} + \varepsilon)u + (p - \sigma)u + q] = 0 , \tag{3}$$

with the pressure p and specific internal energy ε given by

$$p = \frac{\rho kT}{m}, \qquad \varepsilon = \frac{3kT}{2m}.$$

In these equations ρ is the mass density, u is the flow velocity, k is Boltzmann's constant, T is the absolute temperature and m is the mass per atom. In addition, σ is the zz-component of the viscous stress tensor and q is the z-component of the heat flow vector; σ and q both vanish when the flow is uniform.

Equations (1)-(3) may be integrated once immediately to give

$$\rho u = \rho_i u_i, \qquad (4)$$

$$\rho u^2 + \frac{\rho kT}{m} - \sigma = \rho_i u_i^2 + \frac{\rho_i kT_i}{m}, \qquad (5)$$

$$\rho(\frac{u^2}{2} + \frac{3kT}{2m})u + (\frac{\rho kT}{m} - \sigma)u + q = \rho_i(\frac{u_i^2}{2} + \frac{3kT_i}{2m})u_i + \frac{\rho_i kT_i}{m}, \qquad (6)$$

where the index i is either 1 (upstream) or 2 (downstream), depending on whether we have integrated from the uniform conditions far upstream to within the shock wave, or from within the shock wave to the uniform conditions far downstream. It is convenient to use Eq.(5) to eliminate σ from Eq.(6):

$$\rho(-\frac{u^2}{2} + \frac{3kT}{2m})u + (\rho_i u_i^2 + \frac{\rho_i kT_i}{m})u + q$$
$$= \rho_i(\frac{u_i^2}{2} + \frac{3kT_i}{2m})u_i + \frac{\rho_i kT_i}{m} u_i. \qquad (7)$$

Equation (4) shows the mass density ρ is determined by the flow velocity u and allows us to eliminate ρ from Eqs.(5) and (7). Carrying out the elimination and introducing a nondimensional flow velocity v and a nondimensional temperature T defined by

$$v = \frac{u}{u_i}, \qquad T = \frac{T}{T_i},$$

we have

$$v + \frac{3T}{5M_i^2 v} - \frac{\sigma}{\rho_i u_i^2} = 1 + \frac{3}{5M_i^2}, \qquad (8)$$

$$-\frac{v^2}{2} + \frac{9T}{10M_i^2} + (1 + \frac{3}{5M_i^2})v + \frac{q}{\rho_i u_i^3} = \frac{1}{2} + \frac{3}{2M_i^2}. \qquad (9)$$

In Eqs. (8) and (9) the parameter M_i is the Mach number defined by

$$M_i = \frac{u_i}{a_i},$$

with a_i the low frequency speed of sound,

$$a_i = \sqrt{\frac{5kT_i}{3m}}.$$

Equations (8) and (9) may be looked upon as the basic equations governing the structure of a shock wave. According to the Chapman-Enskog "solution" of the Boltzmann equation, σ and q are given in successive approximation by expressions of the form

$$\sigma = \sigma^{(1)} + \sigma^{(2)} + \sigma^{(3)} + \ldots,$$

$$q = q^{(1)} + q^{(2)} + q^{(3)} + \ldots,$$

with $\sigma^{(1)}$ and $q^{(1)}$ of first order in $\frac{d}{dz}$, $\sigma^{(2)}$ and $q^{(2)}$ of second order in $\frac{d}{dz}$ (second derivatives and products of first derivatives), etc. Each of the successive approximations engenders its own hydrodynamic theory of shock structure.

Euler equations. If we neglect σ and q altogether, Eqs. (8) and (9) constitute two algebraic equations in two unknowns with M_i as a parameter. These equations admit two solutions:

$$\frac{u_2}{u_1} = 1 \, , \tag{10a}$$

$$\frac{T_2}{T_1} = 1 \, ; \tag{10b}$$

$$\frac{u_2}{u_1} = \frac{M_1^2 + 3}{4M_1^2} \, , \tag{11a}$$

$$\frac{T_2}{T_1} = \frac{(5M_1^2 - 1)(M_1^2 + 3)}{16M_1^2} \, . \tag{11b}$$

In the present context the first solution is trivial; the second solution determines u_2 and T_2 in terms of u_1 and T_1, and then Eq. (4) determines ρ_2 in terms of ρ_1, u_1 and T_1. In this approximation the shock wave is idealized as a surface; fluid particles are decelerated, heated and compressed as they cross the surface. Despite the idealization, the formulas relating conditions far downstream to conditions far upstream are of general validity. This follows from the fact that the general first integrals (4)-(6) may be effected all the way through the shock wave, thus linking conditions in two regions where σ and q are certainly zero.

Navier-Stokes equations. The first order contributions to σ and q are given by

$$\sigma^{(1)} = \frac{4\mu}{3} \frac{du}{dz} ,$$

$$q^{(1)} = - \kappa \frac{dT}{dz} ,$$

with the viscosity $\mu(T)$ and thermal conductivity $\kappa(T)$ given by

$$\mu = \mu_i T ,$$

$$\kappa = \frac{15k\mu_i}{4m} T .$$

In this approximation the basic Eqs.(8) and (9) become coupled nonlinear first order differential equations for $v(z)$ and $T(z)$.

Although our objective is to find v and T as functions of z, it is instructive to think of $v(z)$ and $T(z)$ as parametric representations of a point in an auxiliary vT-plane. In the auxiliary 2-space a solution point moves along a path or trajectory as z varies from $-\infty$ to $+\infty$, and what we seek is a trajectory connecting the two points whose coordinates are given by Eqs.(10) and (11). Since these points are singular points for the system of differential equations, the question arises as to whether such a trajectory exists. This question has been answered in the affirmative by Gilbarg and Paolucci [4].

The same authors have demonstrated that the Navier-Stokes equations are amenable to numerical integration as an initial value problem if the integration is begun near the downstream singular point, but not if the integration is begun near the upstream singular point. This difference can be understood if we investigate the trajectories near the singular points by means of the linearized Navier-Stokes equations.

Far upstream or far downstream the flow velocity and temperature obey coupled linear first order differential equations with constant coefficients (the coefficients depend on Mach number). The deviations from the asymptotic values are therefore proportional to $e^{\lambda_i z}$, where λ_i is either of the two roots of a characteristic polynomial of second degree. The characteristic roots are shown as functions of M_1 in Fig. 1.

Far upstream both characteristic roots are positive, while far downstream one root is positive and the other is negative. For z increasing (initial value problem from upstream) this means every trajectory in the neighborhood of the upstream singular point is leaving it, but only one [5] trajectory hits the downstream singular point. This presents a formidable "aiming" problem. On the other hand, for z decreasing (initial value problem from downstream) only one [5] trajectory leaves the downstream singular point, while every trajectory which approaches the upstream singular point is attracted into it. This presents a very favorable "aiming" opportunity.

The difference between forward and reverse integrations would not be removed even if the numerical integration were perfect, because the initial values cannot be located precisely on the desired trajectory without an analytic solution to the problem.

Burnett equations. The second order contributions to σ and q are given by

$$\sigma^{(2)} = -\frac{\mu^2}{p}[\frac{8}{9}(\frac{du}{dz})^2 - \frac{4k}{3m\rho}\frac{d\rho}{dz}\frac{dT}{dz} + \frac{4kT}{3m\rho^2}(\frac{d\rho}{dz})^2$$

$$-\frac{4kT}{3m\rho}\frac{d^2\rho}{dz^2} + \frac{2k}{3m}\frac{d^2T}{dz^2} + \frac{2k}{mT}(\frac{dT}{dz})^2] \quad ,$$

$$q^{(2)} = \frac{\mu^2}{\rho} \left[\frac{95}{8T} \frac{dT}{dz} \frac{du}{dz} - \frac{7}{4} \frac{d^2u}{dz^2} - \frac{2}{\rho} \frac{d\rho}{dz} \frac{du}{dz} \right] .$$

When the first and second order contributions to σ and q are included in Eqs. (8) and (9), we obtain coupled non-linear <u>second order</u> differential equations for $v(z)$ and $T(z)$.

Let us adopt the standard device of lowering the order of the system of equations by treating $\frac{dv}{dz}$ and $\frac{dT}{dz}$ as unknown functions on the same footing as v and T. This gives a system of four coupled nonlinear <u>first order</u> differential equations, again with two singular points (in an auxiliary 4-space). We seek a trajectory which connects one singular point to the other, since the singular points represent conditions far upstream and far downstream. To the best of my knowledge it is unknown whether such a trajectory exists.

Sherman and Talbot [6] assumed the existence of a solution and integrated the Burnett equations numerically, starting downstream. They found that the Burnett equations were amenable to numerical integration, provided M_1 was less than approximately 2, but that the solutions developed oscillations upstream for M_1 greater than approximately 1.5. As Sherman and Talbot realized quite clearly, these features can be understood from an analysis of the trajectories in the neighborhood of each singular point.

The deviations of v, $\frac{dv}{dz}$, T and $\frac{dT}{dz}$ from their asymptotic values, far upstream or far downstream, are proportional to $e^{\lambda_i z}$, where λ_i is any one of the four roots of a characteristic polynomial of fourth degree. The characteristic roots are shown as functions of M_1 in Fig. 2.

For $M_1 \simeq 1$ all the upstream characteristic roots have positive real part, whereas only one downstream characteristic root is negative. The first qualitative change in the nature of the singular points occurs upstream at $M_1 \simeq 1.5$, when the characteristic polynomial ceases to have real roots. A second, more drastic change occurs at $M_1 \simeq 1.9$, again upstream, when one pair of conjugate roots develops a negative real part.

<u>Super-Burnett equations</u>. The third order contributions to σ and q are given by [7]

$$\sigma^{(3)} = -\frac{\mu^3}{\rho^2}\left[\frac{47k}{3m\rho}\frac{dT}{dz}\frac{d\rho}{dz}\frac{du}{dz} - \frac{64kT}{9m\rho^2}\left(\frac{d\rho}{dz}\right)^2\frac{du}{dz} + \frac{40kT}{9m\rho}\frac{d^2\rho}{dz^2}\frac{du}{dz}\right.$$

$$- \frac{2kT}{3m\rho}\frac{d\rho}{dz}\frac{d^2u}{dz^2} - \frac{21k}{3mT}\left(\frac{dT}{dz}\right)^2\frac{du}{dz} - \frac{47k}{9m}\frac{d^2u}{dz^2}\frac{dT}{dz}$$

$$\left. - \frac{31k}{9m}\frac{du}{dz}\frac{d^2T}{dz^2} + \frac{2kT}{9m}\frac{d^3u}{dz^3} + \frac{16}{27}\left(\frac{du}{dz}\right)^3\right],$$

$$q^{(3)} = \frac{\mu^3}{\rho\rho}\left[-\frac{8035}{336T}\frac{dT}{dz}\left(\frac{du}{dz}\right)^2 + \frac{166}{21\rho}\frac{d\rho}{dz}\left(\frac{du}{dz}\right)^2 + \frac{949}{168}\frac{du}{dz}\frac{d^2u}{dz^2}\right.$$

$$+ \frac{917k}{8m\rho T}\frac{d\rho}{dz}\left(\frac{dT}{dz}\right)^2 - \frac{1137k}{16m\rho^2}\frac{dT}{dz}\left(\frac{d\rho}{dz}\right)^2 + \frac{397k}{16m\rho}\frac{d\rho}{dz}\frac{d^2T}{dz^2}$$

$$+ \frac{701k}{16m\rho}\frac{dT}{dz}\frac{d^2\rho}{dz^2} - \frac{813k}{16mT^2}\left(\frac{dT}{dz}\right)^3 - \frac{1451k}{16mT}\frac{dT}{dz}\frac{d^2T}{dz^2}$$

$$\left. - \frac{157k}{16m}\frac{d^3T}{dz^3} - \frac{41kT}{8m\rho^2}\frac{d\rho}{dz}\frac{d^2\rho}{dz^2} - \frac{5kT}{8m\rho}\frac{d^3\rho}{dz^3} + \frac{23kT}{4m\rho^3}\left(\frac{d\rho}{dz}\right)^3\right].$$

When the first, second and third order contributions to σ and q are included in Eqs. (8) and (9), we obtain coupled nonlinear third order differential equations for $v(z)$ and $T(z)$. The equivalent system of six coupled nonlinear first order differential equations has two singular points, which correspond to conditions far upstream and far downstream.

These super-Burnett equations are not amenable to numerical integration as an initial value problem for any Mach number, for, as Fig. 3 shows, three of the upstream characteristic roots have positive real part and three have negative real part, while four of the downstream characteristic roots have negative real part and two have positive real part.

Results and Remarks

Schmidt [2] has emphasized on experimental grounds that the asymmetry of the density profile in a shock wave can provide a sensitive test of theories of shock structure. He introduced a normalized density R_ρ defined by

$$R_\rho(z) = \frac{\rho(z) - \rho_1}{\rho_2 - \rho_1},$$

and chose the origin of the z-axis such that

$$R_\rho(0) = 0.5.$$

He then investigated the asymmetry quotient Q_ρ defined by

$$Q_\rho = \frac{A_1}{A_2},$$

in which

$$A_1 = \int_{-\infty}^{0} dz\, R_\rho(z),$$

$$A_2 = \int_{0}^{\infty} dz\,[1 - R_\rho(z)].$$

His results (for argon) are compared with the predictions of the Navier-Stokes equations (for a realistic temperature-viscosity law) in Fig. 4. There can be little doubt about the disagreement between theory and experiment, although it would be desirable to have additional measurements at Mach numbers less than 2.8.

No one knows what the Burnett or higher order equations are for a realistic potential, and so it is impossible to attempt to explain Schmidt's experiments. However, as a provisional guide to experiments, it seems reasonable to investigate the differences between the predictions of the Navier-Stokes equations and the Burnett equations, both for Maxwell molecules. This was done previously by Sherman and Talbot [5], but they did not examine the asymmetry.

Figure 5 shows the normalized density profiles for $M_1 = 1.6$ according to the Navier-Stokes equations and the Burnett equations, and Fig. 6 shows the dependence of Q_ρ on M_1 for the two theories. The results in Fig. 6 suggest that the Navier-Stokes equations cannot account for the asymmetry at any Mach number. It might be worthwhile to test this suggestion experimentally with accurate measurements of the density profile for Mach numbers between 1 and 2.

Results for the normalized temperature profile are similar to those for the normalized density profile. However, the point where $R_T = 0.5$ differs from the point where $R_\rho = 0.5$, and the distance between these two points increases as M_1 increases. Measurement of this distance might lead to an additional sensitive test of theories of shock structure.

There is a formal similarity between the theory of sound propagation and the theory of weak shock waves. The latter requires only the linearized hydrodynamic equations and leads to exponential dependence of the hydrodynamic variables upon z far upstream and far downstream. The reciprocal length scale λ_i in the exponential dependence $e^{\lambda_i z}$ can be expressed as a power series of the form

$$\frac{\lambda_i \mu_i M_i}{\rho_i a_i} = \alpha_1 (M_i^2 - 1) + \alpha_2 (M_i^2 - 1)^2 + \alpha_3 (M_i^2 - 1)^3 + \ldots,$$

where α_1 is determined by the linearized Navier-Stokes equations, α_2 by the linearized Burnett equations, etc. This suggests that it might be possible to base a theory of weak shock structure directly on the <u>linearized</u> Boltzmann equation [8].

The last remark leads naturally to the question of whether a satisfactory theory of shock structure for arbitrary M_1 can be based directly on the Boltzmann equation, as Mott-Smith [9] tried to do. This is still a burning question if one wants to regard predictions of shock structure as tests of the Boltzmann equation rather than tests of the Chapman-Enskog development (see next remark).

The most serious question raised by the present work is the possibility that the super-Burnett equations do not admit a shock wave solution for any Mach number. It would be strange indeed if the messenger for this news were the same linearized super-Burnett equations which seem to have physical significance for sound propagation [10].

Finally, one cannot help wondering why the Navier-Stokes equations are so good qualitatively for all Mach numbers, even when quantitatively wrong, while the higher order hydrodynamic equations are quantitatively correct only up to a certain Mach number (at best!), and qualitatively wrong beyond. This behavior is reminiscent of the derivation of the Boltzmann equation from the Liouville equation, especially if the Burnett and super-Burnett difficulties are genuine.

I should like to take this opportunity to record my gratitude to my friend and teacher, Professor George Uhlenbeck, who transmitted to me part of the legacy of the man we celebrate today.

References

[1] This lecture is not a complete review of the subject. The literature is quite extensive and several interesting contributions have been omitted.

[2] B. Schmidt, J. Fluid Mech. __39__, 361 (1969).

[3] E.P. Muntz and L.N. Harnett, Phys. Fluids __12__, 2027 (1969).

[4] D. Gilbarg and D. Paolucci, J. Rat. Mech. Analysis __2__, 617 (1953).

[5] Two trajectories hit the downstream singular point (from opposite directions), but only one is compatible with the expected monotonic change within a shock wave.

[6] F.S. Sherman and L. Talbot, in *Proceedings of the First International Symposium on Rarefied Gas Dynamics*, edited by F.M. Devienne (Pergamon, New York, 1960), p. 161. The equations used by Sherman and Talbot contain an error, but the error has negligible effect on their results. In the expression for p_{xx} on p. 164 the coefficient of $\frac{\mu^2}{p}(\frac{du}{dx})^2$ is 40/27; it should be 8/9.

[7] J.D. Foch, Jr. and C.E. Simon, to be published.

[8] A theory of weak shock structure based on the linearized Boltzmann equation would probably not lead to such exacting comparisons with experiment as the theory of sound propagation. Its putative value would be in suggesting now to formulate a theory for arbitrary Mach number.

[9] H.M. Mott-Smith, Phys. Rev. __82__, 885 (1951).

[10] Although the predictions of the linearized Burnett equations for sound propagation are in quantitative agreement with experiment, the same cannot quite be said for the linearized super-Burnett equations. We hope to settle this uncertainty soon with absorption measurements at the University of Colorado.

List of Figure Captions

Fig. 1 Characteristic roots for the Navier-Stokes equations.

Fig. 2 Characteristic roots for the Burnett equations, with imaginary parts represented by dotted lines.

Fig. 3 Characteristic roots for the super-Burnett equations, with imaginary parts represented by dotted lines. The imaginary parts which persist in the far right portion of the figure are associated with the real part just under the real axis.

Fig. 4 Comparison between experimental and theoretical values for the symmetry quotient Q_ρ. Experimental values from Ref. 2; theoretical values for the Navier-Stokes equations with $\mu(T) = \mu_i (T/T_i)^{0.68}$.

Fig. 5 Density profiles in a Maxwell gas according to the Navier-Stokes equations and the Burnett equations. Distances in terms of the upstream Maxwell mean free path
$$\ell_1 = \frac{16}{\sqrt{30\pi}} \frac{\mu_1}{\rho_1 a_1} .$$

Fig. 6 Asymmetry quotient Q_ρ in a Maxwell gas according to the Navier-Stokes equations and the Burnett equations.

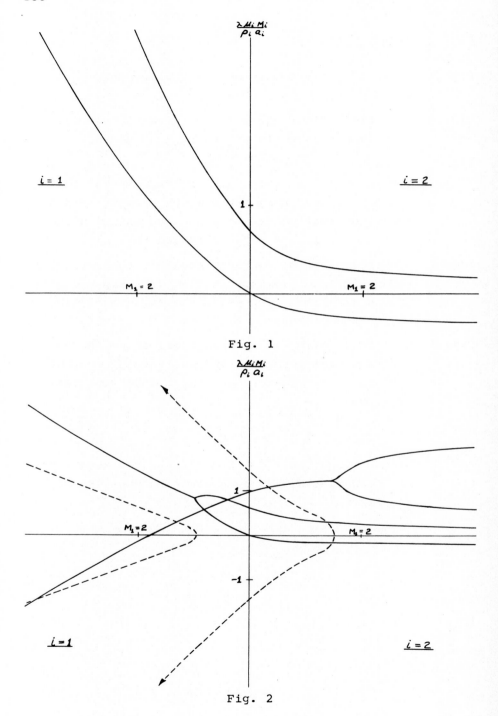

Fig. 1

Fig. 2

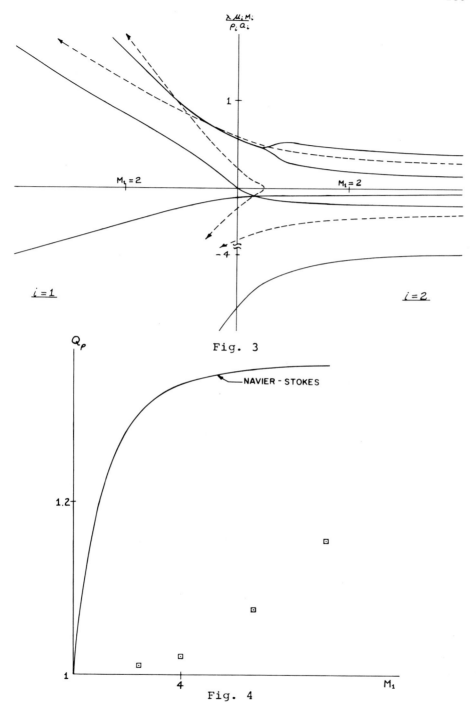

Fig. 3

Fig. 4

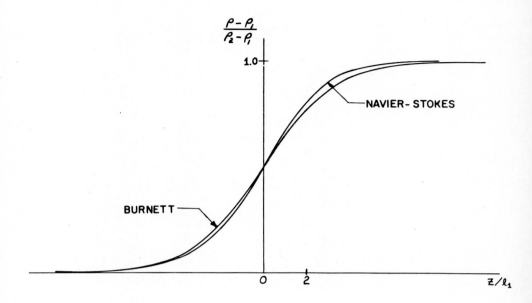

Fig. 5

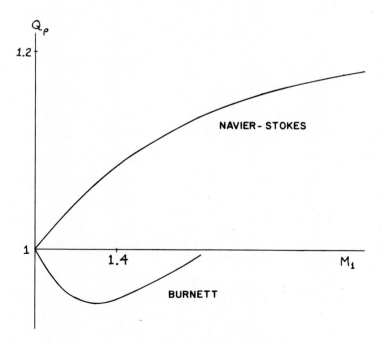

Fig. 6

SOUND FROM THE BOLTZMANN EQUATION

G.W. FORD

Department of Physics, The University of Michigan
Ann Arbor, Michigan

1. Introduction

It is remarkable that after a hundred years we have still so much to learn about the Boltzmann equation. Professor Uhlenbeck has already mentioned some mathematical questions about the H-theorem for the spatially non-uniform case [1]. I would like to draw your attention to some recent results of E. Hauge, who shows that certain solutions of the non-linear Boltzmann equation show an approach to equilibrium which for long times is like an inverse power of the time, not exponential as one would expect [2]. This strange and surprising result is an example showing how much we have yet to learn about the solutions of the Boltzmann equation.

Here, however, I want to tell a little of what is known. More precisely, I will describe some of the mathematical properties of the linearized Boltzmann equation; I will do this in the context of the question of how the collective mode of motion of the gas which we call sound arises directly from the Boltzmann equation. Most of what I have to say is not new and because of limitations of space I can tell only a little of what is known. For a more detailed discussion especially of the question of comparisons with experiment I refer to an article by J. Foch and me in the series Studies in Statistical Mechanics [3].

2. The Linearized Boltzmann Equation

We begin with the Boltzmann equation which governs the temporal development of the distribution function $f(\vec{r},\vec{v},t)$ of the molecules in a monoatomic gas. In more or less standard notation [4],

$$\frac{\partial f}{\partial t} + \vec{v}\cdot\frac{\partial f}{\partial \vec{r}} = \int d\vec{v}_1 \int d\Omega |\vec{v}-\vec{v}_1| \; I(|\vec{v}-\vec{v}_1|,\theta)[f'f'_1 - ff_1] \qquad (1)$$

where the prime and the index 1 refer to the four velocity variables in a binary collision $(\vec{v},\vec{v}_1) \rightleftarrows (\vec{v}',\vec{v}'_1)$ and $I(|\vec{v}-\vec{v}_1|,\theta)$ is the differential cross section for a collision in which the relative velocity $\vec{v}-\vec{v}_1$ is turned through angle θ. Since sound is a <u>small amplitude</u> disturbance from equilibrium it is appropriate to seek solutions of this equation of the form:

$$f(\vec{r},\vec{v},t) = f^\circ(\vec{v})[1 + h(\vec{r},\vec{v},t)] \qquad (2)$$

where

$$f^\circ(\vec{v}) = n(m/2\pi kT)^{3/2} \exp\{-\frac{mv^2}{2kT}\} \qquad (3)$$

is the equilibrium (spatially uniform) Maxwell distribution and h is small. Inserting (2) in (1) and neglecting terms of second order in h we get the <u>linearized Boltzmann equation</u>;

$$\frac{\partial h}{\partial t} + \vec{v}\cdot\frac{\partial h}{\partial \vec{r}} = \int d\vec{v}_1 f^\circ(\vec{v}_1) \int d\Omega |\vec{v}-\vec{v}_1| I(|\vec{v}-\vec{v}_1|,\theta)[h'_1+h'-h_1-h] . \qquad (4)$$

It is convenient to introduce dimensionless velocity variables:

$$\vec{c} = (m/2kT)^{1/2} \vec{v} , \qquad (5)$$

and a time variable:

$$\tau = (2kT/m)^{1/2} t . \qquad (6)$$

The linearized Boltzmann equation then takes the form:

$$\frac{\partial h}{\partial \tau} + \vec{c} \cdot \frac{\partial h}{\partial \vec{r}} = J h , \qquad (7)$$

where

$$J h = \pi^{-3/2} \int d\vec{c}_1 e^{-c_1^2} \int d\Omega \, F(g,\theta) [h(\vec{c}_1')+h(\vec{c}')-h(\vec{c}_1)-h(\vec{c})] \qquad (8)$$

with

$$g = |\vec{c}-\vec{c}_1| , \qquad F(g,\theta) = ngI(|\vec{v}-\vec{v}_1|,\theta) . \qquad (9)$$

The linearized collision operator J is a formally self-adjoint linear operator. For such an operator one first asks about its eigenvalues and eigenfunctions:

$$J \Psi_i = \lambda_i \Psi_i . \qquad (10)$$

Much is known about this question and much remains to be discovered. In general we know:
- i) There are exactly five eigenfunctions with eigenvalue zero: $1, \vec{c}, c^2$. These clearly correspond to the five additive constants of the motion in a binary collision: number of particles, three components of momentum, and energy.

ii) All other eigenvalues are negative. I won't show this here, it isn't difficult, but will remark that the solution of the initial value problem in the spatially homogeneous cases is of the form:

$$h(\vec{c},t) = \sum_i a_i \, e^{\lambda_i \tau} \, \psi_i(\vec{c}) \quad , \qquad (11)$$

where the a_i are determined from the initial value of h. The fact that the λ_i are non-positive means that for long times h goes to zero (exponentially) except for permanent changes in density, average velocity, and temperature. The quantities $|\lambda_i|^{-1}$ clearly represent relaxation times for the approach to equilibrium.

iii) The operator J is a scalar operator. This means that the eigenfunctions of J must be of the form:

$$\psi_{r\ell m}(\vec{c}) = R_{r\ell}(c) \, Y_{\ell m}(\hat{c}) \qquad (12)$$

where $Y_{\ell m}(\hat{c})$ is the spherical harmonic and the "radial eigenfunction" $R_{r\ell}(c)$ is a function only of the magnitude of $\vec{c}$.

More detailed properties of the spectrum of J depend upon the form of the collision cross section and, hence, on the form of the intermolecular force law. For the case of Maxwell molecules, in which the force is repulsive varying as the inverse fifth power of the separation, all the eigenfunctions and eigenvalues are known. The spectrum is purely discrete and unbounded from below.

For the case of elastic spheres $I(g,\theta) = d^2/4$ and we can write

$$J \psi = -m(c)\psi + K\psi , \qquad (13)$$

where

$$m(c) = \frac{nd^2}{\pi^{1/2}} \int d\vec{c}_1 \, e^{-c_1^2} g , \qquad (14)$$

and K, called the Hilbert operator, is given by:

$$K\psi = \frac{nd^2}{4\pi^{3/2}} \int d\vec{c}_1 \, e^{-c_1^2} \int d\Omega \, g[\psi(\vec{c}_1') + \psi(\vec{c}') - \psi(\vec{c}_1)]. \quad (15)$$

The function $m(c)$ has the form shown in Fig. 1. Much is known about the Hilbert operator for the hard sphere case; it has only positive eigenvalues and it is what the mathematicians call a completely continuous operator [5]. A completely continuous operator is one whose spectrum of eigenvalues is bounded, purely discrete and has zero as its only limit point. Now there is a well known theorem of H. Weyl and J. von Neumann which states that when any self-adjoint operator is perturbed by adding to it a completely continuous operator, the continuous spectrum of the first operator is unchanged. Since multiplication by a function is an operator whose spectrum is just the set of values of the function, we conclude that J has a continuous spectrum ranging from $-m(0) = -2nd^2\pi^{1/2}$ to $-\infty$. In addition it has been shown that there are an infinite number of discrete eigenvalues in the gap between zero and $-m(0)$ [6], and we even know something about the distribution of eigenvalues in the gap [7].

What about the spectrum of J for other force laws? Here very little is known. I would speculate that for repulsive power law forces falling off more rapidly than that for Maxwell molecules the spectrum is purely discrete, as in the Maxwell case [8]. For more realistic force laws, such as the Lennard-Jones, I think it fair to say that nothing is known.

3. The Mutilated Collision Operator

As we have heard in Professor Klein's paper, Boltzmann made what he called physically unrealizable assumptions about the mathematical form of his theory in order to make the exposition of the physics clear [9]. I trust I am not being presumptuous in saying that it is in this spirit that I introduce a simplified model for the collision operator. What I will do is replace the Hilbert operator K for elastic spheres by a still simpler operator given by

$$K_m h = m(c)[\rho + 2\vec{c}\cdot\vec{\phi} + \frac{2}{3}(c^2-\frac{3}{2})(\epsilon-\frac{3}{2})] \qquad (16)$$

where

$$\{\rho,\phi,\epsilon\} = \pi^{-3/2}\int d\vec{c}\, e^{-c^2}\{1,\vec{c},c^2\}h(\vec{r},\vec{c},t) \qquad (17)$$

are the moments of h associated with the five constants of the motion. The operator K_m is called the mutilated Hilbert operator, a picturesque terminology due, I believe, to M. Kac, from whom I first heard of this sort of model [10].

It is easy to see that when K is replaced by K_m in (13) the resulting operator still has eigenvalue zero for the five eigenfunctions 1, $\vec{c}$, c^2. For any function h which is orthogonal to these eigenfunctions, i.e., for which the five moments (17) vanish, the mutilated Hilbert operator gives zero and the corresponding mutilated collision operator simply multiplies the function by m(c). Hence the spectrum of the mutilated collision operator consists of a five-fold discrete eigenvalue zero and a continuous spectrum from $-m(0)$ to $-\infty$. This is shown schematically in Fig. 2 where the spectrum of the mutilated collision operator is compared with that for hard spheres.

4. The Normal Modes

Sound is a normal mode of oscillation of the gas, corresponding to a solution of the linearized Boltzmann equation (7) of the form:

$$h(\vec{r},\vec{c},\tau) = h(\vec{c})\, e^{i(\vec{k}\cdot\vec{r} - \omega\tau)} \,. \tag{18}$$

Here we consider the wave vector k as given and we expect that the complex frequency ω of the normal mode will be determined as a condition of existence of such solutions. For the sound mode the frequency ω as a function of k is the dispersion law.

When the form (18) is inserted in the linearized Boltzmann equation (7) we get

$$(\omega - \vec{k}\cdot\vec{c})h = iJh \,. \tag{19}$$

Replacing K by K_m, given by (16), in the expression (13) for J, this equation becomes after a little rearrangement:

$$[\omega+im(c)-\vec{k}\cdot\vec{c}]h = im(c)[\rho+2\vec{c}\cdot\vec{\phi}+\tfrac{2}{3}(c^2-\tfrac{3}{2})(\epsilon-\tfrac{3}{2}\rho)] \,. \tag{20}$$

Here the five moments ρ, $\vec{\phi}$, ε are just five constants, so this is a trivial algebraic equation for $h(\vec{c})$. However, the factor multiplying h can sometimes vanish and we have all learned long ago to beware of dividing by zero! What we must do is distinguish two cases.

Case 1. $[\omega+im(c)-\vec{k}\cdot\vec{c}] = 0$ for some $\vec{c}$. This will be the case if ω is anywhere within the shaded region in the complex ω-plane shown in Fig. 3. The solution of (20) is then

$$h(\vec{c}) = P \frac{im}{\omega+im+\vec{k}\cdot\vec{c}} [\rho+2\vec{c}\cdot\vec{\phi}+\frac{2}{3}(c^2-\frac{3}{2})(\varepsilon-\frac{3}{2}\rho)] + A\delta(\omega+im-\vec{k}\cdot\vec{c}) \quad (21)$$

where P denotes the principal value, $\delta(x)$ is the Dirac δ-function, and A is an undetermined constant. Note that in this case the normal mode solution is not a function but a distribution. We now determine the five moments, ρ, $\vec{\phi}$, ε by inserting this solution in the expression (17). The result is a set of five linear <u>inhomogeneous</u> equations for ρ, $\vec{\phi}$, ε. These can always be solved to express the moments in terms of A, which is the amplitude of the normal mode. When these expressions are inserted in (21) we have the explicit normal mode solution for this case. However, for these modes there is no functional relation between ω and k as a condition of existence, i.e. there is no dispersion law.

Case 2. $[\omega+(m(c)-\vec{k}\cdot\vec{c}] \neq 0$ for any $\vec{c}$. This will be the case whenever ω is outside the shaded region indicated in Fig. 3. Here there is no trouble about dividing by zero, so the solution of (20) is

$$h(\vec{c}) = \frac{im}{\omega+im-\vec{k}\cdot\vec{c}}[\rho+2\vec{c}\cdot\vec{\phi}+\frac{2}{3}(c^2-\frac{3}{2})(\varepsilon-\frac{3}{2}\rho)] \quad . \quad (22)$$

Again we determine ρ, $\vec{\phi}$, ε by inserting this solution in the expressions (17). This time, however, the result is a set of five linear <u>homogeneous</u> equations for ρ, $\vec{\phi}$, ε. These have a non-trivial solution if and only if the determinant of the coefficients vanishes. Setting this determinant equal to zero we get a relation of the form

$$F(\omega,k) = 0 \quad , \qquad (23)$$

which is the dispersion law for these modes. For each satisfying this relation, we obtain an explicit solution for the corresponding normal mode when we insert the corresponding ρ, $\vec{\phi}$, ε in (22). Note that in this case the normal mode solution is a function.

We see that the normal mode solutions fall into two classes. Those corresponding to case 1 are distributions, and there are infinitely many of them, corresponding to a two-dimensional continuum of values of ω. For these modes the imaginary part of ω is always less than $-m(0)$, so they are all strongly damped with a characteristic time of the order of the mean free time of the gas molecules. I call these modes Knudsen modes since they are related to the normal modes in the Knudsen regime where the mean free path is long compared with the wavelength.

The remaining class of normal mode solutions are those corresponding to case 2. These correspond to values of ω which are roots of (23) for fixed k. What we find, at least for small k (long wavelengths), is that there are exactly five such roots, growing out of the five-fold zero eigenvalue of J. The corresponding normal modes can be identified as:

i) Two longitudinal sound modes. The duality arises from the two directions in which a sound wave can propagate.
ii) Two transverse sound modes. The duality here arises from the two senses of polarization of a transverse wave.
iii) A heat conduction mode.

For obvious reasons we call these the hydrodynamic modes. Thus we see how it is that sound, and the other hydrodynamic modes, are singled out of the infinitely many normal mode solutions of the linearized Boltzmann equation.

There is an interesting point concerning what happens to the hydrodynamic modes as we approach the Knudsen regime, i.e., as k increases. What we find is that the roots of (23) corresponding to the longitudinal sound mode move along the path indicated by the dashed curve in Fig. 3 while those corresponding to the other modes move downward along the negative imaginary axis. When these roots reach the boundary of the shaded region they disappear; equation (23) has no roots for larger values of k. Thus, in the Knudsen regime there are only the normal modes corresponding to case 1.

5. Concluding Remarks

Most of the essential features of the normal mode solutions found in the preceding section can be shown, qualitatively at least, to carry over to the elastic sphere case, i.e., when the exact Hilbert operator (15) is used. Unfortunately the discussion must necessarily be couched in rather abstract terms and is not particularly instructive. There remain however many interesting questions, of which I will mention only one: How do the hydrodynamic modes disappear and the Knudsen modes arise in the case of Maxwell molecules for which the separation (13) of the collision operator cannot be made?

Notes

[1] G.E. Uhlenbeck, (this volume).
[2] E.H. Hauge, Phys. Rev. Letters, $\underline{28}$, 1501 (1972).
[3] J.D. Foch, Jr. and G.W. Ford, Studies in Statistical Mechanics, ed. J. de Boer and G.E. Uhlenbeck (North Holland, Amsterdam 1970), Vol. V.
[4] See, e.g., G.E. Uhlenbeck and G.W. Ford, Lectures in Statistical Mechanics (American Mathematical Society, Providence, 1963) Ch. IV.
[5] J.R. Dorfman, Proc. Nat. Acad. Sci. $\underline{50}$, 804 (1963); H. Grad in Rarefied Gas Dynamics, ed. J.A. Laurmann (Academic Press, New York, 1963) pp. 26-59.
[6] I. Kuscer and M.M.R. Williams, Phys. Fluids $\underline{10}$, 1922 (1967).
[7] C.C. Yan and G.H. Wannier, Bull. Am. Phys. Soc. $\underline{13}$, 899 (1968).
[8] For a partial result see C.H. Su and Young-ping Pao, Phys. Fluids $\underline{12}$, 552 (1969).
[9] M.J. Klein, (this volume).
[10] The mutilated collision operator is of course closely related to the so-called Krook models. See, e.g. E.P. Gross and E.A. Jackson, Phys. Fluids $\underline{2}$, 432 (1959).

Figure Captions

Fig. 1. The function m(c) for elastic spheres.

Fig. 2. Comparison of the spectrum of the elastic sphere collision operator, on the left, and that of the mutilated collision operator, on the right.

Fig. 3. The complex frequency plane. Values of ω in the shaded region correspond to Knudsen modes. The dashed curve indicates the path of the sound mode roots.

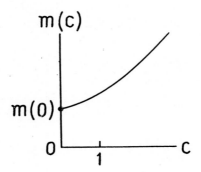

Fig. 1

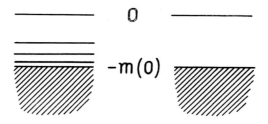

Fig. 2

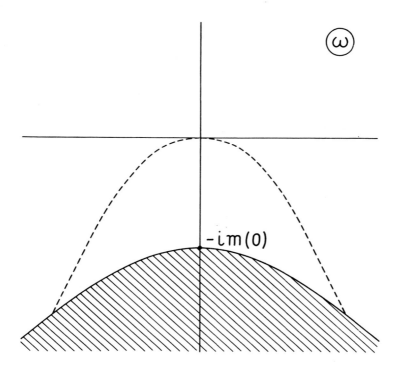

Fig. 3

THE GENERALIZATION OF THE BOLTZMANN EQUATION TO HIGHER DENSITIES

E.G.D. COHEN

The Rockefeller University, New York

ABSTRACT

The problem of the generalization of the Boltzmann equation to higher densities is discussed. The difficulties encountered in including in a systematic way multiple particle collisions are described. The present status of the problem is indicated.

1. Introduction

As was pointed out already by Professor Uhlenbeck [1], the Boltzmann equation is valid for dilute gases, where only binary collisions between particles occur. Up till the present day it is not clear how to generalize Boltzmann's intuitive arguments in a systematic way to include the effect of three - and more particle collisions in the equation. Therefore one has had to go back to the basic equation of statistical mechanics, the Liouville equation, and by expanding it somehow in the density, obtain in first approximation the Boltzmann equation and then correction terms to it due to multiple particle collisions.

The main purpose of my talk is to make it clear to you why such a straight forward generalization of the Boltzmann equation to higher densities is not possible. As a consequence of this, also a virial expansion of the transport coefficients does not seem to exist. Further consequences will be discussed by Dr. Dorfman [2].

2. Bogolubov's Results

Bogolubov was the first to succeed in deriving a generalized Boltzmann equation from the Liouville equation [3]. Let me briefly sketch for you, how Bogolubov's results can be obtained. In order to avoid unnecessary complications, I restrict myself, with Bogolubov, to gases with additive spherically symmetric, purely repulsive intermolecular potentials $\phi(r)$ (r is the interparticle distance), with finite range d.

Starting from the Liouville equation, one can obtain by integration over the coordinates and momenta of all particles but one, for appropriate boundary conditions and in the thermodynamic limit, the following equation for the single particle distribution function $F_1(\vec{r}_1,\vec{p}_1;t)$:

$$\frac{\partial F_1(x_1;t)}{\partial t} + \frac{\vec{p}_1}{m} \cdot \frac{\partial F_1}{\partial \vec{r}_1} = n \int dx_2 \, \theta_{12} \, F_2(x_1 x_2;t) \quad . \tag{1}$$

Here $F_1(x_1;t)$ is the probability density to find particle 1 in the phase $x_1 \equiv \vec{r}_1,\vec{p}_1$, n is the overall density of the system,

$$\theta_{12} = \frac{\partial \phi(r_{12})}{\partial \vec{r}_1} \cdot \frac{\partial}{\partial \vec{p}_1} + \frac{\partial \phi(r_{12})}{\partial \vec{r}_2} \cdot \frac{\partial}{\partial \vec{p}_2}$$

and F_2 is the pair distribution function. F_1 is, but for a change of variables, directly related to Boltzmann's f:
$n\, F_1(\vec{r}_1,\vec{p}_1;t) \rightarrow f(\vec{r},\vec{v};t)$.

The equation (1) is similar in form to the Boltzmann equation for f, except that the collision term on the right hand side contains the unknown function F_2. An equation for F_2 can be obtained from the Liouville equation in a similar fashion as for F_1, by integrating over the phase of all particles but two, expressing $\partial F_2/\partial t$ in terms of F_3 etc.

Thus an infinite set of coupled equations can be generated from the Liouville equation, generally called the B-B-G-K-Y-hierarchy, after a number of investigators who independently derived it from the Liouville equation.

To derive a generalized Boltzmann equation i.e. an equation of the form $\partial F_1/\partial t = O(F_1)$, where O is a time-independent operator, from this hierarchy, one has to express F_2 somehow in terms of F_1. Before I indicate how this can be done, let me remark that there are three basic lengths in the problem: the range of the intermolecular forces d, the mean free path λ and a macroscopic length L. For a moderately dense gas, one has $d \ll \ell \ll L$ or alternatively for the 3 basic times: $t_c = d/<v> \ll t_o = \lambda/<v> \ll T = L/<v>$, where $<v>$ is an average velocity (for instance the velocity of sound).

Bogolubov obtained an expression for F_2 in terms of F_1, by looking for a special solution of the hierarchy, using an expansion in powers of $(t_c/t_o) \sim nd^3$, not unlike the way in which Chapman and Enskog obtained a special solution of the Boltzmann equation by expanding in powers of (t_o/T). I shall follow here a different but equivalent procedure, which lends itself better to a critical discussion of the above mentioned difficulties.

F_2 can be expressed in a systematic way in terms of F_1 by using appropriate non-equilibrium generalizations of the cluster expansions introduced by Uhlenbeck, Kahn and de Boer in the theory of moderately dense gases in equilibrium i.e. by a generalization of the virial expansion known from equilibrium statistical mechanics [4].

Assuming that at the initial instant of time t = 0, the F_s factorize into products of F_1, one obtains in the thermodynamic limit the following formal expansion for F_2 in terms of F_1:

$$F_2(x_1 x_2;t) = F_2^{(0)}(x_1 x_2;t) + nF_2^{(1)}(x_1 x_2;t) + \ldots \quad (2)$$

where

$$F_2^{(0)}(x_1 x_2;t) = S_t(x_1 x_2) F_1(x_1,t) F_1(x_2;t)$$

and

$$F_2^{(1)}(x_1 x_2; t) = \int dx_3 [S_t(x_1 x_2 x_3) - S_t(x_1 x_2) S_t(x_1 x_3) -$$
$$S_t(x_1 x_2) S_t(x_2 x_3) + S_t(x_1 x_2)] \cdot$$
$$\cdot F_1(x_1; t) F_1(x_2; t) F_1(x_3; t)$$

etc.

Here

$$S_t(x_1 \ldots x_s) = S_{-t}(x_1 \ldots x_s) \prod_{i=1}^{s} S_t(x_1)$$

with

$$S_{-t}(x_1 \ldots x_s) = \exp(-t H_s(x_1 \ldots x_s)) \quad .$$

The streaming operator $S_{-t}(x_1 \ldots x_s)$, when acting on the phases of the particles $1 \ldots s$ transforms them into those a time t earlier, if the particles move under their mutual interaction only. The expansion (2) can be considered to be an expansion in the ratio of the first two basic relaxation times $t_c/t_o \sim nd^3$, where $nd^3 \ll 1$. It has a structure very similar to the virial expansion of the pair distribution function in equilibrium, which can be written in the form:

$$f_2(x_1 x_2) = f_2^{(0)}(x_1 x_2) + f_2^{(1)}(x_1 x_2) + \ldots \qquad (2a)$$

where

$$f_2^{(0)}(x_1 x_2) = e^{-\beta \phi(r_{12})} f_{eq}(p_1) f_{eq}(p_2)$$

and

$$f_2^{(1)}(x_1 x_2) = \int dx_3 [e^{-\beta\{\phi(r_{12})+\phi(r_{13})+\phi(r_{23})\}} -$$

$$e^{-\beta\phi(r_{12})} e^{-\beta\phi(r_{13})} - e^{-\beta\phi(r_{12})} e^{-\beta\phi(r_{23})} +$$

$$e^{-\beta\phi(r_{12})}] f_{eq}(p_1) f_{eq}(p_2) f_{eq}(p_3)$$

etc.

Here $f_{eq}(p)$ is the equilibrium momentum distribution function and $\beta = 1/kT$. Clearly the S_t-operators in (2) correspond to the Boltzmann factors in (2a), so that the expansion (2) can be considered as a formal dynamical generalization of the equilibrium virial expansion (2a). Also the cluster property of the integrands of the equilibrium $f_2^{(\ell)}$ ($\ell \geq 1$), which ensures the existence of the $f_2^{(\ell)}$, is supposed to carry over to the $F_2^{(\ell)}$ ($\ell \geq 1$) in the non-equilibrium expansion. The argument for this is that the multiple particle collisions that contribute to the $F_2^{(\ell)}$ are expected to take place effectively in a region of space of $O(r_0^3)$ only.

Bogolubov's generalized Boltzmann equation can be obtained from (2) for $t \gg t_c$, by taking the limit $t \to \infty$ in all the S_t-operators. For, in view of the above, the S_t-operators should have effectively attained their asymptotic values for such times.

Then one obtains, with equation (1), a closed equation for F_1 of the desired form:

$$\frac{\partial F_1}{\partial t} + \frac{\vec{p}_1}{m} \frac{\partial F}{\partial \vec{r}_1} = n\bar{J}(F_1 F_1) + n^2 \bar{K}(F_1 F_1 F_1) + \ldots \qquad (3)$$

where

$$\bar{J}(F_1 F_1) = \int dx_2 \, \theta_{12} \, S_\infty(x_1 x_2) \, F_1(x_1;t) \, F_1(x_2;t)$$

and

$$\bar{K}(F_1 F_1 F_1) = \int dx_2 \theta_{12} \int dx_3 [S_\infty(x_1 x_2 x_3) - S_\infty(x_1 x_2) S_\infty(x_1 x_3) -$$

$$S_\infty(x_1 x_2) S_\infty(x_2 x_3) + S_\infty(x_1 x_2)] F_1(x_1;t) F_1(x_2;t) F_1(x_3;t)$$

etc.

The first term on the right hand side of (3) can be shown to reduce to the usual Boltzmann collision term, if one neglects the difference in position of the two colliding molecules 1 and 2 and takes both F_1 at the same position $\vec{r}_1$. The second term on the right hand side of (3) contains the contributions of the particle collisions (through $S_\infty(x_1 x_2 x_3)$). These do not only include what one would call genuine triple collisions (cf. fig. 1a), but also sequences of at least 3 successive binary collisions (cf. fig. 1b). The reason that at least 3 binary collisions must occur is that the combination of S_∞-operators that appears in $\bar{K}$, is such that this term vanishes if only 1 or 2 binary collisions occur.

The equation (3) was obtained in a different form by Bogolubov.

3. Divergences

The equation (3) has all appearances of being the correct generalization of the Boltzmann equation. Not only does the form look right, the basic expansion (2) reduces term by term to the equilibrium virial expansion when one takes the limit $t \to \infty$ everywhere i.e. replaces S_t by S_∞ and assumes that $nF_1(t \to \infty)$ is the equilibrium distribution $f_{eq}(p)$.

Yet the equation (3) is **not** correct since a closer analysis of the collision terms on the right hand side has shown that the crucial cluster property of the integrands does **not** obtain. In fact, all terms on the right hand side of (3) diverge from a certain term onwards. While in 3-dimensions the four-body – and all higher order terms diverge, in 2-dimensions already the three-body and all higher terms diverge. This was discovered independently by Weinstock [5], Frieman and Goldman [6] and Dorfman and Cohen [7]. In order to understand the origin of these divergences one can best go back to the basic expansion (2) and investigate in more detail the replacement of S_t by S_∞. I shall illustrate the argument on the three-body term in 2-dimensions, since this is the simplest case.

If the three-body collision term in (3) is to exist for a general class of F_1, the volume in the phase space of the 3 particles 1, 2, 3 of all collisional events that contribute to $F_2^{(1)}(t)$ should remain finite, if S_t is replaced by S_∞. To investigate this, a complete dynamical analysis of all collisional events that contribute to $F_2^{(1)}$ in time t should be made. I remark that up till now a complete analysis has only been made for hard disks in 2-dimensions and for hard spheres in 3-dimensions [8]. In figure 1 two typical events are given. Of course, the genuine triple collisions (cf. fig. 1a) give a finite contribution to $F_2^{(1)}$. For the sequence of 3 successive binary collisions of fig. 1b, I shall estimate the phase volume [9].

To do this, the crucial question is: what is the volume $d\Gamma_3(t_1)$ of the phase space of particle 3, for which a recollision of the particles 1 and 2 occurs between time t_1 and $t_1 + dt_1$, when these same particles collided at $t = 0$. Now the recollision is determined by the direction of the momentum of particle 2 at t', which in turn is determined by the direction of the momentum of particle 3 at $t = 0$. For $t \gg t_c$, the freedom of the direction of the momentum of particle 2 will be

$$\sim \frac{r_o}{\ell} d\ell \sim \frac{t_c}{t_1 - t'} dt_1 \sim \frac{t_c}{t_1} dt_1$$

since $t' \sim t_1$, for t', $t_1 \gg t_c$. Thus $d\Gamma_3(t_1) \sim dt_1/t_1$. Therefore the total volume $\Gamma_3(t)$ of the phase space of particle 3 for a recollision to occur within the time t will be

$$\sim \int^t d\Gamma_3(t_1) \sim \int^t \frac{dt_1}{t_1} \sim \log t/t_c \ .$$

Consequently, when S_t is replaced by S_∞, the number of contributions to $F_2^{(1)}(t)$ of collision sequences of the type sketched in fig. 1b, will grow beyond bounds and therefore the three-body collision term $\bar{K}(F_1 F_1 F_1)$ in equation (3) will diverge logarithmically [10].

Similar arguments hold for the higher order collision terms. In fig. 2 some collisional events are sketched that contribute to the four-body collision term. In general it is believed, but no proof exists, that for the interparticle potential considered here, sequences of ℓ successive binary collisions constitute the class of contributions to the ℓ body collision term, that grow fastest with time.

In table I, I have summarized the long time behaviour of the $F_2^{(\ell)}(t)$ ($\ell \geq 1$); this also indicates the nature of the worst divergences that occur in the various collision terms in the generalized Boltzmann equation (3).

Table I

Collision Term	Behaviour for $t \gg t_c$	
	$d=2$	$d=3$
3-body	$\log(t/t_c)$	(t_c/t)
4-body	(t/t_c)	$\log(t/t_c)$
ℓ-body	$(t/t_c)^{\ell-3}$	$(t/t_c)^{\ell-4}$

Thus with the exception of the three-particle correction term in 3-dimensions, all correction terms to the Boltzmann equation appearing in equation (3) diverge.

I remark that one has not been able to generalize Boltzmann's proof of the approach to thermal equilibrium in any way. Not even when one restricts oneself to binary and three-particle collisions (in 3-dimensions) and assumes in addition a spatially homogeneous system, where F_1 only depends on $\vec{p}_1$, but not on $\vec{r}_1$, has a proof been given.

4. Hydrodynamical Consequences

The convergent part of the generalized Boltzmann equation (3) i.e. the equation

$$\frac{\partial f}{\partial t} + \vec{v} \frac{\partial f}{\partial \vec{r}} = \bar{J}(ff) + \bar{K}(fff) \qquad (4)$$

has been investigated (in 3-dimensions) for times $t \gg t_o$ in great detail by Choh and Uhlenbeck [11]. For such time the system is in a hydrodynamical state and Bogolubov already remarked [3a] that the connection of the generalized Boltzmann equation with hydrodynamics can be obtained by application of the Chapman-Enskog development:

$$f = f_o + \mu f_1 + \ldots \qquad (5)$$

Here the order of magnitude of the terms is characterized by the parameter μ, the relative variation of the macroscopic quantities (local density $n(\vec{r},t)$, local velocity $\vec{u}(\vec{r},t)$ and local temperature $T(\vec{r},t)$) over a mean free path, so that for instance $\mu \sim \frac{\lambda}{n}\nabla n \sim \frac{\lambda}{L} \sim \frac{t_o}{T}$. The expansion (5) can therefore also be seen as an expansion in the ratio of the last two basic relaxation times (t_o/T).

Choh and Uhlenbeck were able to derive from the equation (4) the Euler- and Navier- Stokes equations with explicit expressions for the thermodynamic - and transport properties of the gas in terms of the interparticle potential. In fact, for the thermodynamic properties they obtained the first two terms in the virial expansion of the thermodynamic properties, well known from equilibrium statistical mechanics. For instance, for the pressure p they found:

$$p = mkT[1+nB_2(T)] \qquad (6)$$

where $B_2(T)$ is the second virial coefficient. The ideal gas law is obtained from the Boltzmann equation, while the second virial coefficient follows from the difference in position of the two colliding molecules in $\bar{J}(ff)$, which allows collisional transfer of momentum between the two colliding particles.

For the transport coefficients, the viscosity η and the thermal conductivity λ, they obtained the first density correction to the Chapman-Enskog results, $\eta_o(T)$ and $\lambda_o(T)$ respectively, derived from the Boltzmann equation:

$$\eta(n,T) = \eta_o(T) + n\eta_1(T)$$
$$\lambda(n,T) = \lambda_o(T) + n\lambda_1(T) \ . \qquad (7)$$

Here $\eta_1(T)$ and $\lambda_1(T)$ contain contributions from collisional transfer in $\bar{J}$, as well as from three-particle collisions in $\bar{K}$. Dr. Sengers has computed $\eta_1(T)$ and $\lambda_1(T)$ for a gas of hard spheres. I refer to his talk for details of this calculation [12].

In general, while one can expect that the full virial expansions for the thermodynamic properties can be obtained from the right hand side of equation (3) [13], the transport coefficients will have divergent contributions due to the divergences of the higher order collision terms in (3). For instance in 3-dimensions the next term $n^2 \eta_2(T)$ in the expansion (7) for η will diverge logarithmically, due to the logarithmic divergence of the four-body collision term:

$$\eta_2(T) = \eta_2'(T) \lim_{t \to \infty} \log(t/t_c) + \eta_2''(T) \qquad (8)$$

and similarly for $\lambda_2(T)$.

In 2-dimensions a result like (8) is obtained for $\eta_1(T)$ and $\lambda_1(T)$.

5. Remarks

1. From a physical point of view the origin of the divergences can be readily understood. They are caused by extended sequences of successive collisions (such as sketched in figs. 1b, 2b, 2c), with no restriction on the free paths of the particles involved. Clearly a cut-off of these paths at the mean free path must somehow be introduced. This implies, however, that a virial-type of expansion of the properties of a moderately dense gas not in equilibrium in terms of the properties of small groups of particles is not possible and that collective effects, due to all particles together, have to be considered as well.

Thus a rearrangement of the expansion (2) should be performed such that, when introduced in equation (1), the collision terms contain a "damping", that leads to an effective mean free path cut-off of the particle paths. Such a rearrangement has been carried out by Weinstock [14] and Kawasaki and Oppenheim [15] for a linearized theory and by Frieman and Goldman [16] and Dorfman and Cohen [17] for the non-linear theory. The form of the rearranged generalized Boltzmann equation appears to be quite different from the equation (3) in that the collision terms not only contain a damping operator but are non-local in space and time. This has been verified in detail for the three-body term only, up till now.

2. In the hydrodynamical stage, the damping operator is expected to effect a mean-free path cut-off of the divergent expressions for the transport coefficients found before. Therefore one would expect to find that the divergent contribution $\sim \lim_{t \to \infty} \log(t/t_c)$ to η in (8), is replaced

by a convergent contribution $\sim \log(t_o/t_c)$, leading to a term that depends logarithmically on the density. Thus one would expect the following density expansion for η in 3-dimensions [18]:

$$\eta(n,T) = \eta_o(T) + n\eta_1(T) - n^2 \log n \eta_2'(T) + n^2 \eta_2''(T) + \ldots \quad (9)$$

and similarly for $\lambda(n,T)$. Note that if the above obtains the coefficient of the logarithmic terms in (9) is identical with that of the logarithmically divergent term in (8). Neither this, nor the correctness of the expansion (9) have been proved in general, but they have been verified for a simplified case: the Lorentz model [19].

A convincing experimental confirmation of the existence of the logarithmic term in (9) has not yet been obtained [20].

3. In 2-dimensions the expected mean free path cut-off of the divergences in the transport coefficients does <u>not</u> seem to occur. I refer to Dr. Dorfman's talk for a further discussion of this point [2].

4. A generalization of the general theory to include the effects of attractive intermolecular forces, has not yet been given.

From the above it will be clear that the systematic generalization of the Boltzmann equation to higher densities is difficult and far from accomplished.

The greater then is our admiration for the achievement of Boltzmann, who derived in a simple and intuitive way, such a rich and accomplished equation.

Acknowledgment

I am greatly indebted to Dr. J.R. Dorfman, Dr. J.V. Sengers and especially to Professor G.E. Uhlenbeck for suggestions and remarks.

References

[1] G.E. Uhlenbeck, these proceedings.

[2] J.R. Dorfman, these proceedings.

[3] See
   a) N.N. Bogolubov in Studies in Statistical Mechanics, Vol. I, (J. de Boer and G.E. Uhlenbeck eds.), North-Holland Publishing Company, Amsterdam, 1962, p. 5;
   b) G.E. Uhlenbeck and G.W. Ford in Lectures in Applied Mathematics, Vol. I, Amer. Math. Soc., Providence, 1963, Ch. 7.

[4] a) E.G.D. Cohen, Physica 28, 1025 (1962); J. Math. Phys. 4, 183 (1963).
   b) M.S. Green and R. Piccirelli, Phys. Rev. 132, 1388 (1963).

[5] J. Weinstock, Phys. Rev. 132, 454 (1963); 140A, 460 (1965).

[6] R. Goldman and E.A. Frieman, Bull. Amer. Phys. Soc., 10, 531 (1965); J. Math. Phys. 7, 2153 (1966); 8, 1410 (1967).

[7] J.R. Dorfman and E.G.D. Cohen, Phys. Letters 16, 124 (1965); J. Math. Phys. 8, 282 (1967).

[8] See
   a) E.G.D. Cohen in Lectures in Theoretical Physics, Vol. VIII A, (W.E. Brittin, ed.), The University of Coloredo Press, Boulder, 1966, Appendix II;
   b) W.R. Hoegy and J.V. Sengers, Phys. Rev. A2, 2461 (1972).

[9] For a more detailed discussion see:
   a) ref. 8a, Appendix III; and especially
   b) J.V. Sengers in Lectures in Theoretical Physics, Vol. IX C, (W.E. Brittin, ed.), Gordon and Breach,

New York, 1967, p. 335 and in Kinetic Equations, (R.L. Liboff and N. Rostoker, eds.), Gordon and Breach, New York, 1971, p. 137.

[10] This argument also shows why there are no difficulties for the three-body term in 3-dimensions. For, in that case, the angle $\sim (t_c/t_1)$ is replaced by a space angle $\sim (t_c/t_1)^2$, so that $\Gamma_3(t)$ remains finite for $t \to \infty$.

[11] a) S.T. Choh and G.E. Uhlenbeck, The Kinetic Theory of Dense Gases, University of Michigan Report, 1958;
b) E.G.D. Cohen in Fundamental Problems in Statistical Mechanics, Vol. I, (E.G.D. Cohen, ed.) North-Holland Publishing Company, Amsterdam, 1962, p. 110;
c) Ref. 3b, Ch. 6.

[12] J.V. Sengers, these proceedings.

[13] This may seem contradictory in view of the divergence difficulties mentioned above. However, the collision sequences that cause the divergences, do not contribute to the thermodynamic properties of the gas.

[14] J. Weinstock, Phys. Rev. Letters, 17, 130 (1966).

[15] K. Kawasaki and J. Oppenheim, Phys. Rev. 139A, 1763 (1965)

[16] a) E.A. Frieman and R. Goldman, J. Math. Phys. 7, 2153 (1966); 8, 1410 (1967);
b) J.R. Dorfman, Renormalized Kinetic Equations, University of Maryland Techn. Note BN 618, University of Maryland, 1969.

[17] J.R. Dorfman and E.G.D. Cohen, to be published.

[18] The coefficients $\eta_2'(T)$ and $\eta_2''(T)$ defined in (9) depend on the units chosen for n. A more meaningful expansion parameter is the dimensionless quantity nd^3.

[19] a) A. Weyland and J.M.J. van Leeuwen, Physica $\underline{36}$, 457 (1967); $\underline{38}$, 35 (1968);
b) C. Bruin, Physica, to be published.
[20] a) H.M. Hanley, R.D. MacCarty and J.V. Sengers, J. Chem. Phys. $\underline{50}$, 857 (1969);
b) J. Kestin, E. Paykos and J.V. Sengers, Physica $\underline{54}$, 1 (1971).

Figure Captions

Fig. 1 Collision events contributing to the three-body collision term
   a) genuine triple collision of particles 1, 2 and 3.
   b) recollision of particles 1 and 2.

Fig. 2 Some collision events contributing to the four-body collision term.
   a) Genuine quadruple collision between the particles 1, 2, 3 and 4
   b), c) recollisions of particles 1 and 2.

Fig. 3 Recollision of particles 1 and 2, due to particle 3.

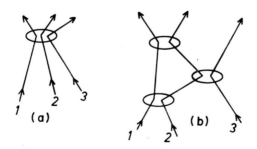

Fig. 1

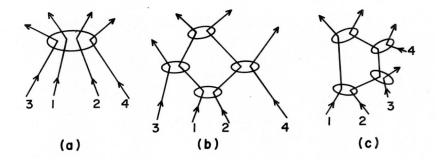

Fig. 2

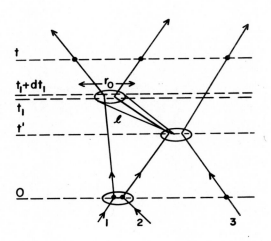

Fig. 3

THE THREE-PARTICLE COLLISION TERM IN THE GENERALIZED BOLTZMANN EQUATION

J.V. SENGERS

Institute for Molecular Physics
University of Maryland, College Park, MD 20742 U.S.A.

ABSTRACT

Generalization of the Boltzmann equation to a moderately dense gas leads to a new collision term which incorporates collisions among three molecules. An analysis of this three-particle collision term is presented for a gas of hard spheres and its effect on the transport properties is discussed.

1. Introduction

The Boltzmann equation describes the rate of change of the single-particle distribution function f for a dilute gas. A systematic procedure for generalizing the Boltzmann equation to higher densities was first proposed by Bogoliubov [1]. Methods for generalizing the Boltzmann equation have subsequently been studied by many authors and the developments are discussed in the paper of Professor Cohen [2].

Retaining only the first few terms, the generalized Boltzmann equation has the form

$$\frac{\partial f(\vec{r}_1,\vec{v}_1;t)}{\partial t} + \vec{v}_1 \cdot \frac{\partial f(\vec{r}_1,\vec{v}_1;t)}{\partial \vec{r}_1} = \bar{J}(ff) + \bar{K}(fff) + \ldots, \quad (1)$$

where $\vec{r}_1$, $\vec{v}_1$ represent the position and velocity of a molecule labeled 1 and where $\bar{J}$, $\bar{K}$, ... are time independent integral operators acting on the function $f(\vec{r}_1,\vec{v}_1;t)$. The operator $\bar{J}$ accounts for the effect of binary collisions and $\bar{K}$ for the effect of collisions that involve <u>three</u> particles. It is the purpose of this paper to elucidate the nature of the three-particle collisions and how they enter into the generalized Boltzmann equation. In doing so, we shall restrict ourselves to a gas of hard sphere molecules. The modifications needed to extend the results to gases with a more realistic potential have not yet been examined adequately.

2. Binary Collision Term

The Boltzmann equation for a gas of hard spheres with mass m and diameter σ reads [3]

$$\frac{\partial f(\vec{r}_1,\vec{v}_1;t)}{\partial t} + \vec{v}_1 \cdot \frac{\partial f(\vec{r}_1,\vec{v}_1;t)}{\partial \vec{r}_1} = J(ff) \qquad (2)$$

where

$$J(ff) = \sigma^2 \int d\vec{v}_2 \int d\vec{k} \; \vec{v}_{21} \cdot \vec{k} \{ f(\vec{r}_1,\vec{v}_1';t) f(\vec{r}_1,\vec{v}_2';t) - f(\vec{r}_1,\vec{v}_1;t) f(\vec{r}_1,\vec{v}_2;t) \} \; . \qquad (3)$$

Here $\vec{v}_{21} = \vec{v}_2 - \vec{v}_1$ is the relative velocity of particles 2 and 1 and $\vec{k}$ the perihelion vector of a binary collision between molecules 1 and 2 (unit vector along the line of centers at time of contact). The effect of collisions on the rate of change of the distribution function is twofold. First, certain molecules with velocity $\vec{v}_1$ will collide with molecules with velocity $\vec{v}_2$; the number of such collisions is proportional to $f(\vec{r}_1,\vec{v}_1;t) f(\vec{r}_1,\vec{v}_2;t)$. Secondly, molecules with velocity $\vec{v}_1'$ will collide with molecules with velocity $\vec{v}_2'$ such that after the collision their velocities become $\vec{v}_1$ and $\vec{v}_2$. Thus $\vec{v}_1'$ and $\vec{v}_2'$ may be interpreted as the <u>initial</u> velocities of the molecules in a binary collision with perihelion vector $\vec{k}$ and <u>final</u> velocities $\vec{v}_1$ and $\vec{v}_2$.[+]

[+] Using the concept of inverse collisions one usually transforms $\vec{v}_1'$, $\vec{v}_2'$ into the final velocities for a binary collision with perihelion vector $-\vec{k}$ and initial velocities $\vec{v}_1$, $\vec{v}_2$. Since a generalization of the concept of inverse collisions to three-particle collisions is not obvious, we use here the Bogoliubov form of the Boltzmann collision term, in which the consideration of inverse collisions is avoided.

We represent such a collision schematically by the diagram of Fig. 1. The lines in this figure indicate particle trajectories and the circles indicate the location of the spheres at the time of contact. The particles traverse the trajectories from top to bottom. However, we read the diagram from bottom to top, that is, we consider the velocities $\vec{v}_1'$, $\vec{v}_2'$ as functions of $\vec{v}_1$, $\vec{v}_2$ (and $\vec{k}$).

The binary collision term $\bar{J}(ff)$ in the generalized Boltzmann equation (1) differs from the Boltzmann collision term $J(ff)$, since it also incorporates the difference in position of the colliding molecules. This term was first considered by Enskog [3,4]

$$\bar{J}(ff) = \sigma^2 \int d\vec{v}_2 \int d\vec{k}\ \vec{v}_{21} \cdot \vec{k} \{f(\vec{r}_1, \vec{v}_1'; t) f(\vec{r}_1 + \sigma\vec{k}, \vec{v}_2'; t) -$$

$$f(\vec{r}_1, \vec{v}_1; t) f(\vec{r}_1 - \sigma\vec{k}, \vec{v}_2; t)\} \quad . \tag{4}$$

This Boltzmann-Enskog collision term is recovered from the Bogoliubov theory, at least up to terms linear in the gradients [5]. It reduces to the Boltzmann collision term, if one neglects variations of the distribution function f over distances of the order of the molecular interaction range σ.

As discussed in the previous lecture [2], the expansion (1) of the generalized Boltzmann equation implies an expansion of the thermal conductivity λ and the viscosity η in terms of the density n

$$\lambda = \lambda_0 + \lambda_1 n + \ldots$$
$$\eta = \eta_0 + \eta_1 n + \ldots \tag{5}$$

The dilute gas values λ_o and η_o are determined by the solutions of the Boltzmann equation (2). The first density corrections λ_1 and η_1

$$\lambda_1 = \lambda_1^{(2)} + \lambda_1^{(3)}$$
$$\eta_1 = \eta_1^{(2)} + \eta_1^{(3)} \tag{6}$$

contain contributions $\lambda_1^{(2)}$, $\eta_1^{(2)}$ from binary collisions, as well as contributions $\lambda_1^{(3)}$, $\eta_1^{(3)}$ from three-particle collisions. The binary collisions enter via the difference in position in $\bar{J}(ff)$ and yield [3]

$$\lambda_1^{(2)} = \frac{4}{5} \pi \sigma^3 \lambda_o$$
$$\eta_1^{(2)} = \frac{8}{15} \pi \sigma^3 \eta_o \tag{7}$$

These terms incorporate the effect of collisional transfer of momentum and energy.

For a discussion of the three-particle collision contributions $\lambda_1^{(3)}$, $\eta_1^{(3)}$ to the first density corrections λ_1, η_1, it will be sufficient to approximate $\bar{K}(fff)$ in the generalized Boltzmann equation (1) by $K(fff)$; $K(fff)$ follows from $\bar{K}(fff)$, if we neglect the spatial dependence of the distribution function f. Just as the difference between $J(ff)$ and $\bar{J}(ff)$ does not affect the dilute gas values λ_o, η_o, so will the difference between $K(fff)$ and $\bar{K}(fff)$ not affect the first density corrections λ_1, η_1. Therefore, in discussing the three-particle collision term we shall assume spatial homogeneity $f(\vec{r},\vec{v};t) = f(\vec{v};t)$ and treat the position $\vec{r}$ as a parameter not considered explicitly.

3. Enskog Theory

In the Boltzmann equation it is assumed that both the positions and the velocities of two molecules which are about to collide, are uncorrelated. The assumption, that the velocities prior to a collision are uncorrelated, is often called the assumption of molecular chaos [3,6].

An intuitive generalization of the Boltzmann equation to a dense gas of hard spheres was proposed by Enskog [4]. For our present purposes it is sufficient to discuss the Enskog theory for the spatially homogeneous case. Enskog estimated the correlations in configuration space by assuming that they are independent of the velocities and that they are the same as for a dense gas in <u>equilibrium</u>. For the probability distribution in velocity space he retained the assumption of molecular chaos. For a spatially homogeneous gas the Boltzmann collision term $J(ff)$ is thus replaced with

$$J_E(ff) = g(\sigma) \, J(ff) \quad , \tag{8}$$

where $g(\sigma)$ is the value of the equilibrium radial distribution function at the distance $r_{12} = \sigma$.

The radial distribution function $g(\sigma)$ can be represented by a virial series [7]

$$g(\sigma) = 1 + n\int d\vec{r}_3 \, f_{13} \, f_{23} + \ldots = 1 + \frac{5}{12} \pi \, n \, \sigma^3 + \ldots, \tag{9}$$

where f_{ij} is the Mayer function: $f_{ij} = -1$ for $r_{ij} < \sigma$ and $f_{ij} = 0$ for $r_{ij} > \sigma$. If we substitute this virial series into (8), we obtain an expansion for the collision term that can be compared with the expansion (1) for the generalized Boltzmann equation. We thus conclude that the Enskog theory approximates the three-particle collision term K(fff) by

$$K_E(fff) = \frac{5}{12} \pi n \sigma^3 J(ff) \quad . \tag{10}$$

In order to deduce the transport properties it is sufficient to consider a perturbation linear in the gradients. One then finds for a spatially homogeneous gas that $g(\sigma)\lambda$ and $g(\sigma)\eta$ in the Enskog theory may be identified with the values λ_o and η_o, respectively, for the dilute gas

$$\lambda_E = \frac{\lambda_o}{g(\sigma)} \quad , \quad \eta_E = \frac{\eta_o}{g(\sigma)} \quad . \tag{11}$$

Again using the expansion (9) for $g(\sigma)$, we conclude that the Enskog theory estimates the three-particle contributions $\lambda_1^{(3)}$ and $\eta_1^{(3)}$ by

$$\lambda_E^{(3)} = -\frac{5}{12}\pi\sigma^3\lambda_o \quad , \quad \eta_E^{(3)} = -\frac{5}{12}\pi\sigma^3\eta_o \quad . \tag{12}$$

4. Three-particle Collision Term

The three-particle collision term K(fff) can be written as [8,9]

$$K(fff) = \int dx_2 dx_3 \; \theta_{12} \{S_\infty(x_1 x_2 x_3) - S_\infty(x_1 x_2)S_\infty(x_1 x_3)$$
$$- S_\infty(x_1 x_2)S_\infty(x_2 x_3) + S_\infty(x_1 x_2)\} \tag{13}$$
$$\cdot f(\vec{v}_1;t) \; f(\vec{v}_2;t) \; f(\vec{v}_3;t) \quad .$$

Here θ_{12} is a differential operator

$$\theta_{12} = \frac{1}{m} \frac{\partial \phi(r_{12})}{\partial \vec{r}_1} \cdot \frac{\partial}{\partial \vec{v}_1} + \frac{1}{m} \frac{\partial \phi(r_{12})}{\partial \vec{r}_2} \cdot \frac{\partial}{\partial \vec{v}_2} \quad ,$$

where $\phi(r_{12})$ is the intermolecular potential, $x_i = (\vec{v}_i, \vec{r}_i)$ represents the velocity $\vec{v}_i$ and position $\vec{r}_i$ of particle i, and

$$S_\infty(x_1 \ldots x_s) = \lim_{t \to \infty} S_t(x_1 \ldots x_s)$$

are streaming operators discussed in the paper of Professor Cohen [2]. The potential $\phi(r_{12})$ for hard spheres is a discontinuous function of r_{12}, but one can nevertheless give a well defined meaning to the operator θ_{12} [10].

The differences between the new collision term (13) and the expression (3) for the Boltzmann collision term J(ff) are two fold. First, J(ff) contains the dynamics of two particles, while K(fff) contains the dynamics of three particles. Secondly, J(ff) contains a two-dimensional <u>surface</u> integral ($\int d\vec{k}$) which depends on the asymptotic velocities $\vec{v}_1'$, $\vec{v}_2'$ and $\vec{v}_1$, $\vec{v}_2$ before and after the collision. On the other hand, the expression (13) for K(fff) contains a six-dimensional <u>volume</u> integral ($\int d\vec{r}_2 d\vec{r}_3$) with contributions from those configurations where $\theta_{12} \neq 0$, i.e. where particle 1 and 2 are actually in the process of colliding. The question arises, whether, in analogy to (3), K(fff) can be transformed into a five-dimensional surface integral which depends on the asymptotic velocities before (and between) and after collisions. For a gas of hard spheres this question can be answered in the affirmative, as pointed out by Green [11,12]. As a consequence, the three-particle collision term K(fff) can be related to a set of well defined collision sequences among three particles in very much the same way as the Boltzmann collision term J(ff) is related to the collision shown in Fig. 1.

In order to understand the structure of K(fff) it is advantageous to make a distinction between genuine triple collisions and successive binary collisions. I define a genuine triple collision as a collision in which more than one pair of molecules are inside each others interaction range. Two hard spheres whose centers are separated by a distance less than (but not equal to) the diameter σ, may be said to be overlapping. Genuine triple collisions among three hard spheres are collisions in which two particles are colliding, while, in addition, at least one of the two colliding particles is overlapping with the third particle. For this reason we refer to the genuine triple collisions as overlap collisions. We may further distinguish between double-overlap and single-overlap collisions. In a <u>double-overlap</u> collision both colliding molecules are overlapping with the third particle, while in a <u>single</u>-<u>overlap</u> collision only one of the two colliding particles is overlapping with the third particle [13,14].

A special feature of a gas of hard spheres is, that the duration of an individual binary collision may be neglected relative the time between successive collisions. Nevertheless, genuine triple collisions, in the sense defined above, do enter into the generalized Boltzmann equation as a consequence of the finite size of the molecules. This may, perhaps, be best understood by first realizing that the density dependence of the equilibrium properties, such as $g(\sigma)$ in (9), is determined by integrals over excluded volume configurations. These excluded volume configurations are not encountered in reality, but need to be

considered to correct the ideal gas law. Similarly, the
Boltzmann equation considers all binary collisions, irrespective of whether the region is already excluded by the
presence of a third particle. Thus a consideration of the
overlap collisions is needed to correct the Boltzmann
equation. The overlap collisions may be interpreted as
accounting for the excluded volume effects in the generalized Boltzmann equation.

It turns out that the three-particle collision term
can be decomposed into a series of four terms [13,14]

$$K(fff) = \sum_{\mu=1}^{4} K_\mu(fff) \tag{14}$$

related to the dynamics of one, two, three and four successive collisions among three particles. We shall represent the various collision sequences by diagrams in Figs.
2 - 5, just as the diagram of Fig. 1 was used to represent
the collision sequence of the Boltzmann collision term. The
lines in Figs. 2 - 5 indicate again particle trajectories,
while the circles indicate schematically the location of
the particles at the time of a collision. The explicit formulas for the collision terms $K_\mu(fff)$ are presented in the
Appendix. In this paragraph we restrict ourselves to an
explanation of the various collision sequences that enter
into $K_\mu(fff)$.

The first term $K_1(fff)$ accounts for binary collisions in which both colliding particles are overlapping
with a third particle (double-overlap collisions). Such a
double-overlap collision is shown schematically in Fig. 2.
The first collision term $K_1(fff)$ is the same as the Boltzmann collision term $J(ff)$, represented by Fig. 1, except
that we now have to integrate in addition over all positions

of particle 3 for which $f_{13} \neq 0$ and $f_{23} \neq 0$. This integration gives the first density correction to the radial distribution function (9), so that $K_1(fff)$ is precisely the three-particle collision term (10) of the Enskog theory

$$K_1(fff) = K_E(fff) \quad . \tag{15}$$

This term $K_1(fff)$ contains the dynamics of only one binary collision and presents a correction to the Boltzmann collision integral solely due to the excluded volume effect.

The terms $K_2(fff)$, $K_3(fff)$ and $K_4(fff)$ in (14) are corrections to the Enskog theory due to sequences of, respectively, two, three and four successive collisions, shown in Figs. 3 - 5. In each case, we consider a collision between particles 1 and 2 at the bottom of the diagram, just as in Fig. 1. However, in contrast to Fig. 1, we now consider all possible trajectories along which particles 1 and 2 may have arrived at this collision due to the interaction with a third particle 3.

The second term $K_2(fff)$ is related to sequences of __two__ successive collisions of which at least one is a single-overlap collision. The three possible events of this type are shown in Fig. 3. The three events differ in that in Fig. 3a both collisions are single-overlap collisions, while in Fig. 3b only the later collision and in Fig. 3c only the earlier collision is a single-overlap collision. This term $K_2(fff)$, therefore, accounts for a combination of excluded volume (overlap collisions) and dynamical effects.

The term $K_3(fff)$ corresponds to sequences of __three__ successive collisions, shown in Fig. 4. In earlier papers [13,14] we have referred to these sequences as recollisions (R), cyclic collisions (C) and hypothetical collisions (H).

Note that in the diagram of Fig. 4c the intermediate collision is a noninteracting collision. That is, particles 1 and 3 are indeed aimed to collide, but they pass through each others interaction sphere and continue along the direction of their original trajectories. This diagram accounts for the possibility that the collision between particles 1 and 2 cannot occur due to the interference of particle 3. It, therefore, represents a dynamical screening effect of particle 3. In the sequences of Fig. 4 particle 2 never overlaps with either particle 1 or 3. Thus the sequences of Fig. 4 do not contain any overlap collisions and $K_3(fff)$ accounts for correlations that are of a purely dynamical nature.

Finally, the term $K_4(fff)$ corresponds to sequences of <u>four</u> successive collisions, shown in Fig. 5. The collision sequence of Fig. 5a has been considered earlier by Sandri et al. [15] and by Murphy and Cohen [16]. In addition, we need to consider the collision sequence of Fig. 5b, in which one of the intermediate collisions is a noninteracting collision. Any other possible sequence of four successive collisions that gives a non-vanishing contribution to $K_4(fff)$ can be obtained from those shown in Fig. 5 by time reversal and/or suitable permutation of the particle numbers [17].

For a gas of hard spheres the three-particle collision term is described completely by the collision sequences presented above. In particular, it can be shown that the expansion (14) terminates after sequences of four successive collisions [17].

5. Transport Properties

As a next step we consider the contribution of the three-particle collision term to the transport properties. As discussed in Section 2, the thermal conductivity λ and the viscosity η depend on the density n as

$$\lambda = \lambda_o + \{\tfrac{4}{5}\pi\sigma^3\lambda_o + \lambda_1^{(3)}\}n + \ldots$$
$$\eta = \eta_o + \{\tfrac{8}{15}\pi\sigma^3\eta_o + \eta_1^{(3)}\}n + \ldots \qquad (16a)$$

where $\lambda_1^{(3)}$ and $\eta_1^{(3)}$ represent the effects of three-particle collisions. The corresponding formula for the self-diffusion D reads

$$nD = D_o + D_1 n + \ldots \qquad (16b)$$

The first density correction D_1 to the self-diffusion is completely determined by three-particle collisions, since there is no collisional transfer of mass.

The expansion (14) for the three-particle collision term K(fff) implies a similar expansion for the transport coefficients $\lambda_1^{(3)}$, $\eta_1^{(3)}$ and D_1

$$\lambda_1^{(3)} = \sum_{\mu=1}^{4} \lambda_{1\mu} \quad ; \quad \eta_1^{(3)} = \sum_{\mu=1}^{4} \eta_{1\mu} \quad ; \quad D_1 = \sum_{\mu=1}^{4} D_{1\mu} \ . \qquad (17)$$

The first term of this expansion is the value predicted by the theory of Enskog as discussed in the previous Section. Thus

$$\lambda_{11} = -\tfrac{5}{12}\pi\sigma^3\lambda_o \ , \quad \eta_{11} = -\tfrac{5}{12}\pi\sigma^3\eta_o \ , \quad D_{11} = -\tfrac{5}{12}\pi\sigma^3 D_o \ .$$
$$(18)$$

For convenience we consider dimensionless numbers by taking the coefficients relative to the Enskog values. We thus define

$$\lambda_{1\mu}^{**} = \frac{\lambda_{1\mu}}{\lambda_{11}}, \qquad \eta_{1\mu}^{**} = \frac{\eta_{1\mu}}{\eta_{11}}, \qquad D_{1\mu}^{**} = \frac{D_{1\mu}}{D_{11}}. \qquad (19)$$

Some time ago I made an attempt to obtain estimates for the three-particle collision integrals [18]. An extensive numerical study of these integrals has now been made by Gillespie. This work will be published shortly; in Table I we list the results obtained for $\lambda_{1\mu}^{**}$, $\eta_{1\mu}^{**}$ and $D_{1\mu}^{**}$ [19]. The values in the Table correspond to the first Sonine approximation; that is, in solving the integral equation for the first density correction we have approximated the solution of the Boltzmann equation by one Sonine polynomial [12]. The results confirm an earlier observation [18] that the theory of Enskog accounts for over 90% of the total three-particle collision contribution. We repeat that the theory of Enskog ($\mu = 1$) incorporates only excluded volume effects. The next term ($\mu = 2$) contains some excluded volume effects via the single-overlap collisions and some dynamical effects via sequences of two successive collisions; the magnitude of this term is of the order of 10% or less compared to the Enskog value. The contribution from sequences of three successive collisions ($\mu = 3$) is even slightly smaller, while the contribution from sequences of four successive collisions ($\mu = 4$) appears to be negligible. An independent calculation of the three-particle collision integrals for $\mu \leq 3$ was made recently by Condiff and co-workers [20]; their results also support the picture presented here.

The density dependence of the equilibrium properties

of a gas of hard spheres is completely determined by excluded volume effects. We conclude that the density dependence of non equilibrium properties, such as the transport properties, is dominated by the same excluded volume effects. The terms accounting for dynamical correlations due to sequences of successive collisions can profitably be treated as correction terms to the theory of Enskog.

6. Discussion Remarks

1. As discussed by Professor Cohen [2] the higher order terms in the generalized Boltzmann equation (1) are divergent. As a consequence the coefficients of a power series expansion for the transport coefficients in terms of the density do not exist beyond the linear term. Upon resummation of the most divergent terms one expects that a transport property, such as the viscosity, depends on the density as

$$\eta = \eta_o + \eta_1 n + \eta_2' n^2 \log n + \eta_2 n^2 + \ldots \qquad (20)$$

Nevertheless, we can extract from the higher order terms of the generalized Boltzmann equation those terms due to excluded volume effects. Using the same arguments as for the three-particle collision term K(fff), we then expect to recover as a first approximation the theory of Enskog in all orders of the density. These terms thus lead to an expression for the transport properties (11) which is a power series in the density. The logarithmic term, on the other hand, originates from sequences of successive binary collision and is therefore of a dynamical nature. From our experience with the three-particle collision integrals that the dynamical effects are small compared to the excluded volume effects, we speculate that the logarithmic term is small compared to the terms of the power series. This may explain why we have not yet been able to demonstrate the existence of the logarithmic term conclusively from experimental transport coefficients [21].

2. The overlap collisions in Figs. 2 and 3 are a direct consequence of the finite size of the molecules and are not contained in the ordinary Boltzmann equation. It is interesting to note, however, that the sequences of successive binary collisions, such as those shown in Figs. 4 and 5, are quite analogous to the collision sequences encountered in the dynamics of rarefied gases in the nearly free molecular flow regime.

Consider as an example the force on an object in a rarefied gas stream. This problem can be handled by solving the ordinary Boltzmann equation in which the object is treated as a boundary [22], but also by the method of the generalized Boltzmann equation in which the object is treated as a heavy particle, as pointed out by Dorfman et al. [23]. One then finds for the drag coefficient C of a sphere with radius R

$$C = C_o + C_1 \alpha + C_2' \alpha^2 \log \alpha + C_2 \alpha^2 + \ldots \quad . \tag{21}$$

The expansion parameter α is the inverse Knudsen number $\alpha = nR\sigma^2$, to be compared with the parameter $n\sigma^3$ in the expansion (20) for the transport properties.

In the free molecular flow limit $\alpha \to 0$ the force on the object is determined by the transfer of momentum of molecules that strike the object individually. The first correction term in α is determined by "three-particle" collisions, that is, by sequences of successive collisions among two molecules and the object. At low densities we may treat the molecules as point molecules and neglect the overlap collisions contained in the diagrams of Figs. 2 and 3. A molecule emitted by the object (a) may either be bounced back to the object (recollision), (b) it may cause an oncoming molecule to collide with the object (cyclic collision) or (c) it may prevent an oncoming molecule from colliding with the object (hypothetical collision). These collision sequences are shown schematically in Fig. 6; they are obtained from the collision sequences shown in Fig. 4, if particle 2 is identified with the object. Just as η_1 in (20), the coefficient C_1 in (21) contains in addition a small contribution due to sequences of four successive collisions among two molecules and the object.

The coefficient C_2 of the logarithmic term can be related to sequences of four successive collisions among three molecules and the object [24]. These sequences are again similar to the four-particle collision sequences that determine the coefficient η_2' in the expansion (20), if one of the molecules is replaced with the object.

We conclude that there exists a close connection between many collision sequences in the generalized Boltzmann equation for a moderately dense gas and those encountered in the solution of the ordinary Boltzmann equation for gas flows in the nearly free molecular regime.

Acknowledgments

The three-particle collision integrals, discussed in this paper, were evaluated by Dr. D.T. Gillespie.

The author is indebted to Dr. D.T. Gillespie for his contributions to the ideas presented in this paper. He is also greatly indebted to Professors J.R. Dorfman and E.G.D. Cohen for many stimulating discussions and a critical reading of the manuscript.

The research was supported by the Arnold Engineering Development Center and, in part, by the Office of Naval Research. The author acknowledges the hospitality of the Cryogenics Division, National Bureau of Standards, Boulder, Colo., where part of the manuscript was prepared.

Appendix

In this Appendix we list explicit expressions for the three-particle collision terms $K_\mu(fff)$ and the corresponding collision integrals for the transport properties. For this purpose it is convenient to introduce binary collision operators most precisely formulated by Ernst, Dorfman, Hoegy and van Leeuwen [10].

The Boltzmann collision term can be written as

$$J(ff) = \int dx_2 \, T_{12} \, f(\vec{r}_1,\vec{v}_1;t) f(\vec{r}_1,\vec{v}_2;t) \quad . \tag{A.1}$$

The operator T_{12} is the sum of two terms

$$T_{12} = T_{12}^i + T_{12}^n \tag{A.2}$$

corresponding to an interacting collision and a noninteracting collision. These collisions are shown schematically in Fig. 7; the solid line in this figure represents the trajectory of particle 1 in the reference frame of particle 2. In an interacting collision (Fig. 1a) the molecules exchange the components of their velocities along the line $\vec{\sigma}_{12}$ of closest approach. In a noninteracting collision (Fig. 1b or 1c) the molecules are also aimed to collide, but pass through each others interaction spheres. The operator T_{12}^i corresponds to an interacting collision (Fig. 1a) and is defined as [10,17]

$$T_{12}^i = \sigma^2 \int_{\vec{v}_{12}\hat{\sigma}_{12}>0} d\hat{\sigma}_{12} |\vec{v}_{12}\hat{\sigma}_{12}| \delta(\vec{r}_{12}-\vec{\sigma}_{12}) R_{12} \quad . \tag{A.3}$$

The symbol $\hat{\sigma}_{12} = \vec{\sigma}_{12}/\sigma$ indicates a unit vector and the operator R_{12} transforms the velocities $\vec{v}_1$ and $\vec{v}_2$ <u>after</u> the collision into the velocities $\vec{v}_1'$ and $\vec{v}_2'$ <u>before</u> the collision

$$R_{12}\vec{v}_1 = \vec{v}_1' = \vec{v}_1 - (\vec{v}_{12} \cdot \hat{\sigma}_{12})\hat{\sigma}_{12}$$

$$R_{12}\vec{v}_2 = \vec{v}_2' = \vec{v}_2 + (\vec{v}_{12} \cdot \hat{\sigma}_{12})\hat{\sigma}_{12} \quad .$$

The operator T_{12}^n corresponds to a noninteracting collision (Fig. 1b) and is defined as

$$T_{12}^n = -\sigma^2 \int_{\vec{v}_{12}\hat{\sigma}_{12} > 0} d\hat{\sigma}_{12} |\vec{v}_{12}\hat{\sigma}_{12}| \delta(\vec{r}_{12} - \vec{\sigma}_{12}) \quad . \tag{A.4}$$

A noninteracting collision may either be registered at the point A in Fig. 1b where particle 1 leaves the interaction sphere, or at the point B in Fig. 1c where particle 1 enters the interaction sphere of 2. This difference is irrelevant in the Boltzmann equation, as long as we neglect variations of the distribution function over distances of order σ. However, this difference becomes crucial in the generalized Boltzmann equation in order to account properly for the excluded volume effects. Therefore, we consider also a second operator

$$\bar{T}_{12} = T_{12}^i + \bar{T}_{12}^n \tag{A.5}$$

where $\bar{T}_{12}^n$ corresponds to the noninteracting collision of Fig. 1c and is defined as

$$\bar{T}_{12}^n = -\sigma^2 \int_{\vec{v}_{12}\hat{\sigma}_{12} < 0} d\hat{\sigma}_{12} |\vec{v}_{12}\hat{\sigma}_{12}| \delta(\vec{r}_{12} - \vec{\sigma}_{12}) \quad . \tag{A.6}$$

Thus T_{12}^n registers the noninteracting collision at point A and $\bar{T}_{12}^n$ at point B.

The terms $K_\mu(fff)$ in the expansion (14) of the three-particle collision term can be written as

$$K_\mu(fff) = \tfrac{1}{2}\int dx_2 dx_3 \, T_\mu(123) f(\vec{v}_1;t) f(\vec{v}_2;t) f(\vec{v}_3;t) \quad . \quad (A.7)$$

where $T_\mu(123)$ are now three-particle collision operators. The factor 1/2 accounts for a symmetrization with respect to particles 2 and 3 [11]. The operators $T_\mu(123)$ can be expressed in terms of the binary collision operators T_{ij} and $\bar{T}_{ij}$, the Mayer function f_{ij} and an operator $S^o = S^o(123)$

$$S^o(123) = \exp\{-t \sum_{i=1}^{3} \vec{v}_i \cdot \frac{\partial}{\partial \vec{r}_i}\}$$

that accounts for the free streaming of the three particles between collisions. In particular [14]

$$T_1(123) = \sum_\alpha f_\beta f_\gamma \bar{T}_\alpha \qquad (A.8a)$$

$$T_2(123) = -\sum_{\alpha\beta}[\bar{T}_\alpha f_\gamma S^o f_\gamma T_\beta + \bar{T}_\alpha f_\gamma S^o * \bar{T}_\gamma^n S^o T_\beta + \bar{T}_\alpha S^o * T_\gamma^n S^o f_\gamma T_\beta] \qquad (A.8b)$$

$$T_3(123) = \sum_{\alpha\beta}[\bar{T}_\alpha S^o * T_\beta^i S^o T_\alpha + \bar{T}_\alpha S^o * T_\beta^i S^o T_\gamma - \bar{T}_\alpha S^o * T_\beta^n S^o * \bar{T}_\beta^n S^o T_\gamma] \qquad (A.8c)$$

$$T_4(123) = \sum_{\alpha\beta}[\bar{T}_\alpha S^o * T_\beta^i S^o * T_\alpha^i S^o T_\gamma - \bar{T}_\alpha S^o * T_\beta^i S^o * T_\alpha^n S^o * \bar{T}_\alpha^n S^o T_\gamma +$$
$$+ \bar{T}_\alpha S^o * T_\beta^i S^o * T_\gamma^i S^o T_\beta - \bar{T}_\alpha S^o * T_\beta^n S^o * \bar{T}_\beta^n S^o * T_\gamma^i S^o T_\beta] \qquad (A.8d)$$

The indices $\alpha \neq \beta \neq \gamma \neq \alpha$ refer to the pairs 12, 13 and 23 among the three particles and the summations are to be taken over all three pairs. The asterisk * indicates a convolution product [10]

$$f \ast g = \int_0^t d\tau\, f(\tau)\, g(t-\tau) = \int_0^t d\tau'\, f(t-\tau')\, g(\tau') \quad . \quad (A.9)$$

The operators $T_\mu(123)$ for $\mu \geq 2$ are in principle still a function of the time t via (A.9). However, it can be shown that contributions from times, large compared to the time that the molecules traverse distances is order σ, are negligibly small [25]. The operators $T_\mu(123)$ in (A.8) are symmetric in the three particle numbers. However, since $\int d\vec{v}_2 d\vec{v}_3\, \bar{T}_{23}\, f(x_2,x_3) = 0$ terms starting with $\bar{T}_{23}$ do not actually contribute to $K_\mu(fff)$.

Each operator product in (A.8) corresponds to a particular sequence of collisions. The relationship between these operator products and the diagrams of Figs. 2 - 5 is given in Table II.

The transport coefficients $\lambda_{1\mu}$, $\eta_{1\mu}$ and $D_{1\mu}$ in (17) are given by the following three-particle collision integrals [14]

$$\lambda_{1\mu} = \frac{1}{3kT^2}\lim_{t\to\infty}\int d\vec{v}_1 dx_2 dx_3\, \phi(v_1)\phi(v_2)\phi(v_3)\vec{A}(\vec{v}_1)\cdot\frac{1}{2}T_\mu(123)$$
$$\{\vec{A}(\vec{v}_1)+\vec{A}(\vec{v}_2)+\vec{A}(\vec{v}_3)\}$$

$$\eta_{1\mu} = \frac{1}{10kT}\lim_{t\to\infty}\int d\vec{v}_1 dx_2 dx_3\, \phi(v_1)\phi(v_2)\phi(v_3)\vec{\vec{B}}(\vec{v}_1):\frac{1}{2}T_\mu(123)$$
$$\{\vec{\vec{B}}(\vec{v}_1)+\vec{\vec{B}}(\vec{v}_2)+\vec{\vec{B}}(\vec{v}_3)\}$$

$$D_{1\mu} = \frac{1}{3}\lim_{t\to\infty}\int d\vec{v}_1 dx_2 dx_3\, \phi(v_1)\phi(v_2)\phi(v_3)\vec{C}(\vec{v}_1)\cdot\frac{1}{2}T_\mu(123)\vec{C}(\vec{v}_1)$$

$$(A.10)$$

where

$$\phi(v) = \left(\frac{m}{2\pi kT}\right)^{3/2} \exp\left(-\frac{mv^2}{2kT}\right) \quad .$$

The vector functions $\vec{A}(\vec{v})$, $\vec{C}(\vec{v})$ and the tensor function $\vec{\vec{B}}(\vec{v})$ are solutions of the linearized Boltzmann equations [3,26]

$$\int dx_2 \phi(v_2) \bar{T}_{12} \{\vec{A}(\vec{v}_1) + \vec{A}(\vec{v}_2)\} = -(\tfrac{1}{2}mv_1^2 - \tfrac{5}{2}kT)\vec{v}_1$$

$$\int dx_2 \phi(v_2) \bar{T}_{12} \{\vec{\vec{B}}(\vec{v}_1) + \vec{\vec{B}}(\vec{v}_2)\} = -m\vec{v}_1^{\,\circ}\vec{v}_1 \qquad (A.11)$$

$$\int dx_2 \phi(v_2) \bar{T}_{12} \vec{C}(\vec{v}_1) = -\vec{v}_1 \quad .$$

The numbers in Table I correspond to the first Sonine approximation, that is, the functions $\vec{A}(\vec{v})$, $\vec{\vec{B}}(\vec{v})$ and $\vec{C}(\vec{v})$ were approximated by

$$\vec{A}(\vec{v}) = \frac{15}{32\sigma^2}\left(\frac{m}{\pi kT}\right)^{1/2} (\tfrac{1}{2}mv^2 - \tfrac{5}{2}kT)\vec{v}$$

$$\vec{\vec{B}}(\vec{v}) = \frac{5}{16\sigma^2}\left(\frac{m}{\pi kT}\right)^{1/2} m\, \vec{v}\vec{v} \qquad (A.12)$$

$$\vec{C}(\vec{v}) = \frac{3}{8\sigma^2}\left(\frac{m}{\pi kT}\right)^{1/2} \vec{v} \quad .$$

Expressions for the three-particle collision contribution to the transport coefficients λ_1 and η_1 were first derived by Choh and Uhlenbeck [27]. Their results were written in a more complicated form, but are equivalent to the formulas presented here.

References

[1] N.N. Bogoliubov, "Studies in Statistical Mechanics I", J. de Boer and G.E. Uhlenbeck, eds. (North-Holland Publ. Comp., Amsterdam, 1962), p.5.

[2] E.G.D. Cohen, Proceedings of this Symposium.

[3] S. Chapman and T.G. Cowling, "The Mathematical Theory of Nonuniform Gases" (Cambridge Univ. Press, London, 3rd. ed., 1970).

[4] D. Enskog, K. Svensk. Vet. Akad. Handl. $\underline{63}$, no.4 (1921).

[5] J.V. Sengers and E.G.D. Cohen, Physica $\underline{27}$, 230 (1961).

[6] J.H. Jeans, "Dynamical Theory of Gases", London 1925.

[7] J. de Boer, Reports Progress Phys. $\underline{12}$, 305 (1949)

[8] E.G.D. Cohen, Physica $\underline{28}$, 1025 (1962); J. Math. Phys. $\underline{4}$, 183 (1963)

[9] M.S. Green and R.A. Piccirelli, Phys. Rev. $\underline{132}$, 1388 (1963).

[10] M.H. Ernst, J.R. Dorfman, W.R. Hoegy and J.M.J. van Leeuwen, Physica $\underline{45}$, 127 (1969).

[11] M.S. Green, Phys. Rev. $\underline{136}$, A905 (1964).

[12] J.V. Sengers, Phys. Fluids $\underline{9}$, 1333 (1966).

[13] D.T. Gillespie and J.V. Sengers, in "Proc. 5th Symposium on Thermophysical Properties", C.F. Bonilla, ed. (ASME, New York, 1970), p. 42.

[14] J.V. Sengers, M.H. Ernst and D.T. Gillespie, J. Chem. Phys. $\underline{56}$, 5583 (1972).

[15] G. Sandri, R.D. Sullivan and P. Norem, Phys. Rev. Letters $\underline{13}$, 743 (1964).

[16] E.G.D. Cohen, in "Boulder Lectures in Theoretical Physics", Vol. 8A (Univ. Colorado Press, Boulder, Colo., 1966), Appendix II, p. 170.

[17] W.R. Hoegy and J.V. Sengers, Phys. Rev. $\underline{A2}$, 2461 (1970).
[18] J.V. Sengers, in "Boulder Lectures in Theoretical Physics", Vol. 9C, W.E. Brittin, ed. (Gordon and Breach, New York, 1967), p. 335; J.V. Sengers, in "Kinetic Equations", R.I. Liboff and N. Rostoker, eds. (Gordon and Breach, New York, 1971), p. 137.
[19] D.T. Gillespie and J.V. Sengers, to be published.
[20] W.D. Henline and D.W. Condiff, J. Chem. Phys. $\underline{54}$, 5346 (1971); D.W. Condiff, private communication.
[21] J. Kestin, E. Paykoç and J.V. Sengers, Physica $\underline{54}$, 1 (1971).
[22] D.R. Willis, in "Rarefied Gas Dynamics", F.M. Devienne, ed. (Pergamon Press, New York, 1963), p.1; Y.P. Pao and D.R. Willis, Phys. Fluids $\underline{12}$, 435 (1969).
[23] J.R. Dorfman and J.V. Sengers, Bull. Am. Phys. Soc. $\underline{15}$, 515 (1970); C.F. McClure and J.R. Dorfman, Bull. Am. Phys. Soc. $\underline{17}$, 616 (1972).
[24] G.E. Kelly and J.V. Sengers, J. Chem. Phys. $\underline{57}$, (1972).
[25] J.R. Dorfman and E.G.D. Cohen, J. Math. Phys. $\underline{8}$, 282 (1967).
[26] J.O. Hirschfelder, C.F. Curtiss and R.B. Bird, "Molecular Theory of Gases and Liquids", (Wiley, New York, 1954).
[27] S.T. Choh and G.E. Uhlenbeck, "The Kinetic Theory of Dense Gases", (Univ. Michigan, 1958).

Table I

Three-particle collision integrals for a gas of hard spheres

number of successive collisions μ	thermal conductivity $\lambda_{1\mu}^{*}$	viscosity $\eta_{1\mu}^{*}$	self diffusion $D_{1\mu}^{*}$
1	1	1	1
2	-0.0303 ± 0.0003	-0.0633 ± 0.0004	-0.1195 ± 0.0005
3	-0.0128 ± 0.0005	$+0.0376 \pm 0.0005$	$+0.0345 \pm 0.0013$
4	$-(0.2 \pm 0.6) \cdot 10^{-6}$	$-(1.0 \pm 0.3) \cdot 10^{-6}$	$-(4.4 \pm 0.7) \cdot 10^{-6}$
sum	0.9569 ± 0.0008	0.9743 ± 0.0009	0.9150 ± 0.0018

Table II

Diagrammatic representation of three-particle collision terms.	
Operator product	Diagram
$f_{13} f_{23} \bar{T}_{12}$	Fig. 2
$\bar{T}_{12} f_{23} S^o f_{23} T_{13}$	Fig. 3a
$\bar{T}_{12} f_{23} S^o \!:\! \bar{T}^n_{23} S^o T_{13}$	Fig. 3b
$\bar{T}_{12} S^o \!:\! T^n_{23} S^o f_{23} T_{13}$	Fig. 3c
$\bar{T}_{12} S^o \!:\! T^i_{13} S^o T_{12}$	Fig. 4a
$\bar{T}_{12} S^o \!:\! T^i_{13} S^o T_{23}$	Fig. 4b
$\bar{T}_{12} S^o \!:\! T^n_{13} S^o \!:\! \bar{T}^n_{13} S^o T_{23}$	Fig. 4c
$\bar{T}_{12} S^o \!:\! T^i_{13} S^o \!:\! T^i_{12} S^o T_{23}$	Fig. 5a
$\bar{T}_{12} S^o \!:\! T^i_{13} S^o \!:\! T^n_{12} S^o \!:\! \bar{T}^n_{12} S^o T_{23}$	Fig. 5b

Captions of Figures

Fig. 1 Schematic representation of a binary collision associated with J(ff). The lines indicate particle trajectories and the circles indicate the location of the spheres at the time of contact.

Fig. 2 Collision associated with K_1(fff). When particles 1 and 2 are colliding, particle 3 overlaps with both 1 and 2.

Fig. 3 Sequences of two successive collisions associated with K_2(fff).
 (a) Particles 2 and 3 overlap at both collisions.
 (b) Particles 2 and 3 overlap at the later collision.
 (c) Particles 2 and 3 overlap at the earlier collision.

Fig. 4 Sequences of three successive collisions associated with K_3(fff).

Fig. 5 Sequences of four successive collisions associated with K_4(fff).

Fig. 6 Sequences of three successive collisions among two molecules and an object. The lines indicate the trajectories of the two molecules and the circle represents the object.

Fig. 7 Geometry of a collision between two hard spheres. The circle represents the action sphere of 2 and the solid lines indicate the trajectory of 1 in the rest frame of 2.
 (a) an interacting collision.
 (b) a noninteracting collision registered at point A.
 (c) a noninteracting collision registered at point B.

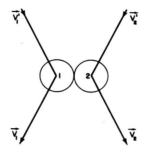

Fig. 1

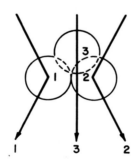

Fig. 2

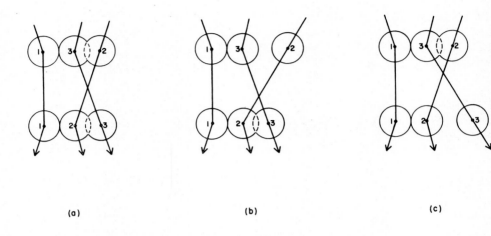

Fig. 3

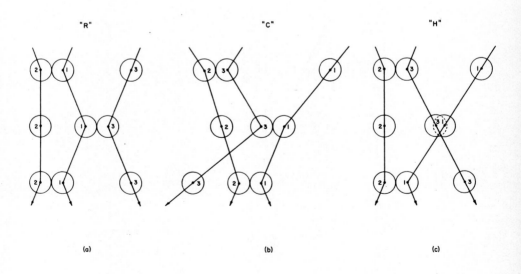

Fig. 4

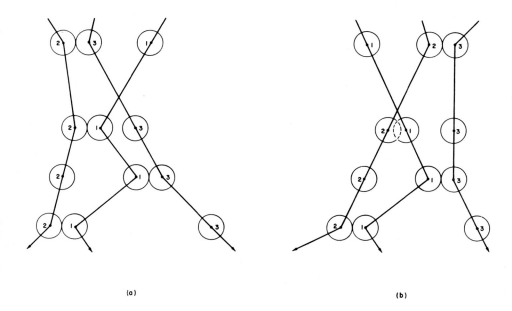

(a) (b)

Fig. 5

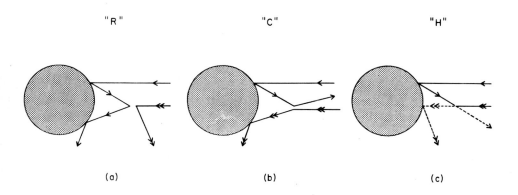

"R" "C" "H"

(a) (b) (c)

Fig. 6

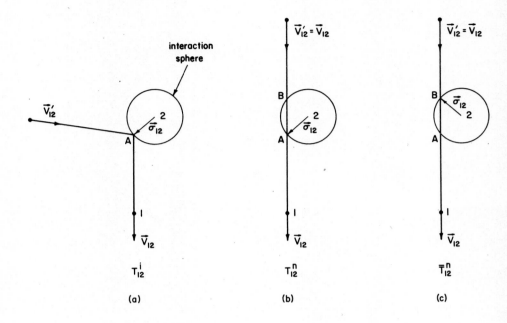

Fig. 7

Acta Physica Austriaca, Suppl.X, 209–221 (1973)
© by Springer-Verlag 1973

VELOCITY CORRELATION FUNCTIONS FOR MODERATELY DENSE GASES

J.R. DORFMAN

Institute for Fluid Dynamics and Applied Mathematics and
Department of Physics and Astronomy
University of Maryland, College Park, Maryland 20742

ABSTRACT

The long-time behavior of the velocity correlation functions characteristic for transport coefficients has been calculated for a gas of hard disks or hard spheres on the basis of an appropriate generalized Boltzmann equation. In d-dimensions one finds that, after several mean free times t_o, these correlation functions exhibit a decay proportional to $(t/t_o)^{-d/2}$. This decay can be understood as the result of the same sequences of binary collisions that are responsible for the divergences encountered in the virial expansion of the transport coefficients. This long-time behavior in two dimensions is consistent with the results of computer studies of velocity correlation functions for hard disk gases. The possible implications for hydrodynamics of the $t^{-d/2}$ behavior of the correlation functions is discussed.

As was discussed by Professor Cohen at this Symposium, the generalized Boltzmann equation leads to an expansion of the transport coefficients for a dense gas. However, general expressions for transport coefficients for a fluid, which are not in the form of expansions, can be derived by another technique, the time correlation function method [1]. This approach has provided a general framework by means of which one can make detailed comparisons between theoretical results, the results of computer-simulated molecular dynamics, and experimental results.

In the time correlation function method, one applies a generalization of the Chapman-Enskog development of the Boltzmann equation to the Liouville equation in order to obtain solutions for a system near a local equilibrium state. Expressions for the hydrodynamic fluxes can be derived which are proportional, under certain circumstances, to the gradients of the hydrodynamic variables. In this way formulas are derived for the transport coefficients in the form of time integrals of correlation functions of microscopic currents. These formulas are related to the expression derived by Einstein for the diffusion coefficient for particles in Brownian motion in a fluid [2]. A transport coefficient, τ, is given by the time correlation function method, as

$$\tau = \int_0^\infty dt <J^{(\tau)}(o) \, J^{(\tau)}(t)> \qquad (1)$$

where $J^{(\tau)}$ represents a microscopic current associated with this transport coefficient, $J^{(\tau)}(o)$ is its value at some time, $J^{(\tau)}(t)$ is its value at a time t later, and the angular brackets denote an average over an equilibrium ensemble. For the coefficient of self-diffusion D, for example, $J^{(D)} = v_x$, the value of the x-component of the velocity of a particular particle.

It is clear from equation (1) that the long-time behavior of the correlation function $<J^{(\tau)}(o) J^{(\tau)}(t)>$ is of considerable interest [3]. I report here some recent theoretical results on the long-time behavior of the velocity correlation functions, $\rho_\tau^{(d)}(t)$, whose time integrals determine the coefficient of self-diffusion and the kinetic parts of the coefficients of shear viscosity and thermal conductivity. The main result [4,5] is that for gases of hard spheres or hard disks, after several mean free times t_o, $\rho_\tau^{(d)}(t)$ decays in time as $(t/t_o)^{-d/2}$, where d is the number of dimensions. In two dimensions, this result is consistent with the available evidence from computer studies of these correlation functions for hard disks [6,7,8]. These studies will be described in more detail by Dr. Wood.

Before discussing the possible implications of this result for transport theory, I will briefly describe how the $t^{-d/2}$ decay is obtained by using the kinetic theory of gases. An analysis based on cluster expansion techniques, similar to those discussed by Professor Cohen but specialized to linear systems, can be used to obtain an expansion of the correlation functions $\rho_\tau^{(d)}(t)$ in terms of the dynamics of small groups of particles in infinite space. The first term in the expansion, the two-body term, provides an exponential decay of $\rho_\tau^{(d)}(t)$ which is damped after a few mean free times. The time integral of this two- body term gives the low-density value for the transport coefficient τ, and it is identical with the result obtained from the Boltzmann equation. If one considers the contributions to $\rho_\tau^{(d)}(t)$ coming from groups

14*

of more than two particles, one encounters difficulties
similar to those that appear in the virial expansion of
the transport coefficients, discussed by Professor Cohen.
In particular, the three-body contribution to $\rho_\tau^{(2)}(t)$ in
two dimensions, and, in three dimensions, the four body
contribution to $\rho_\tau^{(3)}(t)$ appear to decay in time as t^{-1}.
This decay is, in turn, responsible for a logarithmic
divergence of the corresponding three- or four-body term
in the virial expansion of the time correlation function
expressions for the transport coefficients. This divergence is identical to the one that appears in the virial
expansion based on the generalized Boltzmann equation
[9]. The t^{-1} decay in $\rho_\tau^{(d)}(t)$ arises in two dimensions
from the correlated sequences of three binary collisions
among three particles in infinite space, and in three
dimensions from the sequences of four binary collisions
among four particles, which have been discussed by Professor Cohen in his lecture. The higher terms in the series,
which involve the dynamics of five, six, ... particles
in infinite space, also grow with time as a result of correlated sequences of binary collisions taking place among
the particles.

A rearrangement of this series is necessary if
one wants to obtain an expression for $\rho_\tau^{(d)}(t)$ that is
useful for times larger than a mean free time. Such a rearrangement is obtained by summing the contributions from
the binary collision sequences that appear to give the
most rapidly growing contribution to $\rho_\tau^{(d)}(t)$ [10]. This
resummation is schematically illustrated in Figure 1,
where examples of the relevant three- and four-body events
are shown. Using a method originated by Pomeau [11], one
can analyze the contribution coming from the resummed

events. After several mean free times t_o, one finds that in d-dimensions [4,5]

$$\rho_T^{(d)}(t) \sim \alpha_{T,o}^{(d)}(\rho)(t_o/t)^{d/2} \tag{2}$$

where $\rho = na^d$, n is the number density, and a is the hard disk or hard sphere diameter. For example, for $\rho_D^{(d)}(t) = <v_x(o)\, v_x(t)>/<v_x^2(o)>$, the (normalized) velocity correlation function associated with the coefficient of self-diffusion, $\alpha_{D,o}^{(d)}$ is given in two dimensions, by

$$\alpha_{D,o}^{(2)} = [8\pi n\,(D_o + \frac{n_o}{nm})t_o]^{-1} \tag{3}$$

and in three dimensions, $\alpha_{D,o}^{(3)}$ is

$$\alpha_{D,o}^{(3)} = \frac{1}{12n}[\pi(D_o + \frac{n_o}{nm})t_o]^{-3/2} \tag{4}$$

Here D_o and η_o are the coefficients of self-diffusion and shear viscosity, respectively, as determined from the Boltzmann equations appropriate for a gas of hard disks [12] or hard spheres [13], and m is the mass of a particle. Similar results have been obtained for the velocity correlation functions appropriate for the coefficients of shear viscosity and thermal conductivity.

The analysis also shows that the summation of the contributions from the most rapidly growing dynamical events determines the coefficient $\alpha_T^{(d)}$ of the $(t_o/t)^{d/2}$ term only to the lowest order in the density. An attempt has been made to extend these results to higher densities by incorporating into the theory excluded volume effects similar to those that appear in the Enskog theory of dense gases. This method leads to an expression for $\alpha_\tau^{(d)}(\rho)$ denoted by $\alpha_{\tau,E}^{(d)}$ which is applicable to higher densities than is $\alpha_{\tau,o}^{(d)}(\rho)$. For the case of self-diffusion, $\alpha_{D,E}^{(d)}(\rho)$ is identical to $\alpha_{D,o}^{(d)}$ given by equations [3] and [4] above, except that in $\alpha_{D,E}^{(d)}$ the transport coefficients D_o and η_o are replaced by the values D_E and η_E, respectively, as determined by the Enskog theory for hard spheres [13] or disks [14]. Although other dynamical events which might contribute to $\alpha_\tau^{(d)}$ have not been included by this procedure, it is not unreasonable to suppose that their contribution might be small compared to $\alpha_\tau^{(d)}$. This is in view of the fact that for a number of cases [7,12] the Enskog corrections, which incorporate excluded volume effects, are dominant among the contributions of all dynamical events.

Although the theory has not been worked out sufficiently for an estimation of the full range of time for which the $t^{-d/2}$ decay is dominant, it does appear that this decay should hold at least over the time intervals of interest for comparison with the computer results mentioned earlier.

In two dimensions, the results of the computer studies are consistent with a t^{-1} decay for $\rho^{(2)}(t)$ for times in the interval $10t_o \leq t \leq 50t_o$ [6,7,8]. In addition, $\alpha_{D,E}^{(2)}$ is in very good quantitive agreement with the coefficient of (t_o/t) obtained by computer calculations of $\rho_D^{(2)}(t)$ over the entire range of densities for which computer results are available, i.e. up to one-half the density at close packing. The comparison of $\alpha_{D,E}^{(2)}$ with the computer results for $\alpha_D^{(2)}$ is shown in Figure 2. On the same graph we have plotted $\alpha_{D,E}^{(3)}$. In three dimensions, however, computer results are not yet available.

In this connection it is interesting to note that if the resummation of the most divergent events is <u>not</u> carried out, the expansion of $\rho_T^{(d)}(t)$ gives a t^{-1} dependence in both two and three dimensions, coming from the three- and four-body binary collision sequences, respectively. However, in two dimensions, where computer results are available, the coefficient of the un-resummed t^{-1} term is <u>not</u> consistent with the computer results for $\rho_T^{(2)}(t)$ [4]. Thus, the quantitative agreement with the computations can be taken as a vindication of the resummation in two dimensions.

I conclude this discussion with a number of remarks.

1. It should be possible to extend the theory to include a wider class of intermolecular potentials than those for hard spheres or disks. This has not yet been done.

2. There is now some question as to whether the time correlation function expressions for transport coefficients associated with the Navier-Stokes equations exist in two dimensions. For if the t^{-1} behavior persists for asymptotically long times, time integrals like that in equation (1) will not exist. As a result the resummation of the most divergent terms in the virial expansion of the transport coefficients does not seem to provide a mean free path cutoff for the divergent three-body integrals in two dimensions. In three dimensions, although an asymptotic behavior like $t^{-3/2}$ will ensure the existence of the Navier-Stokes transport coefficients, it appears that the transport coefficients associated with

the linear Burnett equations may not exist [4,15]. However we stress that in order to determine the behavior of the correlation functions as $t \to \infty$, and thus to establish whether or not the above mentioned transport coefficients diverge, dynamical events other than those considered here must be taken into account.

3. In view of the slow time decay of the velocity correlation functions, it is not clear to what extent the Navier-Stokes equations with constant transport coefficients can be used even in three dimensions to describe phenomena that are not infinitely slowly varying in space and time.

4. Finally I should like to mention that there is another approach to determining the long-time behavior of the correlation functions, which is based on solutions of the linearized hydrodynamics equation [6,7,15]. This approach also leads to a $t^{-d/2}$ decay for the correlation function, and for $\rho_D^{(d)}(t)$ a result identical to equations (3) and (4) is obtained, except that D_o and η_o are replaced by their full values D and η, respectively. The precise relation of this method to that of kinetic theory has not yet been settled.

In view of the difficulties encountered in trying to extend our knowledge of transport theory beyond that which is predicted by the Boltzmann equation, we can all the more appreciate the importance of Boltzmann's work for present-day kinetic theory.

Acknowledgements

I would like to thank Drs. J.J. Erpenbeck, J.V. Sengers, W.W. Wood, and especially E.G.D. Cohen for their help in the preparation of this paper.

This work was supported in part by the National Science Foundation, under grant NSF GP 29385.

REFERENCES

[1] c.f. R.W. Zwanzig, Ann. Rev. Phys. Chem. <u>16</u>, 67 (1965), and W.A. Steele, in <u>Transport Phenomena in Fluids</u>, H.J.M. Hanley, ed.(Marcel Dekker, New York, 1969), Chapter 8.

[2] A. Einstein, Ann. Phys. <u>17</u>, 549 (1905).

[3] Although we consider here only the relations between the time correlation functions and the transport coefficients, similar time correlation function expressions appear for example in the theory of neutron scattering, of light scattering, and of spectral line shapes.

[4] J.R. Dorfman and E.G.D. Cohen, Phys. Rev. Letters <u>25</u>, 1257 (1970), and Phys. Rev. <u>A6</u>, 776 (1972). A closely related calculation has been made by J. Dufty, Phys. Rev. <u>A5</u>, 2247 (1972).

[5] W.W. Wood, J.J. Erpenbeck, J.R. Dorfman and E.G.D. Cohen (to be published).

[6] B.J. Alder and T.E. Wainwright, Phys. Rev. Letters <u>18</u>, 988 (1967); J. Phys. Soc. Japan Suppl. <u>26</u>, 267 (1969); and Phys. Rev. <u>A1</u>, 18 (1970).

[7] B.J. Alder, T.E. Wainwright, and D.M. Gass, J. Chem. Phys. <u>53</u>, 3813 (1970), and Phys. Rev. <u>A4</u>, 233 (1971).

[8] W.W. Wood and J.J. Erpenbeck (unpublished). See the lecture of Dr. Wood in these Proceedings.

[9] For a bibliography of papers pertaining to the divergence we refer to M.H. Ernst, L.K. Haines and J.R. Dorfman, Rev. Mod. Phys. <u>41</u>, 296 (1969).

[10] K. Kawasaki and I. Oppenheim, Phys. Rev. <u>139</u>, A1763 (1965); and in <u>Statistical Mechanics</u>, T. Bak, ed. (Benjamin, New York, 1967), p. 313, see also J. Weinstock, Phys. Rev. <u>140</u>, A460 (1965).

[11] Y. Pomeau, Phys. Letters, <u>27A</u>, 601 (1968), and Phys. Rev. <u>A3</u>, 1174 (1971).

[12] J.V. Sengers, in <u>Lectures in Theoretical Physics (Boulder)</u>, W.E. Brittin, ed. (Gordon & Breach, New York, 1967) Vol. 9C; p.335. See also the lecture by Prof. Seeger, in these Proceedings.

[13] S. Chapman and T.G. Cowling, <u>The Mathematical Theory of Non Uniform Gases</u>, 3rd ed. (Cambridge University Press, London, 1970).

[14] D.M. Gass, J. Chem. Phys. <u>54</u>, 1898 (1971).

[15] M.H. Ernst, E.H. Hauge, and J.M.J. Van Leeuwen, Phys. Rev. Letters <u>25</u>, 1254 (1970), Phys. Rev. <u>A4</u> 2055 (1971); Y. Pomeau, Phys. Rev. <u>A5</u>, 2569 (1972), see also R. Zwanzig and M. Bixon, Phys. Rev. <u>A2</u>, 2005 (1970), R. Zwanzig in Proceedings of the Sixth IUPAP Conference on Statistical Mechanics, S.A. Rice, K.F. Freed, and J.C. Light, eds., Univ. of Chicago Press, Chicago, 1972, p.241. For an approach based on the mode-mode coupling formulas, see K. Kawasaki, Progr. Theoret. Phys. (Kyoto) <u>45</u>, 1691 (1971) and Phys. Lett. <u>34A</u>, 12 (1971); and for one based on the non-linear Boltzmann equation see E. Hauge, Phys. Rev. Letters <u>28</u>, 1501 (1972).

FIGURE CAPTIONS

Figure 1: Schematic illustration of the resummation of the most divergent events. Only representative three- and four-body events are shown.

Figure 2: $\alpha_{D,E}^{(d)} / (V_o/V)^{d-2}$ plotted as a function of the reduced volume (V_o/V), where V_o is the volume at close packing, for $d=2$ and $d=3$. The crosses indicate the computer results of Alder and Wainwright for $d=2$.

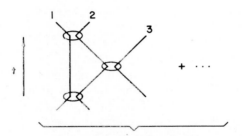

THREE BODY EVENTS

+

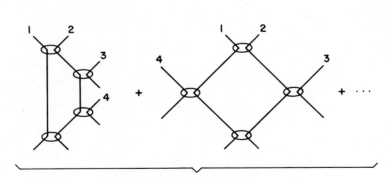

FOUR BODY EVENTS
+
⋮

Fig. 1

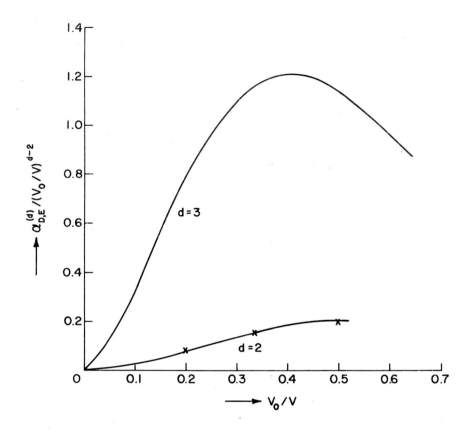

Fig. 2

ON KINETIC EQUATIONS FOR PARTICLES WITH INTERNAL DEGRESS OF FREEDOM

L. WALDMANN

Institut für Theoretische Physik
Universität Erlangen-Nürnberg, Erlangen

ABSTRACT

Already in his famous 1872-paper Boltzmann even included polyatomic dilute gases. More recently this question has been reconsidered from the quantum theoretical point of view. The simplest model in this context is a gas of particles with spin, which has to be described by a one particle distribution matrix with $2S+1$ rows and columns (S = value of spin). The H-theorem can be traced back to the unitarity of the scattering matrix for the binary collision, suitably occurring in the kinetic equation. A special spin-dependent scattering amplitude is considered for which the classical limit can easily be effectuated. In conclusion some remarks on the classical Boltzmann equation for rigid dumb-bells are made.

I. Introduction

During the last fifteen years or so, the kinetics of polyatomic gases has developed from a limited existence into quite a diversified field. This has been due to a beautiful interaction of experiment and theory, the one stimulating the other. For the theory, the prime question is that of an appropriate kinetic equation. Already in his great 1872-paper, which is celebrated today, Boltzmann [1] has included dilute polyatomic gases. More recently this question has been reconsidered from the quantum theoretical point of view [4,5] this avoids some of the difficulties which have been pointed out by H.A. Lorentz [2] and which Boltzmann [3] had to fight with in generalizing his H-theorem to polyatomic gases.

The purpose of this lecture is to give a short survey of quantum kinetic equations now in use and of the relevant H-theorem. Also some considerations on the classical limit will be included. But before starting with this, let me mention with a few words the topics in polyatomic gas kinetics which have been dealt with extensively in the near past. One may distinguish between A) applications to (principally) infinitely large systems which require the knowledge of the kinetic equation alone and B) applications to finite systems (rarefied gases) for which also boundary conditions are vital. Both kinds of applications are listed as follows:

A) Infinite Systems
  1) Alignment of non-spherical molecules by transport processes [6,7,8]
  2) Influence of external fields on transport processes (Senftleben-Beenakker effect) [9,10]

3) Kinetic theory of spin relaxation in gases; diffusion of spin [6,11]; rotational Brownian motion [12]
4) Flow birefrigence in gases [13]
5) Spectrum of depolarized scattered light [14,15]

B) Finite Systems
1) Thermomagnetic torque (Scott effect) [16,17,18] and force [19]
2) Thermomagnetic slip [17,20]

By the way, one method to establish phenomenological boundary conditions is based on non-equilibrium thermodynamics.

Now let us turn to the program envisaged above.

II. Quantum Theoretical Kinetic Equations for Polyatomic Gases

We shall distinguish two cases:
A) molecules with non-degenerate, sufficiently separated levels of internal energy and
B) molecules with degenerate levels (rotating molecules).

Let us first look at the simpler case A). The eigenvalues of the internal molecular energy are denoted by E_V, where V stands for a non-degenerate vibrational quantum number. For each of these states one has a one-particle distribution function

$$f(t,\underline{x},\underline{p}_1)_{V_1} \equiv f_{V_1}(1)$$

with

$$\sum_{V_1} \iint f_{V_1}(1) d^3x d^3p_1 = N \quad ,$$

where N is the total number of particles. The energy of a colliding pair in the c.m. system is (m_{12} = reduced mass of pair)

$$E = \frac{1}{2m_{12}} p_{12}^2 + E_{v_1} + E_{v_2} = \frac{1}{2m_{12}} p_{12}'^2 + E_{v_1'} + E_{v_2'} = E' .$$

The gas is thus pictured as a mixture whose components may undergo inelastic collisions or so to say binary chemical reactions. The relevant cross section

$$\sigma(E, \underline{e}_{12}\underline{e}_{12}')_{v_1 v_2 v_1' v_2'} = |a(\ldots)_{v_1 v_2 v_1' v_2'}|^2$$

will depend on the energy E of the pair, the cosine $\underline{e}_{12}\underline{e}_{12}'$ of the angle of deflection, $\underline{e}_{12}$, V and $\underline{e}_{12}'$, V' being the unit vectors of relative momentum and the internal quantum numbers after and before a gain-collision. More basically, the cross section can be, as indicated, expressed by the scattering amplitude a. With these notations the kinetic equation due to Wang Chang-Uhlenbeck-de Boer (WUB) [21] is written as follows

$$\frac{\partial f_{v_1}(1)}{\partial t} + \underline{c}_1 \cdot \frac{\partial f_{v_1}(1)}{\partial \underline{x}} \tag{1}$$

$$= \sum_{v_2 v_1' v_2'} \iint [f_{v_1'}(1') f_{v_2'}(2') - f_{v_1}(1) f_{v_2}(2)] \sigma_{v_1 v_2 v_1' v_2'} c_{12}' d^2 e_{12}' d^3 p_2 .$$

Here, $d^2 e_{12}'$ means the solid angle element linked with the vector of relative momentum. As this kinetic equation is formally nearly identical with the Boltzmann equation for a monatomic mixture, the proof of the relevant H-theorem is de facto contained in Boltzmann's 1872-paper.

Still a remark is necessary what "sufficiently separated levels" means. The condition of validity of the WUB-equation is [28,25]

$$\frac{|\Delta E_{int}|}{\hbar} t_{free} \gg 1 \; . \tag{2}$$

Here, the first factor is the (smallest) frequency of internal oscillation and t_{free} denotes the time of free flight between two collisions. Indeed, a typical vibrational frequency of molecules is 10^{15} s^{-1}, whereas at 273^0K, 1 at one has $t_{free} \simeq 10^{-10}$ s. So, the above condition is amply fulfilled. This would also be true for the separation of rotational molecular levels but for their degeneracy.

Let us now turn to the case B) of molecules with degenerate levels of internal energy. This case is important with rotating molecules which exhibit the interesting Senftleben-Beenakker effect in external fields. The simplest model for that are particles with spin. From quantum mechanics it is known that for the complete one-particle description of a gas of particles with spin S a (2 S+1) x (2 S+1)-matrix in the magnetic (degeneracy) quantum number is needed:

$$f(t,\underline{x},\underline{p}_1)_{M_1 M_1'} \stackrel{\wedge}{=} f(t,\underline{x},\underline{p}_1,\underline{s}_1) \equiv f(1) \; ,$$

which can equivalently be written as a polynomial in the non-commuting components of the spin vector $\underline{s}_1$. The normalization condition now is

$$tr_1 \int\int f(1) \, d^3x \, d^3p_1 = N$$

where N again is the total number of gas particles and "tr" means the trace on the magnetic quantum numbers. The kinetic equation for this gas is [4,5]

$$\frac{\partial f(1)}{\partial t} + \underline{c}_1 \cdot \frac{\partial f(1)}{\partial \underline{x}} + \frac{1}{i\hbar}[\underline{\mu}_1 \cdot \underline{H}, f(1)]_- \qquad (3)$$

$$= tr_2 \int [\int a f(1') f(2') a^\dagger c'_{12} d^2 e'_{12} - \frac{h}{im_{12}}(a(0) f(1) f(2) - f(1) f(2) a^\dagger(0))]$$

$$\cdot d^3 p_2$$

Even an external magnetic field $\underline{H}$ is included acting on the magnetic moment $\underline{\mu}$ of the particles which again is assumed to be proportional to their spin operator $\underline{s}$. In the collision integral - right hand side - now appears not simply a cross section but the scattering amplitude itself:

$$a = a(E, \underline{e}_{12}, \underline{e}'_{12}, \underline{s}_1, \underline{s}_2) ,$$

in forward direction

$$a(0) = a(E, \underline{e}_{12}, \underline{e}_{12}, \ldots) .$$

In general it does not commute with the distribution operator f. Therefore the order of the factors in the integrand is important. This order results from quite general features of quantum scattering theory. Let us denote the statistical operator of a system before scattering by ρ_i, after scattering by ρ_f. The connection between both in pure quantum mechanics is given by a unitary transformation

$$\rho_f = S \rho_i S^\dagger \quad \text{with} \quad S S^\dagger = S^\dagger S = 1 ,$$

which by introduction of

$$S = 1 - \frac{k}{2\pi i} a \quad , \quad \hbar k = p$$

can be rewritten as

$$\rho_f - \rho_i \propto a \rho\, a^\dagger - \frac{h}{imc'}(a\rho - \rho a^\dagger) \quad .$$

Indeed this matrix form is recognizable in the above collision integral as far as the magnetic quantum numbers are concerned. Of course, the treatment of the relative momentum, especially the $\underline{e}'_{12}$-integration in the collision integral of eq.(3), does not amount to a (mechanical) unitary transformation; it is at this point that the statistical, irreversible behaviour is introduced.

As a matter of fact, the above kinetic equation coincides with the monatomic Boltzmann equation if there is no spin variable at all. This follows immediately from the general identity

$$\frac{h}{im_{12}}(a(0)-a(0)^\dagger) = \int aa^\dagger c'_{12} d^2 e'_{12} = \int a^\dagger a c'_{12} d^2 e'_{12} \quad , \quad (4)$$

called optical theorem, which can be applied to the collision integral if all the factors commute. The optical theorem is nothing but the unitarity of the scattering matrix S expressed in terms of the amplitude a.

In fig. 1 a qualitative plot, a little different from the usual one, is given of the heat conductivities $\lambda_{||}$, $\lambda_\perp$ and $\lambda_{transversal}$ as they result from eq.(3). The influence of a given magnetic field is the smaller the higher the density is. At the low density end, n = 0, and only there, the results of eq.(3) coincide with those of

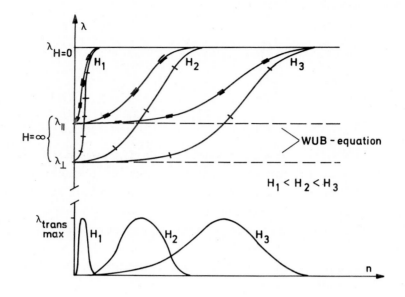

Fig. 1 Heat conductivities λ (H/n), parallel $||$, perpendicular $\perp$ and transversal to magnetic field, as functions of number density n.

eq.(1) if in this the V-quantum numbers are identified with the M's and the correct scattering cross section $\sigma_{M_1 M_2 M_1' M_2'}(\hat{\underline{H}})$ is used. Indeed, condition (2) is fulfilled for any non-zero Zeeman splitting ΔE, if n goes to zero, i.e. t_{free} to infinity.

In conclusion a remark should be made on the restriction to the two cases A and B mentioned at the beginning of this section. What happens in the most general intermediate case, if in condition (2) one has ≈ 1 instead of >>1 or <<1? Several attempts have been made to derive a kinetic equation for this most general situation [22, 23,24]. Scholars are not in perfect agreement in this respect, different philosophies still exist. But in some cases the situation may be better than its appearance. For the Zeeman splitting e.g. it can be shown [25] that eq.(1) is contained in eq.(3) as a limiting case. This makes it plausible that eq.(3) also applies in the intermediate case if the Zeeman frequency is of the same order as the collision frequency (t_{free}^{-1}). As a matter of fact in all the papers on the theory of the Senftleben-Beenakker effect this assumption has been made until now.

III. H-Theorem for a Gas of Particles with Spin

To look at the H-theorem for a polyatomic gas seems appropriate at this memorial meeting. The H-theorem for monatomic gases was one of Boltzmann's outstanding results in his 1872-paper, a result which he immediately tried to generalize for polyatomic gases. Many years later, in 1887, H.A. Lorentz [2] pointed out that Boltzmann's assumption of the existence of inverse collisions was not generally true for polyatomic molecules. Boltzmann reconsidered the problem. In the ensuing paper [3], which has also been incorporated in his famous lectures on kinetic theory [26], he carefully distinguished between initial, final, time reversed and corresponding constellations, finally arriving at a proof of the H-theorem for polyatomic gases by the assumption of "closed cycles of collisions". This part of Boltzmann's argument has frequently been felt to be a little difficult. But its main feature, namely that all the collisions have to be lumped together, remains untouched in the more recent proof of the H-theorem. It has been recognized that there is a direct way leading from the unitarity of the polyatomic scattering matrix in the relevant kinetic equation to the H-theorem [27,28]. The corresponding property of a classical collision probability is the equality of its left and right normalization. Some remarks about that will follow in part V. Now the gas of particles with spin will be considered.

Let us first investigate what becomes of eq.(3) if one chooses a new basis of angular momentum states, a basis which may even be different for different values of t, $\underline{x}$, $\underline{p}_1$, i.e. the time, space and momentum coordinates of a particle "1". Choice of such a new basis means that the distribution matrix f(1) undergoes a unitary transformation U(1) into a new distribution matrix F(1):

$$f(1) = U(1) \; F(1) \; U^\dagger(1) \quad .$$

Afterwards, U will be chosen such that F becomes diagonal; but still let us wait a little with that. The scattering amplitude of eq.(3) will transform analogously:

$$a = U(1) \; U(2) \; A \; U^\dagger(2') \; U^\dagger(1')$$

$$a^\dagger = U(1') \; U(2') \; A^\dagger \; U^\dagger(2) \; U^\dagger(1) \quad .$$

Multiplying both sides of eq.(3) with $U^\dagger(1)$ and $U(1)$ respectively from the left and the right side, we obtain the transformed eq.(3)

$$\mathcal{D}_1 F(1) + [U^\dagger \mathcal{D}_1 U, F]_- + \frac{1}{i\hbar}[\underline{\mu}_1 \cdot \underline{H}, F(1)]_- \qquad (5)$$

$$= \mathrm{tr}_2 \int [\int AF(1')F(2')A^\dagger c'_{12} d^2 e'_{12} \frac{h}{im_{12}}(A(0)F(1)F(2) - F(1)F(2)A^\dagger(0))]$$

$$\cdot d^3 p_2 \quad .$$

The abbreviation

$$\mathcal{D}_1 \equiv \frac{\partial}{\partial t} + \underline{c}_1 \cdot \frac{\partial}{\partial \underline{x}}$$

has been used and $\underline{\mu}$ is the magnetic moment operator transformed analogously:

$$\underline{\mu}_1 = U(1) \underline{\mu}_1 U^\dagger(1) \quad .$$

Comparison of eq.(3) and (5) shows that this kinetic equation is almost form-invariant, as it should be, under a change of the basis. The only new term in eq.(5), the first commutator on the left side, is quite natural and comes from the fact that the new basis has been admitted time- and space-dependent. The optical theorem, eq.(4), is strictly form-invariant under the transformation considered:

$$\frac{h}{im_{12}}(A(0) - A(0)^\dagger) = \int AA^\dagger c'_{12} d^2 e'_{12} = \int A^\dagger A c'_{12} d^2 e'_{12} \quad . \qquad (6)$$

Now, the transformation $U(t,\underline{x},\underline{p})_{MM'}$ is specified in such a way that the transformed distribution matrix

$$F(t,\underline{x},\underline{p})_{MM'} = F_M(\ldots)\delta_{MM'}$$

is diagonal for any time, space point and momentum. There are no physical doubts that the initial value problem of eq.(3) has an existing unique solution f, which can be diagonalized at any time, space point and momentum. This diagonalization can be achieved only when the solution f is already known. Insofar, this procedure is not helpful to solve eq.(3). But it nevertheless leads to considerable insight, namely to the H-theorem. To this end, we now insert the diagonalized F into eq.(5) and select the diagonal element M_1 of it, obtaining

$$\mathcal{D}_1 F_{M_1} \tag{7}$$

$$= \sum_{M_2} \int [\int \sum_{M_1'M_2'} \sum F_{M_1'} F_{M_2'} |A_{M_1M_2M_1'M_2'}|^2 c_{12}' d^2 e_{12}' -$$

$$- F_{M_1} F_{M_2} \frac{h}{im_{12}}(A(0)-A(0)^\dagger)_{M_1M_2M_1M_2}] d^3p_2 .$$

All the commutators on the left side of (5) have dropped out. The diagonal element of the optical theorem (6) is

$$\frac{h}{im_{12}}(A(0)-A(0)^\dagger)_{M_1M_2M_1M_2} \tag{8}$$

$$= \sum_{M_1'M_2'} \int |A_{M_1M_2M_1'M_2'}|^2 c_{12}' d^2 e_{12}' = \sum_{M_1'M_2'} \int |A_{M_1'M_2'M_1M_2}|^2 c_{12}' d^2 e_{12}' .$$

The second line of eq.(8) expresses the equality of the right and left normalization of the total cross section, an identity emerging directly from the unitarity of the scattering matrix. Let us call it the <u>normalization property</u>. With the help of eq.(8) one can rewrite eq.(7) in the equivalent forms

$$D_1 F_{M_1} =$$

$$\sum_{M_2 M_1' M_2'} \int\int (F_{M_1'} F_{M_2'} |A_{M_1 M_2 M_1' M_2'}|^2 - F_{M_1} F_{M_2} |A_{M_1' M_2' M_1 M_2}|^2) c_{12}' d^2 e_{12}' d^3 p_2$$

$$= \sum_{M_2 M_1' M_2'} \int\int (F_{M_1'} F_{M_2'} - F_{M_1} F_{M_2}) |A_{M_1 M_2 M_1' M_2'}|^2 c_{12}' d^2 e_{12}' d^3 p_2 \quad . \quad (9)$$

This looks like the WUB-eq.(1), but the present "σ", taken from the elements of A, is time, space and momentum dependent in an unknown way as long as eq.(3) has not been solved. Eq.(9), first reading, immediately yields the particle conservation equation:

$$\frac{d}{dt} \sum_{M_1} \int F_{M_1} d^3 p_1 = 0 \quad .$$

By means of the normalization property, second reading of (9), one also has

$$\sum_{M_1} \cdots \sum_{M_2'} \int\int\int (F_{M_1'} F_{M_2'} - F_{M_1} F_{M_2}) |A_{M_1 M_2 M_1' M_2'}|^2 c_{12}' d^2 e_{12}' d^3 p_1 d^3 p_2 = 0 \quad .$$
(10)

Herewith, we are ready to look at the entropy density (k_B = Boltzmann's constant, n number density)

$$ns = -k_B \, tr_1 \int f(1) \log f(1) d^3 p_1 = -k_B \sum_{M_1} \int F_{M_1} \log F_{M_1} d^3 p_1 \quad .$$
(11)

For a homogeneous state one obtains with eq.(9), first reading,

$$\frac{dns}{dt} = -k_B \sum_{M_1} \cdots \sum_{M_2'} \int\int\int \log F_{M_1} (F_{M_1'} F_{M_2'} |A_{M_1 M_2 M_1' M_2'}|^2$$

$$- F_{M_1} F_{M_2} |A_{M_1' M_2' M_1 M_2}|^2) c_{12}' d^2 e_{12}' d^3 p_1 d^3 p_2 \quad ,$$

or, by renaming of momenta und subscripts,

$$\frac{dns}{dt} = -\frac{1}{2} k_B \sum_{M_1} \ldots \sum_{M_2'} \iiint \log \frac{F_{M_1} F_{M_2}}{F_{M_1'} F_{M_2'}} F_{M_1'} F_{M_2'} |A_{M_1 M_2 M_1' M_2'}|^2 \cdot$$

$$\cdot c_{12}' d^2 e_{12}' d^3 p_1 d^3 p_2 \cdot$$

By use of the identity (10) which originated from the normalization property this can be rewritten again:

$$\frac{dns}{dt} = \frac{1}{2} k_B \sum_{M_1} \ldots \sum_{M_2'} \iiint \left(\frac{F_{M_1} F_{M_2}}{F_{M_1'} F_{M_2'}} - 1 - \log \frac{F_{M_1} F_{M_2}}{F_{M_1'} F_{M_2'}} \right) F_{M_1'} F_{M_2'} \cdot$$

$$\cdot |A_{M_1 M_2 M_1' M_2'}|^2 c_{12}' d^2 e_{12}' d^3 p_1 d^3 p_2 \cdot \qquad (12)$$

The integrand is ≥ 0 and vanishes, for $\sigma \neq 0$, only if

$$F_{M_1} F_{M_2} = F_{M_1'} F_{M_2'}$$

holds. So, one has

$$\frac{dns}{dt} \geq 0 \quad,$$

which is the H-theorem.

IV. The Classical Limit of the Quantum Kinetic Equation in a Special Case

Let us consider a gas of particles with spin, governed by the kinetic equation (3) with the special scattering amplitude linear in spins

$$a(E, \underline{e}_{12}, \underline{e}'_{12}, \underline{s}_1, \underline{s}_2) \qquad (13)$$

$$= a_o(E,\theta)[1+i\epsilon(E,\theta)\underline{n}\cdot(\underline{s}_1+\underline{s}_2)/\hbar] \quad .$$

Here, θ means the scattering angle:

$$\cos\theta = \underline{e}_{12}\,\underline{e}'_{12}$$

and $\underline{n}$ is a vector normal to the scattering plane:

$$\underline{n} = \underline{e}_{12} \times \underline{e}'_{12} \quad .$$

The spin operators $\underline{s}$ shall have the dimension of an angular momentum so that they become dimensionless if divided by Planck's constant $\hbar$. A scattering amplitude like (13) is caused by spin-orbit coupling and is not at all typical for molecular interaction. We choose it nevertheless because for this model the following considerations are so simple. The function ϵ in (13) is dimensionless; moreover

$$\epsilon \ll 1 \quad , \text{ real}$$

is assumed so that terms of order ϵ^2 etc. can and will be neglected everywhere.

The optical theorem for the amplitude (13) is especially simple. Because of $\underline{n}(0) = 0$ the forward amplitude is

$$a(0) = a_o(0) \quad . \qquad (14)$$

Abbreviating

$$a_o a_o^* = \sigma_o(E,\theta)$$

one has from eq.(4), up to terms linear in ε,

$$\frac{\hbar}{im_{12}}(a_o(0)-a_o(0)^*)$$

$$= \int \sigma_o [1+i\varepsilon \underline{n}\cdot(\underline{s}_1+\underline{s}_2)/\hbar][1-i\varepsilon \underline{n}\cdot(\underline{s}_1+\underline{s}_2)/\hbar]c'_{12}\, d^2e'_{12}$$

$$\simeq \int \sigma_o\, c'_{12}\, d^2e'_{12}\quad . \tag{15}$$

One has now to insert the amplitude (13) into eq. (3). Apart from a factor σ_o the first integrand becomes

$$[1+i\varepsilon \underline{n}\cdot(\underline{s}_1+\underline{s}_2)/\hbar]f(\underline{p}'_1,\underline{s}_1)f(\underline{p}'_2,\underline{s}_2)[1-i\varepsilon \underline{n}\cdot(\underline{s}_1+\underline{s}_2)/\hbar]\quad .$$

Up to terms linear in ε this is a unitary transformation of the f's. According to the commutation relation $\underline{s}\times\underline{s} = i\hbar\underline{s}$ this transformation is nothing but an infinitesimal rotation of the spins. Indeed, one can also write

$$a\, f(1')\, f(2')\, a^\dagger$$

$$= f(\underline{p}'_1,\underline{s}_1+\varepsilon\underline{n}\times\underline{s}_1)\, f(\underline{p}'_2,\underline{s}_2+\varepsilon\underline{n}\times\underline{s}_2)\sigma_o\quad .$$

So, by use also of (14) and (15), the kinetic equation (3) for our model takes the equivalent form

$$\frac{\partial f(1)}{\partial t} + \ldots \tag{16}$$

$$= tr_2 \int\int [f(\underline{p}'_1,\underline{s}_1+\varepsilon\underline{n}\times\underline{s}_1)f(\underline{p}'_2,\underline{s}_2+\varepsilon\underline{n}\times\underline{s}_2)-f(\underline{p}_1,\underline{s}_1)f(\underline{p}_2,\underline{s}_2)]\cdot$$

$$\cdot\sigma_o c'_{12} d^2 e'_{12} d^3 p_2\quad .$$

This is still a quantum theoretical equation, $\underline{s}_1$, $\underline{s}_2$ are still operators with non-commuting components.

But the transition from (16) to a classical kinetic equation is now almost trivial. One has simply to take $\underline{s}_1$, $\underline{s}_2$ as classical variables with commuting components, to replace the trace over the magnetic quantum numbers according to

$$tr_2 \ldots = (S/h) \int d^2 \hat{s}_2 \ldots$$

and to put

$$f_{class} = (S/h) f .$$

Here, S is the (fixed) magnitude of the classical spin vector, $\hat{\underline{s}}$ its unit vector of direction, $d^2\hat{s}$ the pertaining solid angle element. The ensuing classical equation is

$$\frac{\partial f_{class}^{(1)}}{\partial t} + \ldots \qquad (17)$$

$$= \int\int\int [f(\underline{p}_1',\underline{s}_1') f(\underline{p}_2',\underline{s}_2') - f(\underline{p}_1,\underline{s}_1) f(\underline{p}_2,\underline{s}_2)]_{class} \sigma_o c_{12}' \cdot$$

$$\cdot d^2 e_{12}' d^3 p_2 d^2 \hat{s}_2 .$$

Here, according to (16), it has been abbreviated

$$\underline{s}_1' = \underline{s}_1 + \epsilon \, \underline{n} \times \underline{s}_1 \quad , \quad \text{etc.}$$

But instead of that, one might prefer the implicit, symmetrical definition

$$\underline{s}_1' - \underline{s}_1 = \varepsilon\, \underline{n} \times \tfrac{1}{2}(\underline{s}_1 + \underline{s}_1') \quad , \text{ etc.}, \tag{18}$$

which is equivalent to the order ε with the preceding definition, which however beyond that implies exact conservation of the lengths of the $\underline{s}$-vectors

$$s_1^2 = s_1'^2 \quad , \text{ etc. } .$$

Such is required for "classical spin". The magnetic term on the left side of eq.(3) of course goes over into the classical magnetic torque term; again an infinitesimal rotation is at the bottom, this time with the magnetic field as axis. It might seem quite interesting to study more general scattering amplitudes and their classical limits.

V. Remarks on the Classical Boltzmann Equation for Polyatomic Gases

A classical "atom" or "molecule" is something artificial, postulated ad hoc; classical atoms or molecules composed of point particles with Coulomb interaction cannot exist at all. Something quantum theoretical has tacitly been introduced if one considers classical hard spheres or particles with Lennard-Jones potentials or rigid dumb-bells. However, if such objects are once postulated, kinetic theory can very well work with them.

As an example let us consider a gas consisting of rigid dumb-bells, i.e. classically rotating linear molecules. The canonical variables are

$$\theta \, , \, \phi \, , \, p_\theta \, , \, p_\phi \, ,$$

all four rapidly changing in time. But the proper variables for kinetic theory are

$$\underline{j} \; , \quad \alpha \; ,$$

where $\underline{j}$ is the vector of internal angular momentum, constant during free motion, and only the phase angle α, measured in the plane perpendicular to $\underline{j}$, is rapidly changing in time (fig. 2).

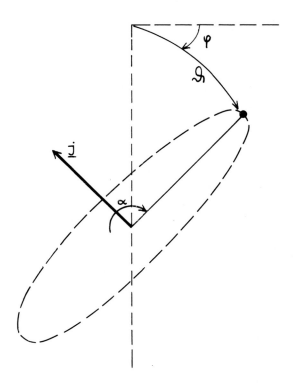

Fig. 2 Rotating rigid dumb-bell.

For the respective distribution functions one has

$$f_{canonical}\, d\theta\, d\phi\, dp_\theta\, dp_\phi = f\, d^3 j\, d\alpha/2\pi$$

or, due to the relation

$$\frac{\partial(\underline{j},\alpha)}{\partial(\theta\phi p_\theta p_\phi)} = j \quad,$$

also

$$f_{canonical} = \frac{j}{2\pi} f \quad.$$

The factor 2π has been inserted in order to have the total particle number given by

$$\iiint f\, d^3x\, d^3p\, d^3j = N \quad,$$

provided that f is independent on α.

This independence shall now be assumed generally:

$$f(1) \equiv f(t,\underline{x},\underline{p}_1,\underline{j}_1) \quad. \qquad (19)$$

Originally this may be taken as an initial condition on f. But it remains then true for all subsequent times thanks to the arbitrariness of the instants of collisions. This is the "Unordnungsannahme" (complete disorder assumption) in the direction of the relative momentum of two colliding particles. With monatomic gases, only complete disorder of positions in the directions perpendicular to the relative momentum has to be considered. - For dipole molecules in an external electric field some modifications of course take place, which are not discussed here.

The Boltzmann equation now must be of the form

$$\frac{\partial f(1)}{\partial t} + \ldots \qquad (20)$$

$$= \iiint [f(1')f(2')j'_1 j'_2 w(121'2') - f(1)f(2)j_1 j_2 w(1'2'12)] d1' d2' d2$$

with the short writing

$$d1' \equiv d^3 p'_1 \, d^3 j'_1 \quad , \quad \text{etc.}$$

The transition "probability" w is a purely (classical) mechanical quantity, as the ordinary monatomic scattering cross section is. One obtains it by evaluating classical orbits; only an averaging over all possible phase angles α before and after collision has to be performed. Again one must get the (purely mechanical) normalization property

$$\iint w(121'2') d1' d2' = \iint w(1'2'12) d1' d2' \quad , \qquad (21)$$

which allows one to rewrite eq.(20) in the form

$$\frac{\partial f(1)}{\partial t} + \ldots \qquad (22)$$

$$= \iiint [f(1')f(2')j'_1 j'_2 - f(1)f(2)j_1 j_2] w(121'2') d1' d2' d2 \quad .$$

In conclusion of these somewhat sketchy remarks let me point out that the normalization property (21) again leads to the H-theorem, in the analogous way as it has been described in section III.

Literature

[1] L. Boltzmann, Wien. Ber. $\underline{66}$, 275-370 (1872).
"Weitere Studien über das Wärmegleichgewicht unter Gasmolekülen".

[2] H.A. Lorentz, Wien. Ber. $\underline{95}$, 115-152 (1887).

[3] L. Boltzmann, Wien. Ber. $\underline{95}$, 153-164 (1887).

[4] L. Waldmann, Z. Naturforsch. $\underline{12a}$, 660 (1957),
$\underline{13a}$, 609 (1958).

[5] R.F. Snider, J. Chem. Phys. $\underline{32}$, 1051 (1960).

[6] L. Waldmann, Nuovo Cimento $\underline{14}$, 898 (1959);
Z. Naturforsch. $\underline{15a}$, 19 (1960).

[7] Y. Kagan and A.M. Afanasew, Zh. Eksp. Teor. Fiz. $\underline{41}$, 1536 (1961), transl. Sov. Phys. JETP $\underline{14}$, 1096 (1962).

[8] L. Waldmann and S. Hess, Z. Naturforsch. $\underline{24a}$, 2010 (1969).

[9] H. Senftleben, Phys. Z. $\underline{31}$, 822, 961 (1930),
J.J.M. Beenakker, G. Scoles, H.F.P. Knaap, and R.M. Jonkman, Phys. Letters $\underline{2}$, 5 (1962).

[10] Y. Kagan and L.A. Maksimov, Zh. Eksp. Teor. Fiz. $\underline{41}$, 842 (1961), transl. Sov. Phys. JETP $\underline{14}$, 604 (1962),
J.J.M. Beenakker and F.R. McCourt, Ann. Rev. Phys. Chem. $\underline{21}$, 47 (1970).
J.S. Dahler and D.K. Hoffmann, in Transfer and Storage of Energy by Molecules, Vol. $\underline{3}$, ed. G.M. Burnett, Wiley, New York 1970.

[11] S. Hess, Z. Naturforsch. $\underline{22a}$, 1871 (1967); $\underline{23a}$, 898 (1968); Physica $\underline{42}$, 633 (1969).
F.M. Chen and R.F. Snider, J. Chem. Phys. $\underline{46}$, 3937 (1967); $\underline{48}$, 3185 (1968); $\underline{50}$, 4082 (1969).
S. Hess and F.R. McCourt, Chem. Phys. Letters $\underline{5}$, 53 (1970).

F.R. McCourt and S. Hess, Z. Naturforsch. 25a, 1169 (1970); 26a, 1234 (1971).

A. Tip and F.R. McCourt, Physica 52, 109 (1971).

[12] S. Hess, Z. Naturforsch. 23a, 597 (1968).

[13] S. Hess, Phys. Letters 30A, 239 (1969).

F. Baas, Phys. Letters 36A, 107 (1971).

A.G.St. Pierre, W.E. Köhler and S. Hess, Z. Naturforsch. 27a, 721 (1972).

[14] V.G. Cooper, A.D. May, E.H. Hara, and H.F.P. Knaap, Can. J. Phys. 46, 2019 (1968).

R.A.J. Keijser, M. Jansen, V.G. Cooper and H.F.P. Knaap, Physica 51, 593 (1971).

[15] S. Hess, Phys. Letters 29A, 108 (1969); Z. Naturforsch. 24a, 1852 (1969); 25a, 350 (1970); Springer Tracts in Mod. Phys. 54, 136 (1970).

S. Hess and H.F.P. Knaap, Z. Naturforsch. 26a, 1639 (1971).

[16] G.G. Scott, H.W. Sturner and R.M. Williamson, Phys. Rev. 158, 117 (1967).

[17] L. Waldmann, Z. Naturforsch. 22a, 1678 (1967).

[18] A.C. Levi and J.J.M. Beenakker, Phys. Letters 25A, 350 (1967).

[19] M.E. Larchez and T.W. Adair, Phys. Rev. A3, 2052 (1971).

S. Hess, Z. Naturforsch. 27a, 366 (1972).

[20] H. Hulsman, F.G. van Kuick, H.F.P. Knaap and J.J.M. Beenakker, Physica 57, 522 (1972).

[21] C.S. Wang Chang and G.E. Uhlenbeck, Eng. Res. Inst. Univ. Michigan Report CM 681 (1951).

C.S. Wang Chang, G.E. Uhlenbeck and J. de Boer in Studies in Statistical Mechanics, ed. J. de Boer and G.E. Uhlenbeck, Vol.II, North Holland Publ. Comp., Amsterdam (1964).

[22] S. Hess, Z. Naturforsch. 22a, 1871 (1967).

[23] A. Tip, Physica 52, (1971).

[24] R.F. Snider and B.C. Sanctuary, J. Chem. Phys. 55, 1555 (1971).

[25] L. Waldmann, Kinetic Theory of Dilute Gases with Internal Molecular Degrees of Freedom. In Fundamental Problems in Statistical Mechanics II, ed. E.G.D. Cohen, p.276-305. Amsterdam, North Holland 1968.

[26] L. Boltzmann, Vorlesungen über Gastheorie II. Teil, VII. Abschnitt; Leipzig 1898.

[27] E.C.G. Stueckelberg, Helvet. Phys. Acta 25, 577 (1952) with a note due to W. Pauli.

[28] L. Waldmann, Transporterscheinungen in Gasen von mittlerem Druck; In Handbuch d. Physik, ed. S. Flügge, 12, p. 484 f., Berlin-Göttingen-Heidelberg 1958.

FLOW-BIREFRINGENCE IN GASES

AN EXAMPLE OF THE KINETIC THEORY BASED ON THE BOLTZMANN-EQUATION FOR ROTATING PARTICLES

S. HESS

Institut für Theoretische Physik
Universität Erlangen-Nürnberg, Erlangen

ABSTRACT

Flow birefringence is an example for the nonequilibrium alignment phenomena which occur in polyatomic gases. In this paper the attention is focussed on gases of linear molecules. Firstly, it is pointed out that the anisotropic part of the electric permeability tensor which is responsible for birefringence is proportional to the (2nd rank) tensor polarization of the rotational molecular angular momentum. Then, the collision-induced tensor polarization set up by a viscous flow is calculated from transport-relaxation equations which, in turn, are derived from the Waldmann-Snider equation, a Boltzmann equation for rotating molecules. Thus the flow birefringence is expressed in terms of properties of single molecules and of their binary scattering amplitude. A relation between flow-birefringence and the Senftleben-Beenakker effect on the viscosity is established. Furthermore, the phenomenon reciprocal to flow birefringence is considered. If a deviation of the tensor polarization from its equilibrium value is maintained externally, the collision-induced coupling between tensor polarization and the friction pressure tensor gives rise to an anisotropy in velocity space. Finally, it is briefly discussed that effects similar to those observed in polyatomic gases should also occur with atomic vapors, provided that the atoms have a nonvanishing electron-orbital angular momentum.

Introduction

One year after Boltzmann's fundamental paper [1] on the "Boltzmann equation" and the H-theorem, Maxwell published a short note on the first observation of a double-refraction caused by a streaming colloidal solution [2](Canada balsam). Later, streaming or flow birefringence has also been found in pure liquids [3,4]. Both in colloidal solutions and in liquids the birefringence is due to an alignment of the figure axis of optically anisotropic particles in a flow field. It has not been realized until the theoretical work of Waldmann [5] and Kagan and coworkers [6] which was based on generalized Boltzmann equations for particles with spin and for classically rotating molecules, respectively, that similar nonequilibrium alignment phenomena should also occur in gases. In this case, however, the internal angular momentum of a particle is aligned in a transport situation.

A strong indirect evidence for the existence of nonequilibrium alignment in polyatomic gases is provided by the influence of a magnetic field on the heat conductivity and the viscosity [7,8]: Senftleben-Beenakker effect. This conclusion is based (i) on the theoretical result that the alignment affects the values of the transport coefficients and (ii) on the fact that the alignment is partially destroyed by an applied magnetic field due to the precessional motion of the rotational angular momenta about the field direction.

Flow birefringence in gases of linear molecules and its relation to the Senftleben-Beenakker effect of the viscosity has been studied theoretically by Hess [9]. Starting point was the Waldmann-Snider equation [10], a generalized Boltzmann equation. An experimental verification of the theory and the first direct proof for the existence of a nonequilibrium alignment in gases of rotating molecules has recently been given by Baas 11 .

This article is divided into 5 sections. Section 1 deals - for gases of linear molecules - with the relation between the anisotropic part of electric permeability tensor which is responsible for the occurrence of birefringence and the tensor polarization of the rotational angular momentum. In section 2, transport-relaxation equations for the friction pressure tensor and the tensor polarization are derived from the Waldmann-Snider equation. From these equations the collision induced tensor polarization set up by a viscous flow and the characteristic flow birefringence coefficient are inferred in section 3. Furthermore, the relation between flow birefringence and the Senftleben-Beenakker effect of the viscosity is discussed. In section 4, the phenomenon reciprocal to flow-birefringence is considered. If a deviation of the tensor polarization from its equilibrium value is produced externally, the collisional coupling between tensor polarization and friction pressure tensor gives rise to an anisotropy in velocity space which might be detectable through the resulting anisotropy of the Doppler width of a spectral line. If the tensor polarization is spatially inhomogeneous, a pressure gradient is built up. Finally, in section 5 it is pointed out that nonequilibrium alignment phenomena which have been observed in polyatomic gases should also occur in vapors of atoms with a nonvanishing electron orbital angular momentum. The appendix contains some details on collision integrals.

1. Formulation of the Problem

A medium is birefringent if its electric permeability tensor $\underline{\varepsilon}$ posseses a nonvanishing anisotropic (symmetric traceless) part* $\overline{\underline{\varepsilon}}$. For a fluid in thermal equilibrium and in the absence of external fields one has $\overline{\underline{\varepsilon}}=0$. This is different for a nonequilibrium situation, in particular for a viscous flow. The constitutive relation for flow birefringence is

$$\overline{\underline{\varepsilon}} = -2\beta \, \overline{\underline{\nabla \, v}} \qquad (1)$$

where $\underline{v}$ is the streaming velocity and β is the "flow-birefringence-coefficient". Sometimes β is written as $\beta = M\eta$ where η is the (shear) viscosity and M is the "Maxwell coefficient". Notice that (1) is very similar to the constitutive relation

$$\overline{\underline{p}} = -2\eta \, \overline{\underline{\nabla \, v}} \qquad (2)$$

for the (traceless) friction pressure tensor $\overline{\underline{p}}$.

To discuss the physical meaning of relation (1) a plane Couette flow in x-direction between parallel plates normal to the y-direction is considered. For this case $\frac{\partial v_x}{\partial y}$ is the only nonvanishing component of $2\overline{\underline{\nabla \, v}}$. Let

---

* Let $A_{\mu\nu}$ ($\mu,\nu=1,2,3$) be the Cartesian components of an arbitrary tensor $\underline{A}$. The components of the tensor $\overline{\underline{A}}$ are given by $\overline{A_{\mu\nu}} = \frac{1}{2}(A_{\mu\nu}+A_{\nu\mu}) - \frac{1}{3}A_{\rho\rho}\delta_{\mu\nu}$. Clearly one has $\overline{A_{\mu\nu}} = \overline{A_{\nu\mu}}$ and $\overline{A_{\mu\mu}} = 0$.

ν_1 (ν_2) be the index of refraction of linearly polarized light propagating in the z-direction with an electric field vector making an angle of 45° (135°) with the x-axis. Then the difference $\delta\nu = \nu_2 - \nu_1$ between the indices of refraction is related to the coefficient β according to

$$\delta\nu = \beta \frac{\partial v_x}{\partial y} \quad . \tag{3}$$

The relation (1) also describes birefringence induced by a sound wave [12] (acoustic birefringence) provided that the sound frequency is small compared with a typical collision frequency. In contrast to flow birefringence, the acoustic birefringence has not yet been observed in gases. In rarefied gases a "heat flow birefringence" characterized by $\overline{\underline{\epsilon}} \sim \overline{\nabla \nabla} T$ may exist [13].

It is the task of the kinetic theory

i   to derive and justify the constitutive relation (1),
ii   to relate β to parameters characteristic for other nonequilibrium phenomena,
iii   to relate β to microscopic properties of single molecules and their mutual interaction.

To accomplish this goal, firstly $\underline{\epsilon}$ is expressed as an average of a molecular quantity. For gases, $\underline{\epsilon}$ is related to the molecular polarizability tensor $\underline{a}$ by

$$\underline{\epsilon} = \underline{\delta} + 4\pi n \langle \underline{a} \rangle \tag{4}$$

where n is the number density and <...> denotes an average evaluated with the one-particle distribution function of the gas. For linear and symmetric top molecules $\underline{a}$ is of the form

$$\underline{\underline{\alpha}} = \alpha \underline{\underline{\delta}} + (\alpha_{||} - \alpha_\perp) \overline{\underline{u}\,\underline{u}} \; , \qquad (5)$$

with $\overline{\underline{u}\,\underline{u}} = \underline{u}\,\underline{u} - \frac{1}{3}\underline{\underline{\delta}}$ where $\underline{u}$ is a unit vector parallel to the molecular axis, $\alpha_{||}$, $\alpha_\perp$ are the polarizabilities for electric fields parallel and perpendicular to $\underline{u}$, and $\alpha = \frac{1}{3}\alpha_{||} + \frac{2}{3}\alpha_\perp$ is an average polarizability. For linear molecules with internal angular momentum $\hbar\underline{J}$ the tensor $\overline{\underline{u}\,\underline{u}}$ can be written as [9,14]

$$\overline{\underline{u}\,\underline{u}} = -\frac{1}{2}(J^2 - \tfrac{3}{4})^{-1} \overline{\underline{J}\,\underline{J}} + \ldots \qquad (6)$$

where the operator J^2 has the eigenvalues $j(j+1)$ with $j = 0,1,2\ldots$ Notice that $\underline{J} \perp \underline{u}$ for linear $^1\Sigma$-molecules. The dots stand for terms which are nondiagonal with respect to the rotational quantum number. These terms are closely associated to the rotational Raman scattering, however, they play no role for the birefringence in gases. Thus Eqs.(4-6) lead to

$$\overline{\underline{\underline{\varepsilon}}} = \varepsilon'\underline{\underline{a}} \qquad (7)$$

where the tensor polarization

$$\underline{\underline{a}} = \langle\underline{\underline{\Phi}}\rangle \text{ with } \underline{\underline{\Phi}} = \sqrt{\tfrac{15}{2}}(\langle\tfrac{J^2}{J^2-3/4}\rangle_o)^{-1/2}(J^2-3/4)^{-1}\overline{\underline{J}\,\underline{J}} \qquad (8)$$

and the abbreviation

$$\varepsilon' = -2\pi n(\alpha_{||}-\alpha_\perp)\sqrt{\tfrac{2}{15}}(\langle\tfrac{J^2}{J^2-3/4}\rangle_o)^{1/2} \qquad (9)$$

has been introduced. The factor occurring in (8) has been chosen such that $\langle\underline{\underline{\Phi}} : \underline{\underline{\Phi}}\rangle_o = 5$ where $\langle\ldots\rangle_o$ refers to an equilibrium average. It has been assumed that $\alpha_{||}-\alpha_\perp$ is approximately independent of the rotational state of a molecule.

According to Eq.(7), the tensor polarization $\underline{a}$ set up by a viscous flow has to be studied in order to derive Eq.(1). This problem is investigated starting from the Waldmann-Snider equation.

2. Boltzmann Equation for Rotating Molecules, Transport-relaxation Equations.

Let $f(\underline{t},\underline{x},\underline{c},\underline{J})$ be the nonequilibrium distribution function operator of the gas. The local average $<A>$ of a quantity $A(\underline{c},\underline{J})$ is given by

$$n <A> = \int d^3c \; Tr \; Af , \qquad (10)$$

where $n = \int d^3c \; Tr \; f$ is the number density of the gas. Clearly, the position $\underline{x}$ and the velocity $\underline{c}$ of a molecule are treated as classical variables, the internal angular momentum $\hbar \underline{J}$ is treated as a quantum mechanical operator. In Eq.(10), "Tr" denotes the trace over magnetic quantum numbers and summation over the rotational states. The distribution operator f obeys a Boltzmann equation which has first been derived by Waldmann [10] and later independently by Snider [10]. The linearized version of this equation can be written as

$$\frac{\partial \Phi}{\partial t} + \underline{c} \cdot \underline{\nabla} \Phi + \omega(\Phi) = 0 , \qquad (11)$$

where $\Phi = \Phi(t,\underline{x},\underline{c},\underline{J})$, defined

$$f = f_o(1 + \Phi) \qquad (12)$$

is a measure for the deviation of f from the equilibrium distribution f_o. In Eq.(11), $\omega(..)$ is the linearized

Waldmann-Snider collision term containing the binary scattering amplitude and its adjoint.

Now it is assumed that the flow velocity $\underline{v} = \langle \underline{c} \rangle$, the friction pressure tensor $\overline{\underline{p}} = nm\langle \overline{\underline{cc}} \rangle$ (where m is the mass of a molecule) and the tensor polarization $\underline{a}$ (cf. Eq.8) are the only macroscopic variables necessary to characterize the nonequilibrium state of the gas for the present problem. In accord with this assumption Φ occurring in Eqs.(11,12) is approximated by

$$\Phi \simeq \frac{m}{kT} \underline{v} \cdot \underline{c} + \frac{m}{kT\, p_o} \overline{\underline{p}} : \overline{\underline{c\,c}} + \underline{a} : \underline{\Phi} \qquad (13)$$

where $p_o = nkT$ is the equilibrium pressure.
Insertion of the ansatz (13) into Eq.(11), its multiplication by $\overline{cc}$ and $\underline{\Phi}$, and the subsequent integration over $\underline{c}$ and performance of the trace "Tr" leads to two coupled linear differential equations (transport-relaxation equations) for $\overline{\underline{p}}$ and $\underline{a}$. These equations are [9,15]

$$\frac{\partial \overline{\underline{p}}}{\partial t} + 2p_o \, \overline{\underline{\nabla\, v}} + \omega_\eta \, \overline{\underline{p}} + \sqrt{2} \, p_o \, \omega_{\eta T} \, \underline{a} = 0 , \qquad (14)$$

$$\frac{\partial \underline{a}}{\partial t} \qquad\quad + \frac{1}{\sqrt{2}\, p_o} \omega_{T\eta} \, \overline{\underline{p}} + \omega_T \, \underline{a} = 0 . \qquad (15)$$

The relaxation coefficients ω_η, $\omega_{\eta T}$, $\omega_{T\eta}$ and ω_T are collision integrals given by [9,16]
($\underline{V} = \sqrt{\frac{m}{2kT}}\, \underline{c}$)

$$\omega_\eta = \frac{2}{5} \langle \overline{\underline{V\,V}} : \omega(\overline{\underline{V\,V}}) \rangle_o , \qquad (16)$$

$$\omega_{\eta T} = \frac{\sqrt{2}}{5} \langle \overline{\underline{V\,V}} : \omega(\underline{\Phi}) \rangle_o, \quad \omega_{T\eta} = \frac{\sqrt{2}}{5} \langle \underline{\Phi} : \omega(\overline{\underline{V\,V}}) \rangle_o, \qquad (17)$$

$$\omega_T = \frac{1}{5} \langle \underline{\Phi} : \omega(\underline{\Phi}) \rangle_o \qquad (18)$$

where $\omega(..)$ is recalled as the linearized Waldmann-Snider collision term. From the time reversal invariance of the binary scattering amplitude follows the Onsager symmetry relation $\omega_{\eta T} = \omega_{T\eta}$. The relaxation coefficients ω_η and ω_T are positive, $\omega_{\eta T}$ may have both signs. Effective cross sections $\sigma_{..}$ pertaining to the relaxation coefficients $\omega_{..}$ may be introduced according to

$$\omega_{..} = n v_{th} \sigma_{..} \qquad (19)$$

where $v_{th} = \sqrt{\frac{8kT}{\pi m_{red}}}$ with $m_{red} = \frac{1}{2}m$ is a thermal velocity.

The relaxation coefficients $\omega_{\eta T} = \omega_{T\eta}$ and ω_T depend in a crucial way on the nonspherical part of the binary scattering amplitude [17] which, in turn, vanishes unless the molecular interaction potential contains a nonspherical part, i.e. a part which depends on the orientation of the molecules. For further details on these relaxation coefficients see the appendix.

3. Tensor Polarization, Flow Birefringence

For a stationary transport situation the time derivatives in Eqs.(14,15) vanish. Then Eq.(15) yields

$$\underline{a} = -\frac{\omega_{T\eta}}{\omega_T} \frac{1}{\sqrt{2} \, p_0} \overline{\underline{p}} \qquad \overline{\underline{p}} = \frac{\omega_{T\eta}}{\omega_T} \frac{2}{\sqrt{2} \, p_0} \eta \, \overline{\nabla \underline{v}} \,, \qquad (20)$$

and Eqs.(14,15) can be solved for $\overline{\underline{p}}$ to yield relation (2) with the viscosity η given by

$$\eta = \eta_{iso}[1 - \frac{\omega_{\eta T}^2}{\omega_\eta \omega_T}]^{-1} \,, \qquad \eta_{iso} = \frac{p_0}{\omega_\eta} \,. \qquad (21)$$

Here η_{iso} is the "isotropic" value of the viscosity which would apply if no alignment would be set up by the viscous flow.

The physical meaning of the tensor polarization $\underline{a}$ as given by Eq.(20) can be elucidated by considering the orientational distribution of the rotating angular momentum of the molecules. Let m be the magnetic quantum number with respect to a quantization axis parallel to the unit vector $\underline{h}$ and N_{jm} the number of particles in the rotational state j with magnetic quantum number m. In thermal equilibrium one has $N_{jm} = \frac{1}{2j+1} N_j^{(o)}$ where $N_j^{(o)}$ is the number of particles in the j-th rotational state. The quantity $\Delta_{jm} = (2j+1)N_{jm}/N_j^{(o)} - 1$ which characterizes the deviation of N_{jm} from its equilibrium value is related to the tensor polarization $\underline{a}$ by

$$\Delta_{jm} = \frac{\sqrt{15}}{2} \frac{m^2 - \frac{1}{3}j(j+1)}{j(j+1) - 3/4} \frac{3}{2} \underline{h} \cdot \underline{a} \cdot \underline{h} \; . \tag{22}$$

Thus Eq.(20) implies

$$\Delta_{jm} = \frac{3}{4} \frac{\sqrt{15}}{2} \frac{m^2 - \frac{1}{3}j(j+1)}{j(j+1) - 3/4} \frac{\omega_{T\eta}}{\omega_T} \frac{n}{p_o} 2 \underline{h} \cdot \overline{\nabla\underline{v}} \cdot \underline{h} \; . \tag{23}$$

For a Couette flow in x-direction between flat plates normal to the y-direction one has $2 \underline{h} \cdot \overline{\nabla\underline{v}} \cdot \underline{h} = \pm \frac{\partial v_x}{\partial y}$ for $\underline{h}$ parallel to $\underline{e}_x \pm \underline{e}_y$ where $\underline{e}_x$, $\underline{e}_y$ are unit vectors parallel to the axes. The ratio $\omega_{T\eta}/\omega_T$ is of crucial importance for the magnitude of Δ_{jm}. Notice that Δ_{jm} as given by Eq.(23) is inversely proportional to the number density. This holds true as long as Knudsen corrections can be disregarded.

In principle, the quantity $\Delta_m = \frac{1}{N} \sum_j N_j^{(o)} \Delta_{jm}$ where $N = \sum_j N_j^{(o)}$ is the total number of particles could be measured by a Stern-Gerlach experiment. So far, however, it proved easier to detect the alignment described by the tensor polarization by means of the birefringence associated with it.

The desired flow birefringence coefficient β defined by Eq.(1) can now be inferred from Eqs.(1,9,20). The result is

$$\beta = -\varepsilon' \frac{n}{\sqrt{2}\,P_o} \frac{\omega_{T\eta}}{\omega_T} = 2\pi n(\alpha_{||} - \alpha_\perp) \frac{1}{\sqrt{15}} \frac{n}{P_o} \frac{\omega_{T\eta}}{\omega_T} \quad . \quad (24)$$

Notice that β as given by Eq.(24) is independent of the number density. For known $\alpha_{||} - \alpha_\perp$, the magnitude of the ratio $\omega_{T\eta}/\omega_T$ and the sign of $\omega_{T\eta}$ can be obtained from a measurement of β. This has been accomplished by Baas [11] for gaseous CO_2.

Now some remarks on the Senftleben-Beenakker effect of the viscosity [7,15,18] are in order. In the presence of a magnetic field $\underline{H} = H\underline{h}$ ($\underline{h} \cdot \underline{h} = 1$) the angular momentum $\underline{J}$ of a diamagnetic molecule precesses about the field direction with the frequency $\omega_H = \gamma H$ where γ is the rotational gyromagnetic ratio. This precession is described by an additional term $i\omega_H[\underline{J} \cdot \underline{h}, \phi]$ on the left hand side of the kinetic equation (11). As a consequence, Eq.(15) contains an additional term due to the precession of the tensor polarization. If the precession frequency ω_H is much larger than the relaxation frequency ω_T the tensor polarization set up by the flow field is essentially destroyed (except for the component $\underline{h} \cdot \underline{a} \cdot \underline{h}$). As a consequence the viscosity $\eta(H)$ approaches its isotropic value η_{iso}. Thus the relative difference $\eta^{-1}\Delta\eta = \eta^{-1}\eta(H) - 1$ between the viscosity with and without a magnetic field approaches, for $\omega_H/\omega_T \to \infty$, the saturation value

$$\left(\frac{\Delta \eta}{\eta}\right)_{sat} = \frac{\eta_{iso} - \eta}{\eta} = -\frac{\omega_{\eta T}^2}{\omega_\eta \omega_T} . \qquad (25)$$

This relative magnetic-field-induced change of the viscosity is of the order 10^{-3} to 10^{-2} for gases of molecules like HD, N_2, CO, CO_2. Thus ω_η is approximately given by $p_0 n^{-1}$. The relaxation coefficient ω_T can be obtained from the magnetic field value for which $\Delta \eta$ reaches half of its saturation value. Furthermore, ω_T can be inferred from the collisional broadening of the depolarized Rayleigh line [9,19,20]. In brief, the magnitude but not the sign of the characteristic quantity $\omega_{\eta T}/\omega_T$ can be determined from the Senftleben-Beenakker effect. The agreement found between the flow birefringence and the viscosity data can be considered satisfactory.

A remark on the orders of magnitude of Δ_{jm} and $\delta \nu$, the difference in index of refraction seems to be useful. For $|\omega_{\eta T}|/\omega_T \approx 10^{-1}$, $\frac{\partial v_x}{\partial y} \approx 100 \text{ s}^{-1}$, and $p \approx 1$ Torr, one has $\Delta_{jm} \approx 10^{-5}$ and $|\delta \nu| \approx 10^{-12}$ for a gas like CO_2. Such small a difference of the occupation number of magnetic substates hardly seems to be detectable in a Stern-Gerlach experiment. On the other hand, differences $\delta \nu$ between the indices of refraction as small as $3 \cdot 10^{-15}$ are measurable.

It seems worth mentioning that the flow-birefringence is also modified by an applied magnetic field [21].

4. The Reciprocal Phenomenon

The flow birefringence which is due to the collision induced coupling between the tensor polarization and the friction pressure tensor is a "cross effect". As is well-known from irreversible thermodynamics, to each cross effect exists a reciprocal phenomenon (e.g. thermal diffusion and diffusio-thermal effect). The relation reciprocal to Eq.(20) can be inferred from Eq.(14). In the absence of a velocity gradient ($\overline{\nabla\,v} = 0$) one finds

$$\underline{p} = n\,m\,\overline{<\underline{c}\,\underline{c}>} = -\sqrt{2}\,p_o\,\omega_{\eta T}/\omega_\eta\,\underline{a}\,, \qquad (26)$$

In the presence of constant external fields which give rise to an equilibrium alignment $<\underline{\phi}>_o$ it is understood that $\underline{a}$ occuring in Eqs.(15,26) is given by $\underline{a} = <\underline{\phi}> - <\underline{\phi}>_o$, i.e. the deviation of the tensor polarization from its equilibrium value. It is assumed that this deviation is produced externally e.g. by radio-frequency induced transitions between the magnetic substates in the presence of constant magnetic or electric fields. Notice that $<\overline{\underline{c}\,\underline{c}}>$ as given by Eq.(26) characterizes an anisotropy in velocity space which is independent of the number density. It gives rise to an anisotropy of the Doppler broadening of spectral lines. More specifically, for small alignment the Doppler width is given by

$$\Gamma_{Dop} = \Gamma_{Dop}^{(o)}(1 - 2\,\frac{\omega_{\eta T}}{\omega_\eta}\,\hat{\underline{k}} \cdot \underline{a} \cdot \hat{\underline{k}}) \qquad (27)$$

where $\Gamma_{Dop}^{(o)}$ is the width for $<\overline{\underline{c}\,\underline{c}}> = 0$ and $\underline{k} = k\,\hat{\underline{k}}$, is the relevant wave vector. It should be stressed that the Doppler width can only be measured if the collisional and diffusional contributions to the broadening of the spectral line are negligible, i.e. if the density is low enough such that $\ell|\underline{k}| \gg 1$ holds true where ℓ is a typical mean free path [9,20,22].

If the tensor polarization $\underline{a}$ is spatially inhomogeneous, Eq.(26) leads to the pressure gradient

$$\underline{\nabla} p = \sqrt{2}\ p_o\ \omega_{\eta T}/\omega_\eta\ \underline{\nabla} \cdot \underline{a}\ . \tag{28}$$

From measurements of the anisotropy of the Doppler width or of the pressure gradient (28) experimental values for $\omega_{\eta T}$ could be obtained which should then be compared with $\omega_{T\eta}$ as inferred from flow birefringence measurements. This would provide a test for the Onsager symmetry relation $\omega_{\eta T} = \omega_{T\eta}$.

5. Atomic Vapors

So far, the Senftleben-Beenakker effect and flow birefringence have only been observed in polyatomic gases [7]. Several attempts to observe an influence of a magnetic field on the transport properties of monatomic gases failed [23]. This is not surprising because all monatomic gases studied so far consisted of atoms without an orbital electron angular momentum $\hbar\underline{L}$. For these atoms, e.g. Na or Cs the interaction potential is essentially spherical if the small magnetic interaction between the electron spins is disregarded. The situation should be different for atomic vapors consisting of atoms with nonvanishing orbital angular momentum $\hbar\underline{L}$ because the "shape" of the atom and consequently the atom-atom interaction potential are nonspherical. The simplest atoms for which a flow birefringence or a Senftleben-Beenakker effect of observable magnitudes can be expected are the ^2P-atoms Al, Ga, In, Tl. These atoms have a $^2P_{1/2}$ ground state. The energy separation to the $^2P_{3/2}$ state is relatively small such that at a temperature of $1000°K$ a significant fraction of the atoms has a total angular momentum $j = 3/2$. A tensor polarization only exists for the atoms with j = 3/2 because $\overline{\underline{J}\,\underline{J}}$ vanishes for j = 1/2. Here $\underline{J} = \underline{L} + \underline{S}$ is the total angular momentum operator (in units of $\hbar$), $\underline{S}$ is the electronic spin operator.

A collision-induced tensor polarization has already been observed accidentally with a beam of Ga-atoms emerging from an oven [24]. Further experiments on flow alignment, flow birefringence and the influence of magnetic fields on the transport properties of such atomics vapors are desirable.

Appendix: Collision Integrals

In this appendix the effective cross sections σ_η, $\sigma_{\eta T} = \sigma_{T\eta}$ and σ_T pertaining to the relaxation coefficients ω_η, $\omega_{\eta T} = \omega_{T\eta}$ and ω_T (cf. Eqs.(16-19)) are studied in more detail. After the integration over the center of mass velocity, these quantities can be written as

$$\sigma_\eta = \tfrac{2}{5}\overline{\{\gamma^4(1-\cos^2\theta)aa^\dagger\}} + \tfrac{2}{5}\overline{\{\gamma^2(\gamma^2-\gamma'^2)(\cos^2\theta-\tfrac{1}{3})aa^\dagger\}}, \quad (A.1)$$

$$\sigma_{T\eta} = \tfrac{2}{5}\sqrt{2}\,\overline{\{\underline{\Phi}_1:(\gamma^2\underline{e}\,\underline{e} - \gamma'^2\underline{e}'\underline{e}')aa^\dagger\}}, \quad (A.2)$$

$$\sigma_{\eta T} = \tfrac{2}{5}\sqrt{2}\,\overline{\{\gamma^2\underline{e}\,\underline{e} : a\,[a^\dagger,(\underline{\Phi}_1 + \underline{\Phi}_2)]\}}, \quad (A.3)$$

$$\sigma_T = \tfrac{2}{5}\overline{\{\underline{\Phi}_1 : a\,[a^\dagger,(\underline{\Phi}_1 + \underline{\Phi}_2)]\}}. \quad (A.4)$$

The colliding molecules are labelled by "1" and "2". Unit vectors parallel to their relative velocity before and after the collision are denoted by $\underline{e}'$ and $\underline{e}$; θ is the scattering angle: $\underline{e}' \cdot \underline{e} = \cos\theta$. As usual, γ'^2 and γ^2 are the relative kinetic energies in units of kT of the colliding pair before and after the collision. The scattering amplitude operator and its adjoint are denoted by a and $a^\dagger$. For any quantities $A = A(\underline{e},\underline{e}',\gamma,\underline{J}_1,\underline{J}_2)$ and $B = B(\underline{e},\underline{e}',\gamma,\underline{J}_1,\underline{J}_2)$ the curly bracket $\{...\}$ is defined by

$$\{A\, a\, B\, a^\dagger\} = \int_0^\infty d\gamma\, e^{-\gamma^2} \gamma^2\, \frac{1}{4\pi} \iint d^2e\, d^2e'\, Q_0^{-2} \cdot$$

$$\cdot \sum_{j_1,j_2,j_1',j_2'} e^{-\varepsilon_{j_1}-\varepsilon_{j_2}} \gamma'\, tr_1 tr_2 A a_{j_1 j_2, j_1' j_2'} B a^\dagger_{j_1' j_2', j_1 j_2},$$

(A.5)

with $Q_0 = \Sigma(2j+1)e^{-\varepsilon_j}$. Here ε_j denotes the internal energy in units of kT of a molecule in the j-th rotational state. The operator $a_{j_1 j_2, j_1' j_2'}$ which depends on $\underline{e},\underline{e}',\gamma$ is defined by

$$a_{j_1 j_2, j_1' j_2'} = P_{j_1} P_{j_2}\, a\, P_{j_1'} P_{j_2'} \qquad (A.6)$$

where a is the scattering amplitude operator and P_j is a projection operator into the j-th rotational state with the property $P_j J^2 = j(j+1) P_j$. The variable γ' occurring within the curly bracket is related to γ

$$\gamma'^2 = \gamma^2 + \varepsilon_{j_1} + \varepsilon_{j_2} - \varepsilon_{j_1'} - \varepsilon_{j_2'}. \qquad (A.7)$$

It is understood that in (A.5) a is put equal to zero for values of γ which would lead to negative values of γ'^2.

Notice that $\sigma_{\eta T} = \sigma_{T\eta}$ and σ_T vanish for a spherical scattering amplitude operator which commutes with $\underline{\Phi}_1 + \underline{\Phi}_2$.

NOTES

[1] L. Boltzmann, Wien Ber. $\underline{66}$, 275 (1872)

[2] J.C. Maxwell, Proc. Roy. Soc. London (A)$\underline{22}$, 46 (1873);
Pogg. Ann. Physik $\underline{151}$, 151 (1874)

[3] D. Vorländer and R. Walter, Z.Phys.Chem.$\underline{118}$, 1 (1925)

[4] G. Szivessy, in Handbuch d. Physik (eds. H. Geiger and K. Scheel) $\underline{21}$, 875 Berlin 1929;
A. Peterlin and H.A. Stuart, in Hand- und Jahrbuch d. Chem. Phys. (ed. Eucken-Wolf) $\underline{8}$, 113 (1943);
J. Frenkel, Kinetic Theory of Liquids, 1943 (Dover, New York, 1955)

[5] L. Waldmann, Nuovo Cimento $\underline{14}$, 898 (1959);
Z. Naturforsch. $\underline{15}$a , 19 (1960)

[6] Y. Kagan and A.M. Afanasev, Soviet Phys. JETP $\underline{14}$, 1096 (1962);
Y. Kagan and L. Maksimov, Soviet Phys. JETP $\underline{14}$, 604 (1962)

[7] H. Senftleben, Phys. Z. $\underline{31}$, 822, 961 (1930);
J.J.M. Beenakker, G. Scoles, H.F.P. Knaap, and R.M. Jonkman, Phys. Letters $\underline{2}$, 5 (1962)

[8] J.J.M. Beenakker and F.R. McCourt, Ann. Rev. Phys. Chem. $\underline{21}$, 47 (1970);
J.S. Dahler and D.K. Hoffmann, in Transfer and storage of energy by molecules, vol.3, ed. G.M. Burnett, Wiley, New York 1970

[9] S. Hess, Phys. Letters $\underline{30}$A, 239 (1969);
Springer Tracts in Mod. Phys. $\underline{54}$, 136 (1970)

[10] L. Waldmann, Z. Naturforsch. 12a, 660 (1957); 13a 609 (1958);
R.F. Snider, J. Chem. Phys. 32, 1051 (1960);
see also
S. Hess, Z. Naturforsch. 22a, 1871 (1967);
A. Tip, Physica 52, 493 (1971);
R.F. Snider and B.C. Sanctuary, J. Chem. Phys. 55, 1555 (1971)

[11] F. Baas, Phys. Letters, 36A, 107 (1971)

[12] M.R. Lucas, Compt. Rend. 206, 827 (1938);
N.C. Hilyard and H.G. Jerrard, J.Appl.Phys. 33, 3470 (1962);
H.G. Jerrard, Ultrasonics 2, 74 (1964);
W.A. Riley and W.R. Klein, J. Acoust. Soc. Am. 45, 587 (1969)

[13] S. Hess, in Proc. 7th Internat. Symp. on Rarefied Gas Dynamics (Pisa 1970), Academic Press, New York 1971

[14] S. Hess, Z. Naturforsch. 24a, 1675 (1969)

[15] S. Hess and L. Waldmann, Z. Naturforsch. 26a, 1057, (1971)

[16] S. Hess and L. Waldmann, Z. Naturforsch. 21a, 1529, (1966); 23a, 1893 (1968)

[17] S. Hess and W.E. Köhler, Z. Naturforsch. 23a, 1903, (1968);
W.E. Köhler, S. Hess, and L. Waldmann, Z. Naturforsch. 25a, 336 (1970)

[18] F.R. McCourt and R.F. Snider, J. Chem. Phys. 47, 4117 (1967)

[19] V.G. Cooper, A.D. May, E.H. Hara, and H.F.P. Knaap, Can. J. Phys. 46, 2019 (1968);
R.A.J. Keijser, M. Jansen, V.G. Cooper, and H.F.P. Knaap, Physica 51, 593 (1971)

[20] S. Hess, Phys. Letters 29A, 108 (1969);
Z. Naturforsch. 24a, 1852 (1969); 25a, 350 (1970)

[21] A.G. St. Pierre, W.E. Köhler, and S. Hess, Z. Naturforsch. 27a, 721 (1972)

[22] S. Hess, Physica 61, 9, 80 (1972); in Festkörperprobleme XII, 649 ed. O. Madelung, Vieweg, Braunschweig, 1972

[23] H. Senftleben, Physik. Z. 32, 550 (1931);
D. Stichtig, Ann. Physik 38, 274 (1940)

[24] P. Toschek, Z. Physik 187, 56 (1965).

NON-EQUILIBRIUM ANGULAR MOMENTUM POLARIZATION IN ROTATING MOLECULES

J.J.M. BEENAKKER

Kamerlingh Onnes Laboratorium der Rijksuniversiteit,
Leiden, Nederland

ABSTRACT

In a non-equilibrium dilute gas of rotating molecules the one particle distribution function can be anisotropic in internal angular momentum $\underline{J}$. The presence of this J "polarization" distinguishes the behaviour of rotating molecules from that of molecules with non-degenerate internal states.
A survey is given of the experimental information that is now available on some of these polarizations, the main source being a study of the influence of external fields on the transport properties. First we consider the type of polarization present in a temperature or velocity gradient. Next we discuss in how far in a solution of the linearized Boltzmann equation the lowest order approximation is sufficient to describe the experimental results. Finally we will briefly indicate the type of information that can be obtained on the interaction between rotating molecules.

Over the last decade an important development has taken place in our understanding of the low density behaviour of systems of molecules with internal degrees of freedom: we learned how to encompass the internal angular momentum of rotating molecules in the frame work of the Boltzmann equation [1,2]. This allowed a systematic treatment of a class of phenomena that have in common their dependence on the molecular orientation. To this group belong among others: the influence of external fields on transport properties, the flow birefringence, depolarized Rayleigh scattering, dielectric non resonant absorption and nuclear magnetic relaxation in polyatomic molecules. Of these the field effect (F.E.) gives the broadest source of information. Since most of the recent developments have been intimately connected to the study of the F.E. we will devote most of our time to this topic.

The most characteristic aspect of the non-equilibrium behaviour of a dilute gas of rotating molecules is the fact that the one particle distribution function, f, is anisotropic both in velocity, $\underline{W}$, and internal angular momentum, $\underline{J}$: $f = f(\underline{W},\underline{J})$. In quantummechanical language the density matrix contains elements $f_{mm'}$, which are off diagonal in the quantumnumber m describing the orientation of the internal angular momentum.

To obtain the actual form of f one starts from the linearized Boltzmann equation appropriate for this case i.e. the Waldmann-Snider equation. In the presence of a temperature gradient the steady state W-S equation for thermal energy transfer is

$$[(W^2 - 5/2) + (J^2 - \langle J^2 \rangle)]\underline{W} \cdot \nabla \ell nT = -n R_o \phi.$$

Here ϕ is the linear deviation from equilibrium given by $f = f^{\circ}(1+\phi)$, R_o is the linearized collision operator which is not degeneracy averaged while the other symbols have their usual meaning. The deviation is of the form

$$\phi = -\underline{A} \cdot \underline{\nabla} \ell n T ,$$

and $\underline{A}$ will depend on the vectors $\underline{W}$ and $\underline{J}$. Following Kagan et al. [3] the deformation of the distribution function is expanded in irreducible tensors $[\underline{W}]^p$ and $[\underline{J}]^q$ of increasing rank p and q:

$$\underline{A} = \Sigma \, \underline{\underline{A}}^{pq} \cdot [\underline{W}]^p \, [\underline{J}]^q .$$

Parity considerations limit the expansion in the case of a temperature gradient to terms odd in $\underline{W}$ but impose no limitation on $\underline{J}$. To express the scalar dependence on W and J, $\underline{\underline{A}}^{pq}$ is further expanded in a set of polynomials in W^2 and J^2:

$$\underline{A} = \Sigma \, \underline{\underline{A}}^{pqrs} \cdot [\underline{W}]^p \, [\underline{J}]^q \, S_r(W^2) \, R_s(J^2)$$

In monatomic gases it is known that in general one term gives sufficient accuracy. Rather optimistically one proceeds here in a similar way:

$$\underline{A} = \underline{\underline{A}}^{1010} \cdot (W^2 - 5/2)\underline{W} + \underline{\underline{A}}^{1001} \cdot (J^2 - \langle J^2 \rangle)\underline{W} + \underline{\underline{A}}^{1100} \cdot \underline{WJ} + \underline{\underline{A}}^{1200} \cdot \underline{W}[\underline{J}]^2 .$$

The first two terms are responsible for the transport of translational and rotational energy. The terms $\underline{W}\,\underline{J}$ and $\underline{W}\,[\underline{J}]^2$ depend explicitly on the orientation of $\underline{J}$ with respect to $\underline{W}$ and correspond to a "polarization" in $\underline{W}$, $\underline{J}$

space induced by the temperature gradient:

$$\langle \underline{W}\, \underline{J} \rangle_{ne} \quad \text{and} \quad \langle \underline{W}[\underline{J}]^2 \rangle_{ne} \neq 0 .$$

The presence of these polarizations distinguishes the behaviour of rotating molecules from that of molecules with non-degenerate internal states.

The most effective tool to study the polarizations is an investigation of the influence of an external field on the transport properties. For a general survey see [4, 5]. Every rotating molecule has associated with its rotation a magnetic moment: $\underline{\mu} = \gamma \underline{J}$. A magnetic field will make $\underline{J}$ precess between collisions with a frequency ω. Hence the direction of $\underline{J}$ is no longer a constant of the motion in free flight. Consequently at high enough fields - $\omega\tau \gg 1$, where τ is of the order of the time between collisions - f will be a function of J_H rather than $\underline{J}$, where J_H is the component of $\underline{J}$ along the field. This results in a destruction of most of the polarization involving $\underline{J}$ and gives rise to a change in the transport properties of the gas.

In this way the F.E. gives direct information on the polarizations in $\underline{J}$. Their destruction by a magnetic field can on a macroscopic level be observed by two types of phenomena:

a. A change of the value of the transport coefficient. This effect is even in the field and saturates at high fields,
b. The appearance of a transverse effect: under the influence of a magnetic field transport takes place perpendicular to both the gradient and the applied field.

This phenomenon is odd in the field and disappears at high fields.

In both cases the field dependence will be a function of the product $\omega\tau$ and consequently of H/p, where p is the pressure. This situation is illustrated in figs.1 and 2 for the case of the heat conduction.

In the following we will give a survey of what we have learned so far from studies of the F.E. on the thermal conductivity and the viscosity. The discussion will be centered on diatomic diamagnetic molecules in a magnetic field. Occasionally a side step will be made to more complicated molecules or to polar molecules in an electric field.

The first question to be answered regards the type of polarization present. An analysis is simplified by the fact that to a good approximation the F.E. is the sum of the contributions of each of the polarizations separately. We know further that terms odd in $\underline{J}$ will give an increase in the transport coefficient in a field while terms even in $\underline{J}$ will give rise to a decrease. To separate the different contributions one can make use of the fact that each polarization has a characteristic dependence on the orientation of the field with respect to the gradient. A study of this angular dependence is an important tool in disentangling the different contributions.

a. <u>The Heat Conductivity</u>.

In the presence of a magnetic field the heat conduction is given by $\underline{q} = - \underline{\lambda} \cdot \nabla T$ where $\underline{\lambda}$ is the second rank tensor given by:

$$\begin{pmatrix} \lambda^{||} & \cdot & \cdot \\ \cdot & \lambda^{\perp} & -\lambda^{tr} \\ \cdot & \lambda^{tr} & \lambda^{\perp} \end{pmatrix}$$

In table I the contributions to the F.E. of the different polarizations are listed

Table I

polarization contribution	$\underline{W}\,\underline{J}$	$\underline{W}[\underline{J}]^2$
$+\Delta\lambda^{||}/\lambda_o$	$+\Psi_{11}\dfrac{\xi_{11}^{2}}{1+\xi_{11}^{2}}$	$-\Psi_{12}\{\dfrac{\xi_{12}^{2}}{1+\xi_{12}^{2}} + 2\dfrac{4\xi_{12}^{2}}{1+4\xi_{12}^{2}}\}$
$+\Delta\lambda^{\perp}/\lambda_o$	$+\Psi_{11}\dfrac{2\xi_{11}^{2}}{1+\xi_{11}^{2}}$	$-\Psi_{12}\dfrac{2\xi_{12}^{2}}{1+\xi_{12}^{2}}$
$+\lambda^{tr}/\lambda_o$	$+\Psi_{11}\dfrac{\xi_{11}}{1+\xi_{11}^{2}}$	$-\Psi_{12}\{\dfrac{\xi_{12}}{1+\xi_{12}^{2}} + 2\dfrac{2\xi_{12}}{1+4\xi_{12}^{2}}\}$
$(\Delta\lambda^{\perp}/\Delta\lambda^{||})_{sat}$	$1/2$	$3/2$

where $\xi_{pq} = \omega\,\tau_{pq} = C_{pq}\dfrac{g\,\mu_N kT}{\hbar}\,H/p$ for diamagnetic molecules, g is the rotational g-factor, μ_N the nuclear-magneton and Ψ_{pq} and C_{pq} are related to collision integrals which depend on the molecular interaction and correspond to the contribution of the $[\underline{W}]^p\,[\underline{J}]^q$ term. Note that the field dependence is a superposition of curves of the type $(\omega\tau)^2/1+(\omega\tau)^2$ for the even effects and $\omega\tau/1+(\omega\tau)^2$ for the transverse effects. Multiples of the precession frequency up to q occur, where q is the rank of the tensor polarization in $\underline{J}$ that is involved.

Experimentally it is found that for all diatomic non-polar molecules the heat conductivity decreases in a magnetic field. This shows that the dominant polarization is even in $\underline{J}$. Extensive measurements were performed on the angular dependence. A schematic diagram of the apparatus that allows a direct determination of $\lambda^{\perp}$ and $\lambda^{||}$ as used by the Leiden group [6,7] is given fig.3. The apparatus has a sensitivity in $\Delta\lambda/\lambda$ of 5 parts in 10^6. The dimensions are such that the measurements can be extended to relatively low pressures without incurring into too large Knudsen corrections. A typical result - for CO at 300K - is given in fig.4. The drawn lines correspond to a contribution from $\underline{W}$ $[\underline{J}]^2$. The shape of the experimental curves agree in general so well with the one predicted by $\underline{W}$ $[\underline{J}]^2$ alone that the deviations hardly exceed the measuring accuracy. Consequently the contributions from other terms are too small to be determined from the shape of the experimental curves. A similar conclusion is reached by comparing $\Delta\lambda^{\perp}$ with λ^{tr}. The results of such a comparison for HD at 85K is given in fig.5. The full lines in each figure represent the set of theoretical curves for the $\underline{W}$ $[\underline{J}]^2$ contribution with the adjustable parameters C_{12} and Ψ_{12} chosen so as to give the best fit.

A more sensitive method is found by studying the ratio $\Delta\lambda^{\perp}/\Delta\lambda^{||}$. This is from the experimental point of view the best determined quantity. Furthermore the value for $(\Delta\lambda^{\perp}/\Delta\lambda^{||})_{sat}$ changes markedly from the value 3/2 corresponding to $\underline{W}$ $[\underline{J}]^2$ with increasing influence of a $\underline{W}$ $\underline{J}$ polarization as is shown in fig.6. The results of this study are given in table II.

Table II

Gas	300K	85K
N_2	1.57 ± 0.01	1.50 ± 0.03
CO	1.52 ± 0.01	1.49 ± 0.03
HD	1.51 ± 0.01	1.49 ± 0.03
pH_2	1.50 ± 0.10	
oD_2	1.60 ± 0.10	
CH_4	1.65 ± 0.015	1.53 ± 0.03
CF_4	1.53 ± 0.025	

For diatomic molecules the deviations from the value 3/2 are hardly significant. One may safely conclude that for diatomic-non polar-molecules $\underline{W}\,[\underline{J}]^2$ is by far dominant. For spherical top molecules like CH_4 the indications for a systematic deviation from 3/2 are more significant, but contributions like $\underline{W}\,\underline{J}$ remain very small.

The absence of odd in $\underline{J}$ polarizations has an interesting consequence for the properties of the linearized collision operator. As was shown by Levi and McCourt [8] such terms will be absent if the collision operator is selfadjoint. In a classical collision model this corresponds to the situation where every collision has its inverse. This approximation seems to hold rather well for linear non-polar molecules.

The situation is different for strongly polar symmetric top molecules. In 1965 Senftleben [9] found indications that in CH_3CN the heat conductivity increased instead of decreased on application of an electric field. This was later confirmed in an elaborate study by De Groot

et al.[10]. Comparing the behaviour of symmetric top
molecules with different dipole moments they found
that the change in an electric field goes with increasing dipole moment from a negative to a positive effect,
see fig.7. There are strong indications that two contributions of opposite sign are present as shown by a
change of sign of the F.E. at low values of E/p. Such
a phenomenon was first observed by Borman et al.[11].
That the presence of a positive contribution is related
to the interaction of a pair of such polar molecules is
shown strikingly by the behaviour of mixtures of CH_3CN
with N_2, see fig.8 [12]. Dilution with N_2 suppresses
clearly the positive contribution.

More detailed conclusions cannot be drawn without further studies of the angular dependence of the F.E.
in polar molecules. These are now in progress. Only then
will one be able to decide unequivocally on the types of
polarization present. Furthermore it is not yet possible
to conclude whether the anomalous change of sign is due
to the presence of a large dipole moment or to the asym-
metric top structure of the molecules in which the effect
has so far been observed. For this reason measurements -
in a magnetic field - are needed on linear strongly polar
molecules such as HCl.

b. The Viscosity

In the presence of a magnetic field the relation
between stress and velocity gradient is conveniently expressed [13,14] in the following way

$$\begin{pmatrix} (3/2)^{1/2} \pi_{zz} \\ \pi_{xz} \\ \pi_{yz} \\ 1/2(\pi_{xx}-\pi_{yy}) \\ \pi_{xy} \end{pmatrix} = -2 \begin{pmatrix} n_o^+ & . & . & . & . \\ . & n_1^+ & -n_1^- & . & . \\ . & n_1^- & n_1^+ & . & . \\ . & . & . & n_2^+ & -n_2^- \\ . & . & . & n_2^- & n_2^+ \end{pmatrix} \begin{pmatrix} (3/2)^{1/2} S_{zz} \\ S_{xz} \\ S_{yz} \\ 1/2(S_{xx}-S_{yy}) \\ S_{xy} \end{pmatrix}$$

where S_{rs} has been written for

$$S_{rs} \equiv (\nabla V_o)_{rs} = \tfrac{1}{2}\{(\tfrac{\partial V_{o_r}}{\partial x_s} + \tfrac{\partial V_{o_s}}{\partial x_r}) - \tfrac{1}{3}(\nabla \cdot V_o)\delta_{rs}\},$$

and where the dots indicate those tensor elements which vanish because of the axial symmetry. The viscosity coefficients $n_\mu^\pm$ relate the real spherical components of the symmetry-2 (traceless, symmetric) parts of the pressure and velocity gradient tensors by:

$$\pi^{(2)\mu\pm} = -2\{n_\mu^+ [\nabla V_o]^{(2)\mu\pm} \mp n_\mu^- [\nabla V_o]^{(2)\mu\mp}\}.$$

The relation between $n_\mu^\pm$ and the coefficients of De Groot and Mazur [15] is given in table III, n_μ^+ corresponds to even in field effects and n_μ^- to transverse effects.

Table III

n_o^+	n_1		
n_1^+	n_3	n_1^-	n_5
n_2^+	$2n_2-n_1$	n_2^-	$-n_4$

As stated above parity considerations restrict the deformation of the distribution function to terms even in $\underline{W}$. The lowest order polarization in $\underline{J}$ that contributes is of the type $[\underline{J}]^2$. The expressions for the behaviour in a magnetic field become very simple; one has

$$\frac{\Delta n_\mu^+}{\eta} = -\Psi_{02}\frac{(\mu\xi_{02})^2}{1+(\mu\xi_0)^2} \quad , \quad \frac{n_\mu^-}{\eta} = \Psi_{02}\frac{\mu\xi_{02}}{1+(\mu\xi_0)^2} \quad \text{for } \mu = 1, 2 \text{ and } \Delta n_0^+ = 0.$$

The fact that $\Delta n_0^+ = 0$ for this type of polarization allows a direct test of the presence of other polarizations for which this is no longer true. To measure the even effects one makes use of a capillary Wheatstone-bridge as shown schematically in fig.9. A high sensitivity differential manometer M serves as a null detector. When a field is applied on C_1 the resulting viscosity change causes an unbalance of the bridge that is measured by M.

By judiciously selecting the geometric conditions Hulsman et al.[16] succeeded to determine Δn_0^+, Δn_1^+ and Δn_2^+ separately. Typical results are given in fig.10 for N_2 at 300K. For all gases studied N_2, CO, HD, CH_4 and CF_4, Δn_0^+ is much smaller than the change in the other two coefficients indicating that $[\underline{J}]^2$ is strongly dominant. A similar conclusion can be reached from a study of the transverse coefficients n_1^- and n_2^- [17].

The transverse effect is measured by observing the pressure difference across a capillary as explained in fig.11. The result for HD at 300K are shown in fig.12. The drawn curve is the best fit for a $[\underline{J}]^2$ contribution. The values obtained from the fit for the two adaptable parameters Ψ_{02} and τ_{02} are within the measuring accuracy equal to those obtained from the even coefficients. (We will come back to this point later). The polar gases [10] do not show a markedly different behaviour, this is in contrast to the case for the heat conductivity. The only gas were sofar a different behaviour has been found is NH_3 [18], where the effect is positive.

The fact that the dominant polarization $[\underline{J}]^2$ does not contain $\underline{W}$ makes it insensitive to collisions in which only $\underline{W}$ and not $\underline{J}$ changes. Consequently it is insensitive to purely elastic collisions which change only $\underline{W}$. At room temperature collisions in which $\underline{J}$ changes in direction and/or magnitude are less frequent than those in which only $\underline{W}$ changes. Consequently the field needed to make $\omega\tau = 1$ is smaller for the viscosity than for the heat conductivity. This is clearly demonstrated by the behaviour of HD where the inelastic collisions are an order of magnitude less frequent than the elastic ones; see fig.13.

For strongly polar molecules the opposite is true. For large enough dipole moment the long range dipolar forces change $\underline{J}$ more effectively than $\underline{W}$. This is demonstrated in fig.14. Here Z_{02} the ratio between effective cross sections for $[\underline{W}]^2$ and $[\underline{J}]^2$ decay is plotted versus dipole moment.

There are other interesting consequences of the structure of the $[\underline{J}]^2$ polarization. For example as the molecular polarizability α_{ik} is not isotropic a non-random distribution of the molecular axis will result in an anisotropic refractive index. This will give rise to flow birefringence. Expressions for this effect were given by Hess [19]. As the refractive index n of dilute gases is only slightly different from 1 and as the amount of $[\underline{J}]^2$ polarization is small the birefringence will remain small. Recently Baas [20] succeeded to construct an apparatus with the desired sensitivity: $\frac{\Delta n}{n} \approx 3.10^{-15}$. A schematic diagram is given in fig.15. The apparatus consists of two concentric cylinders. Rotation of the inner cylinder sets up the velocity gradient in the gas, which makes the medium inside birefringent. The laser light linearly polarized by prism 1, becomes elliptically polarized after passing between the cylinders. A quarter wave plate converts this light into linearly polarized light, but with a direction of polarization rotated over a very small angle. The amount over which it is rotated is detected with the aid of a second polarization prism. A typical result is given in fig.16.

Apart from giving a direct proof of the existence of the $[\underline{J}]^2$ polarization and allowing a check on the theory, this type of measurements gives information on the sign of the $[\underline{J}]^2$ polarization produced by a given flow-field. The direction is found to be perpendicular to the $[\underline{W}]^2$ polarization by which it is produced.

Summarizing we can say that for simple molecules which are not strongly polar in the presence of a temperature or velocity gradient the non-equilibrium distribution function is of the form $f = f^o(1+\phi)$.

$$\phi = -\{\underline{\underline{A}}^{1010} \cdot (\underline{W}^2 - 5/2)\underline{W} + \underline{\underline{A}}^{1001} \cdot (\underline{J}^2 - 3/2)\underline{W} + \underline{\underline{A}}^{1200} \vdots \underline{W}[\underline{J}]^2\} \cdot \underline{\nabla} \ell nT$$

$$-\{\underline{\underline{B}}^{2000}/[\underline{W}]^2 + \underline{\underline{B}}^{0200}/[\underline{J}]^2\} : [\nabla v_o]^2.$$

So far in the analysis only the lowest order terms in the polynomial expansion have been used. Model calculations

for rough spheres and hard ellipsoids seem to justify
this approach [21,22,23]. For these models the higher
order polynomials give very small contributions to the
orientational polarization and its effects. Calculations, however, for monatomic gases indicate that for
soft potentials, where the collision cross sections
are rather sensitive to the velocity of the colliding
particles, higher order Sonines can no longer be neglected [24]. For polyatomic gases cross sections involving changes in $\underline{J}$ can be strongly dependent on $\underline{W}$
or $\underline{J}$. It is therefore an open question whether the
hard core model calculations are a sound enough basis
to discard higher order polynomials. It is thus interesting to consider how far this question can be answered experimentally.

One of the ways in which higher order polynomials could manifest themselves is by changing the shape
of the curves describing the field dependence. These
curves would in fact become a superposition of curves
with slightly different positions at the H/p axis. Looking over the existing experimental material we conclude
that indications of this type are virtually absent in
the case of the heat conductivity. The situation is,
however, different for the viscosity. While the results
for HD - as discussed earlier - agree well with the
shape predicted by the lowest order approximation, this
is not the case for N_2 as shown in figs.17. The curves
for the transverse effect in N_2 show a pronounced broadening. The fact that Δn_o^+ remains very small (see fig.10)
supports the conclusion that the broadening arises from
higher order polynomials. Further indications in this
direction come from a study of depolarized light scattering in dilute gases.

In a gas of non-spherical molecules in equilibrium the orientational distribution of the molecules will fluctuate. If the molecular polarizability is anisotropic the fluctuations in $<[\underline{J}]^2>$ will give rise to a fluctuating refractive index and hence to light scattering. In fact it gives rise to a depolarized contribution to the Rayleigh line [25]. The linewidth of the depolarized component is directly related to the decay time of the $[\underline{J}]^2$ fluctuations. In the same approximation in which the F.E. for the viscosity contains curves corresponding to one value of τ the depolarized Rayleigh line will be a Lorentzian as shown by Hess [26]:

$$I(\omega) \sim \frac{\Delta\omega_{1/2}}{(\omega-\omega_0)^2 + \Delta\omega_{1/2}^2}$$

where $\Delta\omega_{1/2}$ is the halfwidth and ω_0 the frequency of the incoming light. To test the lineshape it is convenient to consider the Fourier transform F(t) of the measured - deconvoluted - line. If the line is a Lorentzian ln $\tilde{F}$ versus t will give a straight line with a slope $\sim \Delta\omega_{1/2}$. While this is indeed found for H_2 gas as shown in fig. 18 it is no longer true for N_2. As shown in fig.19 [28] the results for this gas show marked deviations from the behaviour of a single Lorentzian. This situation has to be expected if higher polynomials have to be included. In conclusion one can state that both the F.E. and depolarized Rayleigh scattering suggest that for molecules where many rotational levels are occupied (e.g. N_2) higher order polynomials have to be included. Rather then use a description in terms of a polynomial expansion one can try another approach in which the molecules in different rotational states are treated separately. Such a line is followed in the uncoupled model as proposed by Coope and Snider [29]. In such a model the field dependence (and also the depolarized Rayleigh line) will be a superposition of contributions of the different rotational states. As the weighting in such a superposition differs from experiment to experiment one has to be very careful in correlating effective cross sections. At the time of writing this paper this problem is still under study.

From the foregoing it will be clear that the field effect is of great help in clarifying questions related to the transport theory of rotating molecules. It helps furthermore to solve another problem. Up till now one is unable to perform ab initio calculations of the interaction between all but the simplest molecules. Rather than to start from a known interaction potential and use kinetic theory to predict the transport properties, we are forced to attempt the opposite. We need measurements of transport properties to arrive at the molecular interaction. Studies in the dilute gas regime are to be preferred because then one is dealing with binary collisions only without the additional complications of more than two body interactions. This has been the way along which most of our knowledge on the interaction between noble gas atoms was obtained. For polyatomic gases this way proved to be more difficult. The best one could obtain was an effective spherical potential. The reason for this is that transport properties as such are in a first approximation insensitive to the non spherical part of the molecular interaction. The F.E., however, is directly related to the non-spherical part in the interaction.

In fig.20 the results of a study of the F.E. on the viscosity of CO as a function of temperature [30] are presented in terms of effective cross sections Σ. $\Sigma(02)$ correspond to the decay of $[\underline{J}]^2$ polarization, $\Sigma\binom{02}{20}$ to its production by the presence of a $[\underline{W}]^2$ polarization. For comparison the effective cross section corresponding to the viscosity $\Sigma(20)$ is also given. Although the calculation of Σ remains even for the simplest models a formidable task one may go one step further without too much computation. For a given type of angle dependent interaction a number of relations between the different Σ exist. An experimental test of such relations can be of great help in selecting a simple potential model. Work along these lines was recently started by Chen, Moraal and Snider [32]. They showed e.g. that under certain simplifying conditions the coupling of $[\underline{J}]^2$ to $[\underline{W}]^2$ is directly related to the cross section describing the bulk viscosity Σ_{rot} (rotational relaxation)

$$\Sigma_{rot} \simeq \sqrt{30}\ \Sigma\binom{02}{20}\ .$$

A test of this relation for N_2 is given in fig.20 [33]. The results look promising. We stand, however, still at the beginning of a long way!

At this point I like to conclude this survey. Many important and interesting aspects of the subject remain untouched. But I hope to have shown how the F.E. has proved to be an important tool in the study of rotating molecules in dilute gases. Its usefulness is not limited to this regime as is proven by its successful application in the near hydrodynamic regime in a study of the Maxwell stresses and in its application to boundary layer phenomena.

At the end of this paper I like to express my gratitude to my collaborators in the Leiden group for molecular physics who made this survey possible.

This work is part of the research program of the "Stichting voor Fundamenteel Onderzoek der Materie (F.O.M)" and has been made possible by financial support from the "Nederlandse Organisatie voor Zuiver Wetenschappelijk Onderzoek (Z.W.O)".

REFERENCES

[1] Waldmann, L., Z. Naturforsch. 12a (1957) 660-662; 13a (1958) 609-620.

[2] Snider, R.F., J.Chem.Phys. 32 (1960) 1051-1060.

[3] Kagan, Yu.M. and Afanas'ev, A.M., Zh.Eksp.Teor.Fiz. 41 (1961) 1536-1545. Transl. Sov.Phys.--JETP 14 (1962) 1096-1101.

[4] Beenakker, J.J.M., Festkörperprobleme 8 (1968) 275-311.

[5] Beenakker, J.J.M. and McCourt, F.R., Ann.Rev.Phys. Chem. 21 (1970) 47-68.

[6] Hermans, L.J.F., Koks, J.M., Knaap, H.F.P. and Beenakker, J.J.M., Phys.Lett. 30A (1969) 139-140.

[7] Hermans, L.J.F., Koks, J.M., Hengeveld, A.F., Knaap, H.F.P., Physica 50 (1970) 410-432.

[8] Levi, A.C. and McCourt, F.R., Physica 38 (1968) 415-437.

[9] Senftleben, H., Ann.Phys. 15 (1965) 273-277.

[10] Tommasini, F., Levi, A.C., Scoles, G., de Groot, J.J., van den Broeke, J.W., van den Meijdenberg, C.J.N. and Beenakker, J.J.M., Physica 49 (1970) 299-341.

[11] Borman, V.D., Gorelik, L.L., Nikolaev, B.I., Sinitsyn, V.V. and Troyan, V.I., Zh.Eksp.Teor.Fiz. 56 (1969) 1788-1795.

[12] De Groot, J.J., van den Broeke, J.W., Martinius, H.J. and van den Meijdenberg, C.J.N., Physica 49, (1970) 342-344.

[13] Beenakker, J.J.M., Coope, J.A.R. and Snider, R.F., Phys.Rev. $\underline{A4}$ (1971) 788-796.

[14] Coope, J.A.R. and Snider, R.F., J.Che.Phys. $\underline{56}$ (1972) 2056-2071.

[15] De Groot, S.R. and Mazur, P., Nonequilibrium Thermodynamics, Amsterdam: North-Holland (1962) 501 pp.

[16] Hulsman, H. and Knaap, H.F.P., Physica $\underline{50}$ (1970) 565-572.

[17] Hulsman, H., van Waasdijk, E.J., Burgmans, A.L.J., Knaap, H.F.P. and Beenakker, J.J.M., Physica $\underline{50}$ (1970) 53-76.

[18] Korving, J., Physica $\underline{46}$ (1970) 619-625.

[19] Hess, S., Phys.Lett. $\underline{30A}$ (1969) 239-240.

[20] Baas, F., Phys.Lett. $\underline{36A}$ (1971) 107-108.

[21] Klein, W.M., Hoffman, D.K. and Dahler, J.S., J.Chem. Phys. $\underline{49}$ (1968) 2321-2333.

[22] McCourt, F.R., Knaap, H.F.P. and Moraal, H., Physica $\underline{43}$ (1969) 485-512.

[23] Moraal, H., McCourt, F.R. and Knaap, H.F.P., Physica $\underline{45}$ (1969) 455-468.

[24] Chapman, S. and Cowling, T.G., The mathematical theory of non uniform gases, Cambridge, chapter 10.

[25] Cooper, V.G., May, A.D., Hara, E. and Knaap, H.F.P. Phys.Lett. $\underline{27A}$ (1968) 52-53.

[26] Hess, S., Z.Naturforsch. $\underline{24a}$ (1969) 1852-1853.

[27] Hess, S., Springer Tracts Mod. Phys. $\underline{54}$ (1970) 136-176.

[28] Keijser, R.A.J. et al., to be published.

[29] Coope, J.A.R. and Snider, R.F., J. Chem. Phys. $\underline{56}$ (1972) 2049-2055.

[30] Burgmans, A.L.J., Thesis Leiden (1972);
Burgmans, A.L.J. et al., Physica, to be published.

[31] Moraal, H. and Snider, R.F., Chem. Phys. Lett. $\underline{9}$ (1971) 401-405.

[32] Chen, F.M., Moraal, H. and Snider, R.F., J. Chem. Phys. $\underline{57}$ (1972) 542-561.

[33] Prangsma, G.J., Thesis Leiden (1971); Prangsma, G.J., Burgmans, A.L.J., Knaap, H.F.P. and Beenakker, J.J.M., Physica, to be published.

FIGURE CAPTIONS

Fig.1   The different heat conductivity coefficients occurring in a magnetic field.

Fig.2   Typical behaviour of $\Delta\lambda^{\perp}$, $\Delta\lambda^{||}$ and λ^{tr} as a function of H/p.

Fig.3   Schematic diagram of apparatus used to measure $\Delta\lambda$ as a function of the orientation of the magnetic field.

Fig.4 $(\frac{\Delta\lambda^{\perp}}{\lambda})$ and $(\frac{\Delta\lambda^{||}}{\lambda})$ versus H/p for CO at 300K. Drawn lines $\underline{W}[\underline{J}]^2$ contribution.

Fig.5 $(\frac{\Delta\lambda^{\perp}}{\lambda_o})$, $(\frac{\Delta\lambda^{||}}{\lambda_o})$ and $\frac{\lambda^{tr}}{\lambda_o}$ versus H/p for HD at 85K. Drawn lines $\underline{W}[\underline{J}]^2$ contribution. Shaded areas estimated experimental uncertainties. See also text.

Fig.6   The influence of the relative strength of the $\underline{W}\,\underline{J}$ contribution on
$$\frac{|\Delta\lambda^{\perp}|_{sat}}{|\lambda_{tr}|_{max}} \quad \text{and on} \quad (\frac{\Delta\lambda^{\perp}}{\Delta\lambda^{||}})_{sat}.$$

Fig.7   The importance of the electric dipole moment for the magnitude of the field effect in the heat conductivity.

Fig.8 $\frac{\Delta\lambda^{||}}{\lambda_o}$ versus E/p for CH_3CN-N_2 mixtures. Drawn curves correspond to a combination of a positive $\underline{W}\ \underline{J}$ and a negative $\underline{W}[\underline{J}]^2$ contribution.

Fig.9 Schematic diagram of the capillary viscosity bridge.

Fig.10 η_o^+, η_1^+ and η_2^+ versus H/p for N_2 at 300K.

Fig.11 Schematic diagram of arrangement to measure the transverse viscomagnetic effect. The pressure difference between A and B is directly related to η_1^- and η_2^-.

Fig.12 η_1^- and η_2^- versus H/p for HD at 300K. Drawn line $[\underline{J}]^2$ contribution.

Fig.13 $\frac{\Delta\eta}{\eta}$ and $\frac{\Delta\lambda}{\lambda}$ versus H/p for HD.

Fig.14 The ratio between the effective cross section for $[\underline{W}]^2$ and $[\underline{J}]^2$ decay, Z_{02}, versus electric dipole moment, d.

Fig.15 Schematic diagram of apparatus to measure flow birefringence.

Fig.16 The birefringence, $\frac{\Delta n}{n}$, in CO as a function of the gradient of the flow velocity.

Fig.17 η_1^- and η_2^- versus H/p for N_2 at 300K. Drawn curves $[\underline{J}]^2$ contribution.

Fig.18   The Fourier transform of the intensity of the depolarized Rayleigh line, $\tilde{F}_I(t)$ plotted as $\ln \tilde{F}_I(t)$ versus t for H_2 at 300K.

Fig.19   The Fourier transform of the intensity of the depolarized Rayleigh line, $\tilde{F}_I(t)$ plotted as $\ln \tilde{F}_I(t)$ versus t for N_2 at 300K.

Fig.20   The effective cross sections $\Sigma(02)$, $\Sigma(20)$ and $\Sigma\binom{20}{02}$ of CO versus temperature.

Fig.21   A test of the relation $\Sigma_{rot} \simeq \sqrt{30}\ \Sigma\binom{20}{02}$ for N_2.

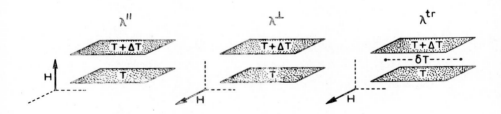

Fig. 1

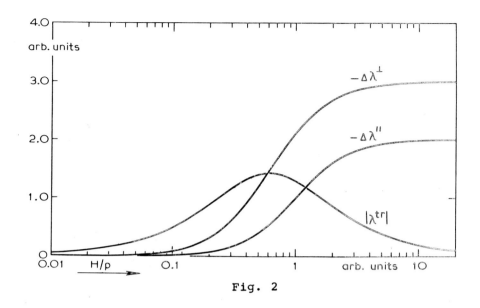

Fig. 2

λ'' and $\lambda^\perp$

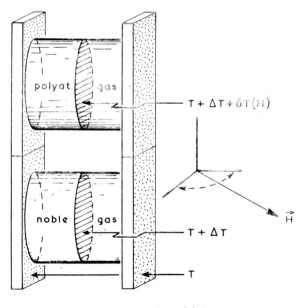

Fig. 3

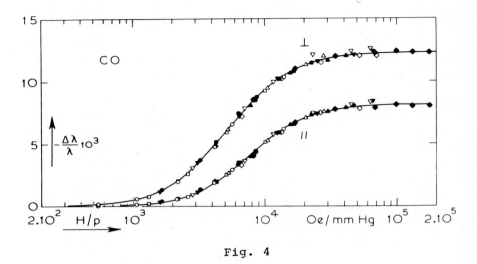

Fig. 4

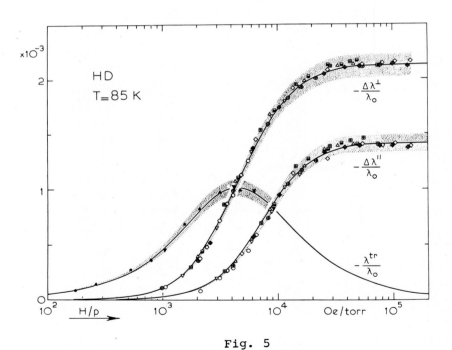

Fig. 5

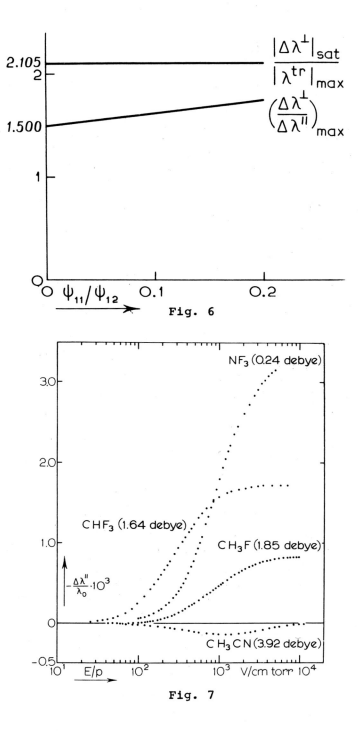

Fig. 6

Fig. 7

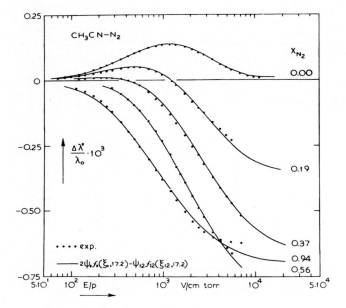

Fig. 8

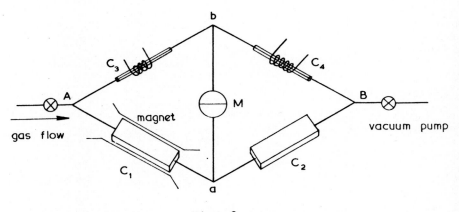

Fig. 9

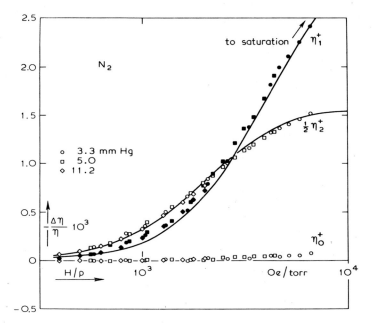

Fig. 10

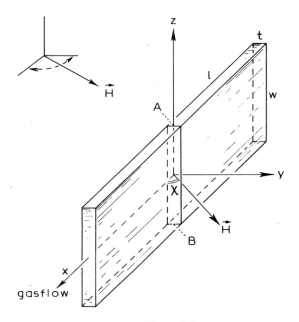

Fig. 11

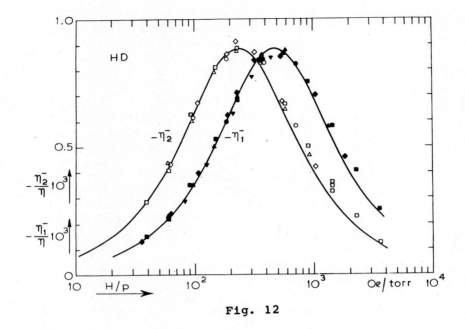

Fig. 12

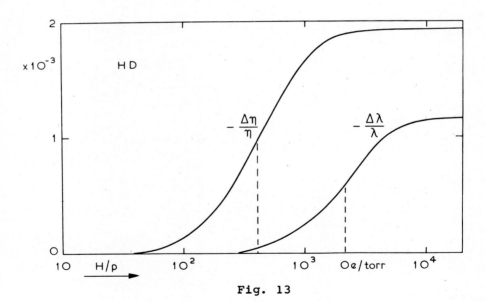

Fig. 13

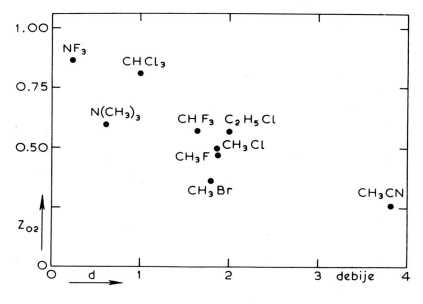

Fig. 14

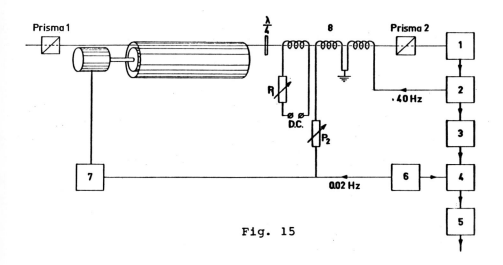

Fig. 15

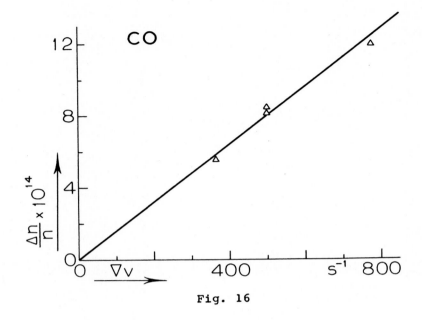

Fig. 16

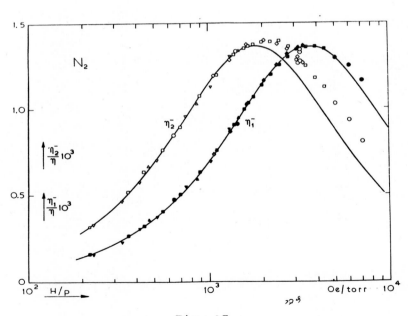

Fig. 17

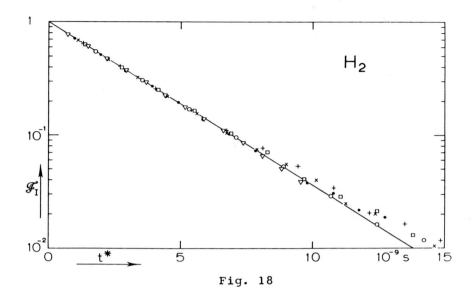

Fig. 18

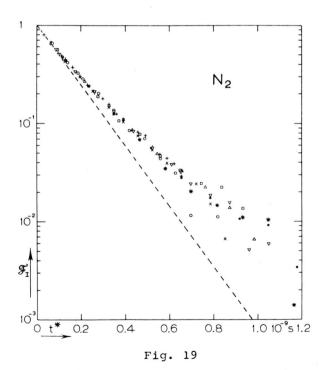

Fig. 19

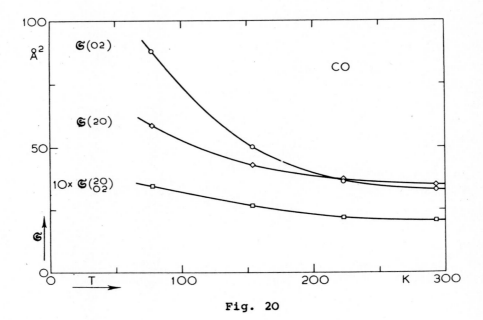

Fig. 20

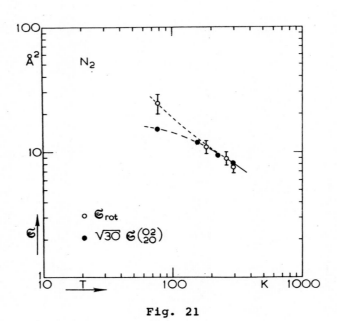

Fig. 21

THE BOLTZMANN EQUATION IN SOLID STATE PHYSICS

RYOGO KUBO

Department of Physics, University of Tokyo

ABSTRACT

The Boltzmann equation has been used successfully in solving a number of problems in solid state physics. This paper presents a historical review of such applications and further a brief review of some recent developments which go beyond the limitations of its applicability. The first use of a Boltzmann equation goes back to Lorentz (1905) who formulated the Drude theory of metallic electrons as a kinetic theory. Sommerfeld (1928) rewrote the Lorentz theory introducing the Fermi statistics. This was immediately followed by the work of Bloch who opened the new era of the quantum theory of solids. The Boltzmann-Bloch equation was a great success to elucidate the problem of metallic conduction. The phonon transport theory was established by Peierls in 1929. Between 1928 and the mid-fifties the progress had been more or less along the line set up by these pioneers. The soil of solid state physics became exceedingly rich after the war mainly because of rapid growth on the experimental side, for example in semiconductor physics. After the mid-fifties, however, there has been a remarkable progress on the theoretical side. Quantum-statistical theories of many-body systems developed various new methods and new concepts, which cover not only many-body systems in thermal equilibrium but also those in non-equilibrium. Thus they necessarily have some connection with the Boltzmann equation approach. They should be able to provide a satisfactory foundation of a Boltzmann equation if it is really to be justified, or they should provide us with a better method to approach reality when a naive kinetic approach is not legitimate. It should be remembered that the gas-like particles in solid states are only quasi-particles representing some sort of excitation modes, which are by no means elementary objects like real gas particles. Thus, we need a more sophisticated approach to understand transport phenomena in a condensed matter like a solid.

1. Introduction

For many years since the end of the last century the kinetic theory of matter has been a very important subject of physics. Looking back from the viewpoint of theoretical physics of today, its importance may be said to lie more in the new concepts and new methods it stimulated at each stage of its development rather than the vast accumulation of physical facts it has disclosed.

The Boltzmann equation is the oldest of such contributions from the kinetic theory of matter. It was given by L. Boltzmann with the purpose of establishing a statistical theory of dilute gases. It can be, and in fact has been, used for studying a great variety of statistical problems, not necessarily problems in physics. Even though its limitations are now clearly recognized and more advanced methods may be available, the simplicity and transparency of the Boltzmann equation approach makes it remain the most useful standard method of statistical physics.

Here I would like to try a short historical survey of the Boltzmann equation used in solid state physics. Before doing that, a few remarks may be appropriate as an introduction. Solids are of course very much different from gases to which the Boltzmann equation originally was devised to apply. A solid is a condensed matter in which material particles are strongly interacting. In a sense it is rather surprising to find the Boltzmann equation so useful in physics of solids. Its usefulness is by no means obvious. Indeed, there can exist something gas-like in solids, - electrons and phonons -, as had been recognized already at the early beginning of this century. In the modern language, low-lying excitations in a solid would behave like a collection of quasiparticles. However it had been a rather long way to reach this modern concept, for example, to reach the so-called quasi-particle picture of metallic electrons from the naive Drude picture of free electrons. Gas-like quasi-particles in a solid are in fact very complex objects not easily understandable by intuition. Thus a Boltzmann equation approach to study physics of a given solid is subject to very basic questions which would not be asked when we work on a Boltzmann equation for a gas. In the first place, what are the objects to which you want to apply the Boltzmann equation? Is it really legitimate to write Boltzmann equations for those objects? What are the collision terms to describe statistical change of those objects? What are the consequences of such equations which are relevant to the physics of solids? If a Boltzmann equation turns out to be inapplicable to the problem, how do you go beyond the limitations of Boltzmann equations? The kinetic theory of solids, or more generally physics of solids, has been struggling many years with these questions. Thus the history of Boltzmann equations in solids is necessarily connected to our deeper understanding of the physics of solids itself.

One characteristic feature of the kinetic theory of solids is that it is more quantum-mechanical than classical-mechanical. Thus the above-mentioned questions have to be answered on the basis of quantum-mechanics. This situation is somewhat different from that in physics of gaseous systems. Fortunately, quantum mechanics often makes life easier. This is an advantage. Recent success of quantum theory of many-particle systems does heavily rely upon this luck. It should be kept in mind however that this is, to some extent, merely from a formal point of view. It does not necessarily really help to understand nature. From time to time we are obliged to go back to the simplest method such as Boltzmann equation approach to seek our way through thick mist of complexity.

2. From 1900 to 1928.

In 1900 P. Drude [1] published two monumental papers entitled "Zur Elektronentheorie der Metalle" in which he gave a thorough study of transport properties of metals on the basis of a free electron model. This work appeared perhaps too early, many years earlier than the proper time. Assuming the Maxwellian law for electron velocity and the existence of a definite mean free path at a given temperature, he was able almost to exhaust all points of metallic conduction. In particular, he derived the Wiedemann-Franz law

$$\frac{\kappa}{\sigma} = 3(\frac{k}{e})^2 T$$

or the Lorentz number

$$L = \kappa/\sigma T = 3(k/e)^2 \quad . \tag{1}$$

Drude's treatment corresponds to the elementary theory of gas kinetics. An important step forward was taken in 1905 by H.A. Lorentz [2] who wrote down the Boltzmann equation

$$X\frac{\partial f}{\partial \xi} + \xi\frac{\partial f}{\partial x} + \eta\frac{\partial f}{\partial y} + \zeta\frac{\partial f}{\partial z} + \frac{\partial f}{\partial t} = b - a \qquad (2)$$

for the distribution function f of the electron gas. Here X is the external force, (ξ, η, ζ) is the velocity, and b-a on the right hand side is the collision term. This is the first use of the Boltzmann equation in the kinetic theory of solids. It was assumed that the mutual collision of electrons can be ignored and that electrons are scattered by atoms in the bulk of metal so that the mean free path is nearly constant, of about the order of the interatomic distance. Assuming f in the form

$$f = f_o + g \qquad (3)$$

where f_o is the Maxwellian distribution, Lorentz showed how the Boltzmann equation is solved for g in the presence of an external field or of non-uniformity of temperature. Thereby electric or heat current was calculated. The results were of course largely the same as those from the elementary theory of Drude, but with some difference; for example the value of the Lorentz number now turned out to be

$$L = 2(k/e)^2 \qquad (4)$$

instead of the Drude value, which is actually closer to empirical values.

Lorentz had to use a classical naive picture of free electrons which led him to some wrong results. Nevertheless his version of the Drude theory was a very essential improvement because it became the prototype of the real theory of electron transport in metals, which was established only after a quarter of century when new quantum mechanics suddenly opened the door to the new era of modern physics.

Perhaps, this period must be considered in a certain sense like the period before Planck's discovery of quantum. The Drude-Lorentz theory was certainly on the right track but it lacked something very essential because it definitely contradicted the experience.

It could not explain the temperature dependence of electrical and thermal resistivities. It predicted a wrong sign for the pressure effect of the electrical resistance. The lack of electronic specific heat was also a great difficulty of the free electron model. These serious difficulties caused some speculations. J.J. Thomson [3], stimulated by the discovery of superconductivity by Kamerlingh Onnes, proposed a dipolar model. F.A. Lindemann imagined metallic electrons forming a rigid lattice. Earlier than Thomson and Lindemann [4], Wien [5] clearly recognized that the electron theory of metals required an entirely new basis. His argument was phenomenological, reminding us of another great contribution of his own to the radiation theory. But he knew what must be right. He pointed out that the electron velocity must be essentially independent of temperature and the conductivity would become infinite at the absolute zero of temperature if the metal structure is completely regular. Assuming that electrons are scattered by lattice vibrations, he even gave a resistivity formula which predicted the resistivity to be a function of T/θ. This was the first attempt to associate the electrical resistance with the Debye temperature. The real breakthrough came in 1927 - 1928 when A. Sommerfeld [6] removed

at a stroke the fundamental difficulties which had beset
the free-electron gas theory [7] by applying the concept
of Fermi-Dirac statistics which was discovered by Fermi
and Dirac in 1926. Sommerfeld rewrote the Drude-Lorentz
theory replacing the Maxwellian distribution function by
the Fermi-Dirac distribution function and showed that the
new theory of a degenerate electron gas could account for
all the essential features of metallic properties. As is
well known, the Sommerfeld theory recovered a good agreement of the theoretical Lorentz number,

$$L = \frac{\pi^2}{3}(\frac{k}{e})^2 \quad , \tag{5}$$

with the experimental values. However he did not consider
the problem of determining the mean free path, which was
left for Bloch. Before going to the new era of solid state
physics started by the initiation of a quantum-mechanical
theory of transport by Bloch, we should not forget a word
about phonons, which are nearly as old as electrons in
metals.

Debye invented the Debye-model for the lattice
vibrations. This is not merely a simple improvement replacing Einstein's atomic oscillators with single frequency
by a set of oscillators with a continuous frequency spectrum. It was an essential improvement in that the Debye
modes are spread over the whole piece of a crystal. This
lead Debye [8] already in 1914 to see the fact that Debye
waves are indeed the carriers of heat in dielectric solids.
Like x-ray propagating in a crystal, Debye wave would be
scattered by thermal agitations of the lattice vibration.
Thus Debye explained the 1/T dependence of thermal conduc-

tivity which was found already by A. Eucken [9].

The Debye model was a simple model treating lattice vibrations as elastic vibrations. More faithful lattice dynamics was initiated by Max Born and von Kármán [10]. Unlike the free electron model of metals, the lattice dynamics had from the beginning a much safer, much sounder dynamical basis. Quantization and statistics did not present any difficulty. The concept of phonons was almost there. However, there was still a large gap between ideal independent phonons and non-ideal interacting phonons. In order to proceed from the stage of an equilibrium ideal phonon gas to a non-equilibrium phonon gas with weak interaction, one had to wait nearly twenty years before a quantum-mechanical theory of phonon transport was first established by R. Peierls [11] in 1929.

3. From 1928 to the Mid-fifties

The year 1928 should be remembered as the beginning of the modern theory of solids. The Sommerfeld theory was still based on the free electron model and left open the question how the model can be justified on the basis of quantum-mechanics. Nor did it answer the question how the electron mean free paths are determined. Several attempts immediately followed Sommerfeld's paper. Houston [12], and Frenkel and Mirolubow [13] discussed the scattering of electron waves in a crystal, but the decisive work was done in this year by F. Bloch.

In a monumental paper [14] entitled "Über die Quantenmechanik der Elektronen in Kristallgittern", Bloch established the basis of modern theory of metallic electrons. First, he proved the Bloch theorem which shows that electron waves are only modulated by the periodic potential of a perfect lattice but are never scattered by lattice atoms as long as the atoms stay in their regular positions. Thus the electrical resistance is caused only by lattice irregularities such as lattice vibrations or lattice imperfections. Next he derived the acceleration equation for a Bloch electron, which shows that an external electric field makes the electron wave vector increase in time, but in the real space an electron may be accelerated or decelerated according to its energy-momentum relationship.

Then Bloch took

$$H_{ep} = V(\vec{r}-\vec{u}) - V(\vec{r}) = -(\vec{u} \text{ grad } V) \tag{6}$$

as the interaction potential of an electron with the lattice vibrations. Here V is the periodic potential and $\vec{u}$ is the displacement at the position $\vec{r}$ caused by the lattice vibrations. It should be noted however that the electron-phonon interaction is not a simple problem and in fact Bloch's assumption is by no means well justified. But the essential point here is that the electron-phonon interaction causes an electron scattered from one-Bloch state $\vec{k}$ to another $\vec{k}'$ inelastically by absorbing or emitting a phonon of momentum $\vec{q}$, namely

$$\vec{k}' = \vec{k} \pm \vec{q} , \tag{7}$$

$$E_{k'} = E_k \pm \hbar\omega_q . \tag{8}$$

Probabilities of such transitions were obtained by using Dirac's perturbational method. This enabled him to write down his celebrated equation of transport - the

Boltzmann-Bloch equation – but with an additional assumption that phonons remain in thermal equilibrium. If the problem is further simplified assuming a spherical energy function $E(\vec{k})$ for Bloch electrons and the simplest Debye model for phonons, the collision term of the Boltzmann-Bloch equation is given in the form,

$$\left(\frac{\partial f}{\partial t}\right)_{coll.} = A \int_0^{q_0} q^2 dq \int_0^{2\pi} d\phi$$

$$\{[(1-f(\vec{k}))f(\vec{k}+\vec{q})(N_q+1)-f(\vec{k})(1-f(\vec{k}+\vec{q}))N_q]_{E_{k+q}=E_k+\hbar\omega_q}$$

$$+ [(1-f(\vec{k}))f(\vec{k}+\vec{q})N_q - f(\vec{k})(1-f(\vec{k}+\vec{q}))(N_q+1)]_{E_{k+q}=E_k-\hbar\omega_q}\}$$

(9)

where A is a constant for a given material, q_0 the maximum wave number of phonons, ϕ the angle between $\vec{k}$ and $\vec{q}$, N_q the equilibrium phonon number. For small derivations from equilibrium, this collision term is linearized in the deviation g in Eq.(3) to

$$\left(\frac{\partial g}{\partial t}\right)_{coll.} = A \int_0^{q_0} q^2 dq \int_0^{2\pi} d\phi$$

$$\{[g(\vec{k}+\vec{q})(N_q+1-f_0(\vec{k}))-g(k)(N_q+f_0(\vec{k}+\vec{q}))]_{E_{k+q}=E_k+\hbar\omega_q}$$

$$+ [g(\vec{k}+\vec{q})(N_q+f_0(\vec{k}))-g(\vec{k})(N_q+1-f_0(\vec{k}+\vec{q}))]_{E_{k+q}=E_k-\hbar\omega_q}\}.$$

(10)

In the presence of an electric field F the stationary distribution is determined by

$$\frac{eF}{\hbar}\frac{df_0}{dE}\frac{dE}{dk}\frac{k_x}{k} = \left(\frac{\partial g}{\partial t}\right)_{coll.} \tag{11}$$

which is an integral equation for the function g. It is convenient to write

$$g = -\Phi(E) \frac{df_o}{dE} \qquad (12)$$

and transform the integral equation to that for $\Phi(E)$. No rigorous solutions of this integral equation is known. But in the high and the low temperature [15] limits, $T \gg \theta$ and $T \ll \theta$, the integral kernel becomes degenerate, so that analytical solutions are available. Fortunately, the low temperature solution turns out to give a good interpolation formula for the resistivity, known now as the Grüneisen formula [16] which is expressed as

$$\rho = \text{const.} \ (\tfrac{T}{\theta})^5 \ J_5(\tfrac{\theta}{T}) \ , \qquad (13)$$

where

$$J_5(x) = \int_o^x \frac{x^5 dx}{(e^x-1)(1-e^{-x})} \ .$$

In this way Bloch showed that the electrical resistance in a metal is proportional to T at high temperatures and to T^5 at low temperatures. The last conclusion differs from that by earlier theories, because now the inelasticity of scattering and the Pauli principle were properly considered.

When the temperature is not uniform, Eq.(11) has to be replaced by

$$v_x (\frac{\partial f_o}{\partial T} \frac{dT}{dx} + eF \frac{\partial f_o}{\partial E}) = (\frac{\partial g}{\partial t})_{\text{coll.}} \qquad (14)$$

Generally, electric current and heat current are coupled

together. This thermoelectric effect was studied by Nordheim [17] and further by Kroll [18]. The calculated heat conductivity K is nearly constant at high temperatures and increases in proportion with T^{-2} at low temperatures [19]. The Wiedemann-Franz law is thus followed at high temperatures but not at low temperatures.

Bloch's theory was the first great success of quantum theory of solids. Like many other examples of successful theories the very basis of its success was not so sound. It contained simplifications and assumptions which are not so easily nor necessarily justified. Later we shall come back to these points in connection with more recent progress, but here we should note the following three points made by Peierls [20-22].

First, the momentum conservation, Eq.(7), is not necessarily observed. It is more generally replaced by the law,

$$\vec{k}' = \vec{k} \pm \vec{q} + \vec{K} \tag{15}$$

where K is a reciprocal lattice vector representing the amount of momentum absorbed by the crystal lattice. Scattering with $\vec{K} = 0$ is called a normal process and that with $\vec{K} \neq 0$ an Umklapp process. Umklapp processes can be crucial [20].

Second, the phonons should be more or less in disequilibrium when electrons flow, because the momentum is transferred to phonons from electrons. Bloch assumed implicitly that phonons relax to equilibrium quickly enough, which may be right but needs a careful analysis.

Third, the application of Dirac's perturbation method is questionable. In this calculation there appears the well-known function,

$$D(t) = 2(1-\cos(E'-E)t/\hbar)/(E'-E) \sim (2\pi t/\hbar)\delta(E'-E) \qquad (16)$$

which multiplies the distribution function of electrons. Since the latter function appreciably changes over an energy interval kT, the time t must be long enough to make the corresponding uncertainty of energy smaller than kT. In other words the mean free time τ has to be so long that the condition

$$\hbar/\tau \ll kT \qquad (17)$$

holds, which, however, is not satisfied in ordinary metals [22].

Already in 1929 Peierls [11] worked out a kinetic theory of lattice vibrations both on classical and on quantum mechanical grounds. Classically he treated the problem by the Fokker-Planck equation. Quantum-mechanically he used Dirac's perturbation method to calculate the transition probability of phonons and showed that both treatments are almost parallel. The interaction between lattice vibrations was assumed to be the third order anharmonicity which causes, in a quantum-mechanical picture, split of a single phonon into two phonons and the reverse process. He particularly emphasized that phonons can approach to a unique thermal equilibrium only through the Umklapp processes in which the phonon momenta satisfy the conservation law,

$$\vec{q} - \vec{q}' - \vec{q}'' = \vec{K} \quad , \quad \text{or} \quad \vec{q} + \vec{q}' - \vec{q}'' = \vec{K} \quad , \qquad (18)$$

with $\vec{K} \neq 0$ equal to a reciprocal vector. In other words, the normal processes with $\vec{K} = 0$ alone do not guarantee a unique thermal equilibrium nor a finite heat conductivity of a lattice. This is an important difference between the kinetics of phonons and that of gas molecules, which comes from the fact that phonons can be created or annihilated at the boundaries whereas real gas particles can not. At high temperatures $T \gg \theta$, where oscillators are nearly classical, Peierls gave a heat conductivity inversely proportional to T as Debye [8] had pointed out. At low temperatures $T \ll \theta$, where the long wave phonons are dominant, chances of Umklapp processes become rare and so the heat conductivity grows very large. As the quantum effect it will be like

$$\kappa \propto T^n \exp(\theta/bT)$$

where n and b are certain constants of the order one. This increase of heat conductivity in very pure, perfect non-metallic crystals was verified experimentally only in 1950's [23]. Peierls also pointed out the lattice imperfections are an important factor to limit the conductivity.

The Boltzmann equation for phonons with cubic anharmonicity takes the form

$$-\vec{v}_q \frac{\partial n_q}{\partial T} \nabla T$$

$$= \int\int d\vec{q}' d\vec{q}'' [\{n_q n_{q'}(1+n_{q''}) - (1+n_q)(1+n_{q'})n_{q''}\} Q_{qq'}^{q''}$$

$$+ \frac{1}{2}\{n_q(1+n_{q'})(1+n_{q''}) - (1+n_q)n_{q'}n_{q''}\} Q_q^{q'q''}] \tag{19}$$

for a stationary state in the presence of temperature gradient. Here n_q is the phonon distribution as a function of space, $\vec{v}_q$ the phonon group velocity and Q's are the relevant transition probabilities containing delta functions to secure the equalities (19) and the energy conservation,

$$\hbar\omega_q + \hbar\omega_{q'} = \hbar\omega_{q''} \quad \text{or} \quad \hbar\omega_q = \hbar\omega_{q'} + \hbar\omega_{q''} \quad .$$

If phonons are not far from equilibrium, the distribution function n_q may be put

$$n_q = n_q^o - \phi_q \partial n_q^o / \partial (\hbar\omega_q) \quad . \tag{20}$$

Then the right hand side of Eq.(19) is linearized to the form

$$(\frac{\partial n_q}{\partial t})_{coll.} = \int\int dq' dq'' \{P_{qq'}^{q''}(\phi''-\phi-\phi') + \frac{1}{2} P_q^{q'q''}(\phi'+\phi''-\phi)\} \quad . \tag{21}$$

Peierls gave only a sketch of this Boltzmann equation without detailed consideration of the nature of its solution. Ten years later in 1941, Pomeranchuk [24] pointed out that the cubic anharmonicity is incapable of guaranteeing a finite thermal conductivity if the sound waves lack dispersion or anisotropy, because the mean free path of low frequency phonons will then be large in proportion to q^{-4}. The integral for the conductivity diverges at zero wave number. This means that either quartic anharmonicity must be called for, or the dispersion and anisotropy of lattice waves play rather delicate roles [25]. In the former case, the high temperature thermal resistance will no longer be proportional to T but to a higher power of T. Although the quantum theory of phonon kinetics in nonmetallic crystals seems to be better founded than that of metallic electrons, the real physics is thus by no means simple.

Coming back to metallic electrons, it is now clear that the phonon kinetics should not be simply ignored even if we are primarily interested in electron transport. If it were not for Umklapp processes either in electron-phonon or in phonon-phonon scattering, the momentum gain from the applied electric field would not be dissipated. This would lead to an infinite electrical conductivity and a large thermopower [26]. Depending on the efficiency of Umklapp processes or other processes dissipating the momentum, there may occur the phonon drag effect which affects the electrical conductivity and particularly the thermopower. Thus the Bloch theory has to be reformulated, at least as a matter of principle, in a more rigorous manner as the transport of electrons and phonons considering both the normal and the Umklapp processes. This leads to two formidable integral equations to determine simultaneously the distributions of electrons and phonons.

In a real system there exist many other possible agents causing scattering of electrons and phonons. Conduction electrons interact by Coulomb interaction, which is shielded to a short range interaction by conduction electron themselves. More important is the Pauli principle which almost forbids electron-electron scattering in strongly degenerate electrons. The mean free path of metallic electrons for electron-electron scattering is thus proportional to $(T_0/T)^2$ where T_0 is the degeneracy temperature. Thus the electron-electron scattering is unimportant in metals except at low temperature where it might dominate over electron-phonon scattering.

On the other hand, various kinds of lattice imperfections very often play important roles governing electron or phonon scattering and hence the transport properties. Defects such as vacancies, intersticials, impurity atoms, and dislocations often limit the mean free paths of conduction electrons. In magnetic materials, disordered spins and spin waves may scatter electrons. Lattice imperfections and surfaces are important scatterers for phonons. Thus the transport phenomena of electrons, phonons and other possible sorts of carriers in metals or non-metals make a very colorful field of solid state physics.

A standard theoretical method to analyze a transport phenomenon is to follow the line set up by such pioneers as Bloch and Peierls. One first assumes an appropriate interaction mechanism between the carriers and the scatterers. This defines the model. A Boltzmann equation is written down in terms of the scattering probabilities calculated from the assumed scattering mechanism. Then the equation is solved and relevant transport coefficients are calculated. The result is compared with experiments to examine the validity of the model.

This is not always easy. In the first place, choice of a reasonable model of interaction is in itself a hard problem. For example, the electron-phonon interaction, which Bloch assumed to be of the form (6), has been an outstanding problem, because this is essentially a many-body problem. Bardeen [27] was the first who tried a careful analysis of this problem. The interaction of an impurity atom with conduction electrons is not simple either. An important contribution to this problem was done only in 1952 by Friedel [28]. Very often one has to take a phenomenological standpoint leaving the interaction parameters to be determined only by comparing with experiments. The Fröhlich Hamiltonian of an electron-phonon system [29] is such an example. Even then there is generally the question

of validity of simple perturbational calculation - the
golden rule for transition probabilities. If the interaction is not weak enough, the interaction will dress the
carriers to something different from what was originally
assumed. As a phenomenological theory, however, it would
be possible to regard the model as already dressed or renormalized and use the golden rule.

If one could thus justify the use of a Boltzmann
equation, then comes the task to solve it. In many cases,
the equation turns out to be a very complicated integral
equation, - generally non-linear, but linearized if the
deviation from equilibrium is not large, - which does not
allow a rigorous solution. Then we have to look for approximations. Commonly two standard methods are used, - the
relaxation time approximation and the variational method.

In the relaxation time approximation one writes
the collision term, the expression (10) or (20) for example,
in the form

$$(\frac{\partial g}{\partial t})_{coll.} = -g/\tau \qquad (22)$$

where τ is the relaxation time which may, in general, be
a function of $\vec{k}$ or $\vec{q}$. This relaxation time can be calculated from the transition probabilities. One has to be
careful to use a proper definition of the relaxation time,
because the transport relaxation time and the life time
of a state are generally to be distinguished. Once this
simple assumption (21) is made, the Boltzmann equation is
no longer an integral equation so that its solution is
immediately obtained and the transport coefficients are
calculated as integrals over the momentum space or the
Fermi surface, thus allowing a straight-forward extension

of the Lorentz-Sommerfeld theory of transport. Obviously this approximation cannot be justified in general, because the relaxation times are the eigenvalues of the linearized collision operator

$$\left(\frac{\partial g}{\partial t}\right)_{coll.} \equiv -\Gamma g \qquad (23)$$

and are defined with respect to proper modes of relaxation of the distribution function. It could be correct or a reasonable approximation in some special cases where the scattering is elastic or nearly elastic and isotropic. For example, in semiconductors where charged or neutral impurities are the scatterers, the relaxation time of conduction electrons can be thought of as a function of energy and this energy dependence is the essential point.

The relaxation time approximation has been very commonly used, often beyond its limits of justification, mainly because it is the only tractable method and is useful to obtain at least qualitative or semi-quantitative results. Such examples are numerous particularly in physics of semiconductors. An application of this to phonons in nonmetals was made by Klemens [30]. It should be kept in mind that this approximation could lead to erroneous conclusions because it may not take care of possible existence of conserved quantities.

The variational method [31] is a more systematic approach. This was first used by Kohler [32] and then by Sondheimer [33] for the Boltzmann-Bloch equation. An application to phonon transport in nonmetallic solids was made by Leibfried and Schlömann [34]. The collision operator, (10) or (21) for example, becomes a linear operator for the function Φ in (12) or (20), so that the Boltzmann equation (14) or (19) generally takes the form

$$X = P\Phi \qquad (24)$$

where P is a linear self-adjoint integral operator. If the system should relax to equilibrium in the absence of external perturbation, all of the eigenvalues of P are positive except the one which is zero and corresponds to the equilibrium distribution function. Thus P is positive (non-negative) definite. From this property of P, we have the variational principle which now says that of all functions satisfying the condition,

$$(\Phi,P\Phi) = (\Phi,X) \quad ,$$

the solution of Eq.(24) gives the expression $(\Phi,P\Phi)$ its maximum value. Equivalently, it says that the solution of Eq.(24) gives the expression

$$\rho = \frac{(\Phi,P\Phi)}{(\Phi,X)^2}$$

its minimum value.

It is easily seen in the case of electron or phonon transport, that the value of ρ calculated for unit strength of external perturbation is the relevant resistance. Thus the value of resistance calculated with the use of a trial function is always larger than the exact value. An example of this theorem is the Grüneisen formula, (13), of electrical conductivity which is a lower bound of the conductivity to be obtained from the Boltzmann-Bloch equation.

In the same way as in the case of the ordinary

Boltzmann equation for a gas, this variational principle is related to the H-theorem or the entropy production. The entropy in non-equilibrium states may be defined in terms of one-particle distribution function, for example

$$S = -k_B \int d\vec{k} \{f_k \log f_k - (1-f_k) \log(1-f_k)\}$$

for Fermions. Then the first statement of the variational principle is equivalent to the maximum principle of irreversible production of entropy consistent with its subsequent conservation [31].

This variational principle has proved very useful. It can be extended easily to a system in which several kinds of carriers are coexistent [31]. For example, the phonon drag effect can be treated by this method for an electron-phonon system in a metal or in a semiconductor.

During ten or fifteen years after the last war, the progress in various fields of solid state physics was really remarkable. It was largely supported by very rapid developments on the experimental side - progress of experimental techniques controlling impurities and imperfections in crystals, use of low temperatures, high magnetic fields, high frequency electromagnetic waves and so forth. The richness of our present knowledge of transport phenomena in metals and nonmetals is beyond comparison to that in 1930's. A great variety of interesting new phenomena have been discovered in these years. In many cases, the Boltzmann equation approach has proved to be extremely fruitful. At the same time it has been well recognized that there are certainly many cases which are beyond the limitation of this traditional approach.

4. Beyond a Boltzmann Equation - Recent Developments

Let us now turn to more recent developments since the mid-fifties on the theoretical side in connection with Boltzmann equations in solid state physics, focusing our attention to fundamental aspects. Three subjects will be mainly considered - first, an extension of the Boltzmann equation to Fermion liquids, second, justification of a Boltzmann equation, and third, the linear response theory and the Green's function method.

As already remarked in the Introduction, gas-like particles in a solid are usually complex objects. For example, conduction electrons in a metal are strongly interacting with each other and with lattice ions. Part of these interactions is taken into account when we regard the electrons as Bloch electrons moving in a crystal periodic potential, which is a sort of self-consistent field. Still the remaining interactions are not weak. By some reason, the free electron model or the band electron model seems to work surprisingly well. This observation leads us to the concept of quasi-particles. Imagine that the interactions are switched off so that the electrons are free. The ground state and the excited states of this free electron gas are simply defined as the Fermi sea at zero temperature and as the possible occupations of electrons and holes above and below the Fermi level. When the interactions are switched on adiabatically, the whole system is changed slowly into the real system. We now assume that the scheme of excitation levels is not so much mixed up by this switching so that the excitation levels of a real system keep one-to-one correspondence to those of the ideal system, at least with regard to the low excitations. This assumption is not valid for superconductors, but should be valid for normal metals. In normal metals, electrons and holes are thus quasi-electrons or holes dressed in some way by their interactions. Similarly, real phonons are not pure vibrations of lattice ions but are accompanied by electronic motion.

Liquid He^3 is perhaps an ideal case of such a system consisting of quasi-fermion particles. The interaction between helium atoms is not too strong, but not very weak. Landau [35] assumed that the energy E of such a liquid is a functional of the distribution function $n_\sigma(\vec{p},\vec{r},t)$ of the Fermi particles, where $\vec{p}$ and $\vec{r}$ are the momentum and position of a particle and σ is its spin. The variation δE of the total energy as n varies by δn can be expressed as

$$\delta E = \sum_\sigma \int \epsilon_\sigma \delta n \, d\vec{r} \, d\vec{p}/\hbar^3 \quad . \tag{27}$$

This equation defines the energy $\epsilon_\sigma(\vec{p},\vec{r})$, which corresponds to the change of the whole system due to the addition of a single quasi-particle with $\vec{p}$ at $\vec{r}$ and is regarded as its Hamiltonian function. Now Landau [36] wrote a Boltzmann equation or a kinetic equation,

$$\frac{\partial n}{\partial t} + \frac{\partial n}{\partial \vec{r}} \frac{\partial \epsilon}{\partial \vec{p}} - \frac{\partial n}{\partial \vec{p}} \frac{\partial \epsilon}{\partial \vec{r}} = I(n) \quad , \tag{28}$$

the right hand side being the collision term. It is important to recognize that the energy ϵ is itself a functional of n, which may be expressed by the relation

$$\delta \varepsilon_\sigma = \sum_{\sigma'} \int f_{\sigma\sigma'}(\vec{p},\vec{p}') \delta n_{\sigma'}(\vec{p}') d\vec{p}' \quad . \tag{29}$$

This is an extension of the idea contained in the Vlasov equation [37] for a plasma, in which the electric field acting on a charge particle is self-consistently determined by the particle distribution. In fact, if the collision term is ignored, the Fermi-liquid will perform an oscillation [36], - an analogue to the plasma oscillation, - called the zero sound, at high frequency collisionless regime or at very low temperatures. Predictions of the Landau theory have been confirmed by experiments on He^3 liquid [38], [62].

The Fermi liquid theory of Landau is a bold intuitive extension of the Boltzmann equation approach to an interacting particle system - a liquid. It is hard to prove, not at all surprisingly if one remembers that even the simplest classical Boltzmann equation is pretty hard to prove.

The collision term for a pure Fermi liquid is determined by the scattering processes of two particles. This is not independent of the energy ε, for the function f in Eq.(29) is in fact nothing but the amplitude of forward scattering. This function f plays a central role in Landau's theory. Through the function f, equilibrium and non-equilibrium properties of the system are internally related. It is true that this is a common feature of the kinetic method. In the simplest Boltzmann equation, the structure of the collision term is such that it will assure the Maxwellian distribution. For a system of interacting particles, the kinetic equation must guarantee the true equilibrium and at the same time describe non-equilibrium processes. The complication is that the interaction does affect not only the collision term but also drift terms. In Eq.(28), the dressing effect of interaction on the quasi-particles appears as the energy ε. There have been a great deal of efforts to prove or justify the Landau theory from the many-body theoretical point of view.

Let us now turn to our second subject of this chapter, that is the question whether the Boltzmann equation can really be proved from first principles. This question for a classical system was carefully analyzed by Bogoliubov [39] in 1947. Here we consider a quantum-mechanical problem.

Earlier than Bloch, in 1927 Pauli [40] gave the so-called master equation to describe a stochastic evolution of a quantum system under a weak perturbation. We all believe that this is essentially correct, but the derivation is not quite satisfactory. Both Pauli and Bloch made repeated randomizations of phase of quantum states in order to throw out the off-diagonal elements of the density matrix $\rho(t)$ which evolves in time following the equation of motion,

$$\frac{\partial \rho}{\partial t} = \frac{1}{i\hbar}[H_o + \lambda V, \rho] \qquad (30)$$

where H_o is the unperturbed Hamiltonian and λV is the perturbation. Bloch dealt with many electrons so that he was obliged to make another ansatz to express the diagonal elements of the density matrix in terms of one-particle distribution functions. This is another problem, but here we shall not go further into this point.

Thus they did not prove but assumed rather that a dynamical equation, Eq.(30), is really transformed into a stochastic equation like the Pauli equation or the Boltzmann-Bloch equation. Generally speaking we now believe that this fundamental question of statistical physics can be answered by a careful analysis of temporal and spatial coarse-graining processes from a microscopic stage to a macroscopic stage of observation.

In a series of papers devoted to a systematic analysis of perturbations in a large system, van Hove [41] pointed out that for a large system a dissipative perturbation V may be characterized by a special property which be named as the diagonal singularity. In the representation (α) diagonalizing the unperturbed Hamiltonian H_o, the perturbation V is offdiagonal, but in second order an operator with the form VAV, where A is a diagonal operator, may possess a singularity which appears as

$$<\alpha'|VAV|\alpha''> = W(\alpha')\delta(\alpha'-\alpha'') + F(\alpha',\alpha'') \qquad (31)$$

in the limit of infinite size of the system. The diagonal elements are singular containing the delta function in this limit, whereas the offdiagonal elements are expressed by a non-singular function of α' and α''. For example, if V is an electron-phonon interaction, the forward scattering of an electron by emission and absorption of a single phonon makes this singularity when summed over the intermediate states. The same sort of diagonal singularity may persist to higher order operators in the form of $VA_1V\ldots VA_nV$.

If the perturbation V possesses this diagonal singularity, then it is proved that the evolution of the diagonal part of the density matrix ρ in Eq.(30) is separated from that of the off-diagonal part if the diagonal part is initially coarse-grained in some sense. This evolution can be regarded as a stochastic process, but it is in general non-Markoffian and has some sort of memory effect. If, furthermore, the perturbation is weak, then the process approaches a Markoffian process in its long time behavior.

The condition for this is characterized by two time constants. One is the time τ_c associated with the energy width δE over which the matrix elements of V appreciably vary; namely

$$\tau_c = \hbar/\delta E \ . \tag{32}$$

The other one is the relaxation time τ_r, which is of the order of

$$\tau_r^{-1} \sim (\lambda V)^2/\hbar \delta E \ . \tag{33}$$

If the perturbation is so weak that the condition

$$\tau_r \gg \tau_c \tag{34}$$

or

$$\lambda V \ll \delta E \tag{35}$$

is satisfied, the time t can be scaled by the relaxation time τ_r. In the limit

$$\lambda \to 0 \qquad\qquad \lambda^2 t \sim 1 \tag{36}$$

for a fixed δE, the diagonal part of the density matrix behaves like a Markoffian process, which is exactly the same as that described by a Pauli equation obtained with the use of the golden rule. Here however no use is made of the repeated random-phase procedure, so that van Hove's derivation is far more satisfactory than the conventional one.

The time constant τ_c corresponds to the duration of a single collision process, or the correlation time of the perturbing field acting on the scattered particle. The condition (34) corresponds to the condition of motional narrowing of a spectral line to a Lorentzian line as the present author [42] has stressed on several occasions. It should be remembered that van Hove's conditions for obtaining a master equation are sufficient but not necessary. Two typical cases are conceivable in which a stochastic description is possible in terms of a certain form of master equation. One is a system with a weak perturbation which is modulated so fast that a narrowing condition of the type (35) is satisfied. In this case, the diagonal singularity is fine but not necessary. The other case is where the perturbation is localized and random, but not necessarily weak.

As a model of the second case, we imagine an electron scattered by impurity atoms distributed in space at random. The problem is whether or not the time-evolution of the probability distribution $f(p,t)$ can be described by a stochastic equation of a Boltzmann type. This problem has been investigated by a number of authors. The scattering potential needs not be weak. If the Born approximation is not accurate, it could be replaced by a more rigorous cross section. Only if the potential is short-ranged and the concentration of impurities is sufficiently low, then a Boltzmann equation can be obtained. The condition (34) still holds, but Eq.(33) is no longer correct. Van Hove and Verboven [43] discussed this problem as a simple example of van Hove's theory. Kohn and Luttinger [44] made a somewhat different approach to this problem by observing the fact that the density matrix ρ in Eq.(30) can be expanded in ascending powers of λ starting from a term in λ^{-2}. When the off-diagonal part is eliminated, the diagonal part is seen to obey an equation which is identified with a Boltzmann equation in the low concentration limit. In both of these seemingly different approaches the diagonal singularity makes an essential condition.

A Boltzmann equation may thus be proved under certain ideal conditions, which are not necessarily satisfied in reality. In order to apply it to the real world, we have to seek some sort of generalization. We could proceed from the zero density limit to finite densities, or from the weak interaction limit to finite interactions. There has been a great deal of efforts in these directions. Working on solids or condensed matters however, we feel the need to explore a quite different approach. Isn't there any way, other than the traditional way, to approach non-equilibrium states of a system in which particles are strongly interacting?

Fortunately there exists such a way, which is now widely known as the linear response theory [45]. If a weak external force is applied to a system in thermal equilibrium, it is brought slightly in disequilibrium. This will be observed in a certain physical quantity as a deviation from its equilibrium value linear in the given disturbance. The coefficient in such a linear relationship is the relevant transport coefficient or more generally the admittance, which the linear response theory tells us to be given by an analysis of the system in equilibrium in the absence of external disturbance.

This is an approach to non-equilibrium statistical mechanics in a direction quite different from the traditional one. We start from the limit of weak disequilibrium, not from the limit of weak interaction or zero density. We do not dare to jump directly into strong disequilibrium. A Boltzmann equation, on the other hand, may be able to deal with a state far from equilibrium, in its own limitations. For example, it is at present almost the only method available to treat hot electrons in semiconductors [46]. There is of course a certain area which is accessible in both ways, but in the regime of weak disequilibrium the linear response theory is general and includes the Boltzmann equation approach.

Consider now a system in thermal equilibrium, to which an external perturbation,

$$H_{ex}(t) = - F(t) A \qquad (37)$$

is switched on. Here F is an external force and A is a physical quantity conjugate to it. We observe another physical quantity B which responds to this perturbation. Its deviation from the equilibrium value is generally expressed as

$$\Delta B(t) = \int_{-\infty}^{t} \phi(t-t') F(t') dt' \qquad (38)$$

as a linear response to the force. If the force is periodic

with a frequency ω, the response is expressed in terms of an admittance $\chi(\omega)$ which is the Fourier-Laplace transform of $\phi(t)$.

We shall not discuss here this theory in any greater detail, but note only the following points. By solving Eq. (30) to the first order of $H_{ex}(t)$, we find that the response function $\phi(t)$ is nothing but the statistical average of the Poisson bracket of two relevant quantities represented by the Heisenberg operators, A(0) and B(t), namely

$$\phi(t) = <\frac{1}{i\hbar} [A(0),B(t)]> . \qquad (39)$$

This can also be written as

$$\phi(t) = \beta <A(0);B(t)> \equiv \int_0^\beta <e^{\lambda H} A(0) e^{-\lambda H} B(t)> d\lambda \qquad (40)$$

$$(\beta = 1/kT)$$

which is a quantum definition of correlation of the quantities, A(0) and B(t). For example, the conductivity $\sigma(\omega)$ is given by

$$\sigma(\omega) = \beta \int_0^\infty dt\, e^{-i\omega t} <J_x(0);J_x(t)> \qquad (41)$$

corresponding to Eq. (40), or equivalently by

$$\sigma(\omega) = \int_0^\infty dt\, \frac{e^{-i\omega t}-1}{i\omega} <\frac{1}{i\hbar} [J_x(0),J_x(t)]> \qquad (42)$$

where J_x is the current density operator.

In a case where a Boltzmann equation can be used, Eq. (41) is expressed in terms of the collision operator.

Or starting from a model Hamiltonian the expression (41) can somehow be calculated and gives a result which in effect justifies the use of the Boltzmann equation. Such a calculation was performed for the impurity resistance problem by several authors [48]. It is interesting to note here that Eq.(41) immediately shows that the condition (18) is not actually required for a Boltzmann equation to hold as long as the scattering is elastic or nearly elastic. We only need

$$\frac{h}{\tau} \ll E_F \qquad (43)$$

where E_F is the Fermi energy, because the expression (41) is reduced to a correlation function of electron velocity on the Fermi surface. The condition (43) is usually satisfied in metals. This is just what Landau had conjectured [49].

There are many other cases where we do not know a Boltzmann equation to start with. In liquid or amorphous conductors, the mean free paths are often estimated to be just of the order of atomic distance. Obviously the concept of a mean free path is no longer well defined. Then the formula (41) gives us the starting point. Another typical case is the galvano-magnetic effect in high magnetic field, where the off-diagonal elements of current operators become the essential part [50].

As we see in Eq.(39), a Poisson bracket means physically an effect produced by a disturbance or propagation of a disturbance. Thus a function like (39) in an Einflussfunktion or a Green's function. Equation (40) shows further that it is related to a correlation function. If a system consists of independent free particles, such a

Green's function has a simple structure. When some interactions are introduced it becomes generally very complex, but its structure may keep some resemblance to that for non-interacting particles, which gives the mathematical basis for the concept of the quasi-particle picture. A remarkable progress has been achieved in recent years in the perturbational analysis of Green's functions. Particularly useful is the thermal Green's function method, first introduced by Matsubara [51] and extensively developed by many investigators, which enables us to perform systematic perturbational calculations using the Feynman diagrams, Dyson's equation, analytic continuation and other mathematical tools taking full advantage of the quantum structure of dynamics [52]. There has been an enormous amount of theoretical work on various problems of many-body systems in the last ten years. For example, Langer [53] studied the impurity resistance, considering the electron-electron interaction which shields the impurity potential to an effective one. Holstein [54] examined an electron-phonon gas assuming the Fröhlich Hamiltonian and showed that the computation of the expression (41) for zero frequency reduces to coupled integral equations which correspond to the linearized Boltzmann equation. For high frequencies the result is a generalization of Drude's formula. He concludes that the criterion of a Boltzmann-like equation is again (43), or

$$k\theta/E_F \ll 1 \qquad (44)$$

rather than (18).

The linear response theory also provides us with a mean to derive a generalized Boltzmann equation. We suppose an external potential field $U(\vec{r},t)$ to be applied and ask for the response in the one-particle distribution $n(\vec{r},\vec{p},t)$. Quantum-mechanically, this distribution function is defined as the statistical average of the Wigner density,

$$n_w(\vec{r},\vec{p},t) = \int d\vec{\xi}\, \psi^\dagger(\vec{r} - \frac{\vec{\xi}}{2},t)\psi(\vec{r} + \frac{\vec{\xi}}{2},t)e^{i\vec{\xi}\vec{p}/\hbar} , \quad (45)$$

where $\psi^\dagger$ and ψ are the quantized wave functions of the particles. The response is given in the form, if the system is uniform,

$$\Delta n(\vec{r},\vec{p},t) = \int\int \phi(\vec{r}-\vec{r}_1,\vec{p},t-t_1)U(\vec{r}_1,t_1)d\vec{r}_1 dt_1 , \quad (46)$$

where ϕ is the response function. If there exists a Boltzmann equation for $n(\vec{r},\vec{p},t)$, ϕ must be identical with the fundamental solution or the Green's function of the linearized Boltzmann equation. Conversely, a Boltzmann equation, perhaps in a generalized sense, is obtained from the calculated functional form of ϕ for small wave number and low frequencies. An excellent treatise on this subject was given by Kadanoff and Baym [55] on the basis of the work by Martin and Schwinger [56]. This method was applied by Prange and Kadanoff [57] to an electron-phonon gas in metals and by Kwok and Martin [58] to interacting phonons. The former work takes advantage of the smallness parameter m/M, the ratio of masses of electron and ion, in the same spirit of Migdal's treatment of electron-phonon interaction in equilibrium. Thus the condition (44) is again confirmed. The latter work shows that the phonon gas may, under certain circumstances, allow the existence of wave-like propagation of heat, the second sound, which can also be derived from an elementary Boltzmann equation of phonons [59].

Another useful way to look at the expression (41) is due to Mori [60], who showed that the equation of motion of a dynamical quantity can be transformed into an equation which has a similarity to the Langevin equation of a Brownian motion. This transformation is achieved in a damping theory formalism [61] by defining a proper way of projection and rewriting the equation for the projected part. The integrand of the right hand side of Eq.(41) is the correlation function of the current which makes a stochastic process, generally non-Markoffian, defining this generalized Langevin equation, and is written as

$$\sigma(\omega) = \frac{e^2 n}{m} \frac{1}{i\omega + \gamma(\omega)}$$

where $\gamma(\omega)$ corresponds to a certain correlation function of the force acting on electrons. This expression has also some connection with a generalized Boltzmann equation.

We now conclude this review with one remark. From an operational point of view, we have only to know the relations between inputs and outputs. The linear response theory provides us with a basis to calculate the admittance from the microscopic structure and dynamics of a physical system. The calculation may be carried out with a help of mathematical machinary. A physicist however will not be quite happy just to watch the machine producing the answer. From the very basic level of microscopic physics to the level of everyday macrophysics, there exist various levels of submacro- or quasimicrophysics. At each level, we construct a physical picture to bridge the two sides of that particular level. The Boltzmann equation remains a very valuable, educational example of such a physical picture.

References

[1] P. Drude, Ann. d. Phys. 1, 566 (1900); 3, 369 (1900).
[2] H.A. Lorentz, Proc. Amst. Acad. 7, 438, 585 (1905), The Theory of Electrons, Teubner, Leipzig und Berlin 1909, 2nd ed. Dover Publication, New York, p. 267.
[3] J.J. Thomson, Phil. Mag. 30, 192 (1915).
[4] F.A. Lindemann, Phil. Mag. 29, 127 (1915).
[5] W. Wien, Sitzungsber. Preuss. Akad. Wiss. Berlin 5, 184 (1913).
[6] A. Sommerfeld, Naturw. 15, 825 (1927, Zs. f. Phys. 47, 1, (1928).
[7] Cited from D.K.C. MacDonald, Electrical Conductivity of Metals and Alloys in Handb. d. Phys. Bd XIV ed. S. Flügge, Springer, Berlin Göttingen-Heidelberg 1956.
[8] P. Debye, Vorträge über die kinetische Theorie der Materie und Elektrizität. Teubner, Berlin, 1914.
[9] A. Eucken, Ann.d.Phys. 34, 185 (1911).
[10] M. Born and Th.V. Kármán, Phys. Zeits. 13, 297 (1912).
[11] R. Peierls, Ann. d. Physik. 3, 1055 (1929).
[12] W.V. Houston, Zs. f. Phys. 48, 449 (1928).
[13] J. Frenkel and N. Mirolubow, Zs. f. Phys. 49, 885 (1928).
[14] F. Bloch, Zs. f. Phys. 52, 555 (1928).
[15] F. Bloch, Zs. f. Phys. 59, 208 (1930).
[16] E. Grüneisen, Ann. d. Phys. 16, 530 (1933).
[17] L. Nordheim, Ann. d. Phys. 9, 607, 641 (1931).
[18] W. Kroll, Zs. f. Phys. 77, 322 (1932), 80, 50 (1933), 81, 425 (1933).
[19] A. Sommerfeld and H. Bethe, in Handb. d. Phys. 34/2 (1933).
[20] R. Peierls, Ann. d. Phys. 4, 121 (1930).

[21] R. Peierls, Quantum Theory of Solids, Oxford Clarendon Press 1953.

[22] R. Peierls, Helv. Phys. Act. $\underline{7}$ (Suppl.), 24 (1934), Zeits. f. Phys. $\underline{88}$, 786 (1934).

[23] K.R. Wilkinson and J. Wilks, Proc. Phys. Soc. A $\underline{64}$, 89 (1951), F.J. Webb, K.R. Wilkinson and J. Wilks, Proc. Roy. Soc. A $\underline{214}$, 546 (1952), F.J. Webb and J. Wilks, Phil. Mag. $\underline{44}$, 664 (1953).

[24] I. Pomeranchuk, J. of Physics USSR $\underline{4}$, 259, 357 (1941), $\underline{6}$, 237 (1942), $\underline{7}$, 197 (1943).

[25] C. Herring, Phys. Rev. $\underline{95}$, 954 (1954).

[26] L. Gurevich, J. Phys. USSR $\underline{9}$, 477 (1945), $\underline{10}$, 67 (1946).

[27] J. Bardeen, Phys. Rev. $\underline{52}$, 688 (1937).

[28] J. Friedel, Phil. Mag. $\underline{43}$, 153 (1952).

[29] H. Fröhlich, Proc. Roy. Soc. $\underline{215}$, 291 (1952).

[30] P.G. Klemens, Proc. Roy. Soc. A $\underline{208}$, 108 (1951).

[31] J.M. Ziman, Electrons and Phonons, Oxford Clarendon Press 1960, Chapter VII.

[32] M. Kohler, Zeits. f. Phys. $\underline{124}$, 772 (1948), $\underline{125}$, 679 (1949).

[33] E.H. Sondheimer, Proc. Roy. Soc. A $\underline{203}$, 75 (1950).

[34] G. Leibfried and E. Schlömann, Nachr. Gött. Akad. $\underline{2a}$, 71 (1954).

[35] L.D. Landau, J. Exptl. Theoret. Phys. USSR, $\underline{30}$, 1058 (1956).

[36] L.D. Landau, J. Exptl. Theoret. Phys. USSR, $\underline{32}$, 59 (1957).

[37] A. Vlasov, Zhur. eksp. teor. fiz. $\underline{8}$, 291 (1938), J. Phys. USSR $\underline{9}$, 25 (1945).

[38] For a review, see for example A.C. Anderson, in Physics of Quantum Fluids ed. R. Kubo and F. Takano, Syokabo, Tokyo 1971.

[39] N.N. Bogoliubov, J. Phys. USSR 10, 256, 265 (1946), also in Studies in Statistical Mechanics I. ed. J. de Boer and G.E. Uhlenbeck, North-Holland Pub. Comp., Amsterdam 1962.

[40] W. Pauli, in Festschrift zum 60. Geburtstag A. Sommerfelds, Hirzel, Leipzig (1928).

[41] L. van Hove, Physica 21, 517 (1955); 23, 441 (1957); 25, 268 (1959).

[42] R. Kubo, in Lectures in Theoretical Physics I. ed. W.E. Brittin and L.G. Dunham, Interscience New York (1959), R. Kubo, in Fluctuation, Relaxation and Resonance in Magnetic Systems ed. ter Haar, Oliver and Boyd, Edinburgh 1962, R. Kubo, J. Math. Phys. 4, 174 (1963).

[43] L. van Hove and E. Verboven, Physica 27, 418 (1961).

[44] W. Kohn and J.M. Luttinger, Phys. Rev. 108, 590 (1957), J.M. Luttinger and W. Kohn, Phys. Rev. 109, 1892 (1958).

[45] R. Kubo, J. Phys. Soc. Japan 12, 570 (1957), R. Kubo, in Lectures in Theoretical Physics I. ed. W.E. Brittin and L.G. Dunham, Interscience, New York 1959, and also in "Statistical Mechanics of Equilibrium and Nonequilibrium", ed. J. Meixner, North Holland, Amsterdam, 1965.

[46] See for example, E.M. Conwell, High Field Transport in Semiconductors, Academic Press, New York and London, 1967.

[47] R. Kubo, Canad. J. Phys. 34, 1274 (1956), H. Nakano, Progr. Theor. Phys. Kyoto 15, 77 (1956).

[48] D.A. Greenwood, Proc. Phys. Soc. 71, 585 (1958), S.F. Edwards, Phil. Mag. 3, 1020 (1958), G.V. Chester and A. Thellung, Proc. Phys. Soc. 73, 745 (1959).

[49] L.D. Landau, as quoted by Peierls [22].
[50] R. Kubo, H. Hasegawa and N. Hashitsume, J. Phys. Soc. Japan 14, 56 (1959), R. Kubo, S. Miyake and N. Hashitsume, Solid State Phys. 17 ed. F. Seitz and D. Turnbull, Academic Press, New York and London 1965.
[51] T. Matsubara, Progr. Theor. Phys. 14, 351 (1955).
[52] See as a standard textbook, A.A. Abrikosov, L.P. Gorkov and I.E. Dzyaloshinski, Methods of Quantum Field Theory in Statistical Physics, translated from Russian Prentice-Hall, Englewood Cliffs, N.J. 1963.
[53] J.S. Langer, Phys. Rev. 120, 714 (1960).
[54] T. Holstein, Ann. of Physics, 29, 410 (1964).
[55] L.P. Kadanoff and G. Baym, Quantum Statistical Mechanics, W.A. Benjamin, Inc., New York 1962.
[56] P.C. Martin and J. Schwinger, Phys. Rev. 115, 1342 (1959).
[57] R.E. Prange and L.P. Kadanoff, Phys. Rev. 134A, 566 (1964).
[58] P.C. Kwok and P.C. Martin, Phys. Rev. 142, 495 (1966).
[59] R.A. Guyer and J.A. Krumhansl, Phys. Rev. 148, 766 (1966).
[60] H. Mori, Progr. Theor. Phys. 33, 423 (1965).
[61] A. Messiah, Quantum Mechanics, Vol II. p. 994, North-Holland Publ. Comp., Amsterdam, 1964. R. Zwanzig, Phys. Rev. 124, 983 (1961).
[62] J.C. Wheatley in Progress in Low Temperature Physics, Vol. 6, North-Holland, 1967, Chapter 3.

EXPERIMENTAL AND THEORETICAL INVESTIGATIONS IN SEMICONDUCTORS CONCERNING THE BOLTZMANN EQUATION

K. SEEGER and H. PÖTZL

Ludwig Boltzmann Institute for Solid State Physics,
Vienna

ABSTRACT

Transport phenomena in solids including semiconductors are described by the Boltzmann equation. For obtaining solutions it is essential whether the scattering processes cause a weak or a strong anisotropy of the distribution function. The former case is known as the "diffusion approximation" and has been investigated very carefully while the second case (spiked distribution, streaming effect) occurs mainly at strong interaction of hot carriers with the optical modes of lattice vibration at low lattice temperatures. Essentially four methods have been applied for an investigation of transport phenomena:

(1) The assumption of a drifted Maxwell-Boltzmann distribution in non-degenerate semiconductors and the transformation of the transport equation into balance equations for momentum and energy,
(2) Variational methods,
(3) Monte Carlo methods, and
(4) Methods of iteration.

All four methods have been applied in interpreting experimental data on energy relaxation both in non-polar and polar semiconductors. The Monte Carlo method has been used for static and dynamical calculations which demonstrate the influence of various physical constants on the mechanisms of heating and intervalley transfer of carriers. Finally the relaxation of the high-frequency conductivity of hot electrons in InSb has been calculated by the iterative method.

In this paper we will deal with investigations of the Boltzmann transport equation in connection with experimental results about hot carriers in semiconductors which have been obtained at the Ludwig Boltzmann-Institute for Solid State Physics.

In the first part we will deal with phenomena which can be described either by a drifted Maxwell-Boltzmann distribution for the solution of the transport equation or with a solution obtained by a variational method. We will then apply the more recent and refined Monte Carlo method and the method of iteration, also known as the "path variable method" to polar semiconductors where other methods fail.

According to the Boltzmann transport equation a change in the distribution of carriers in phase space with time due to externally applied electric and magnetic fields and temperature gradients, $(df/dt)_{field}$, equals the change due to collisions with lattice vibrations and imperfections, $(\partial f/\partial t)_{coll}$:

$$\left(\frac{df}{dt}\right)_{field} = \frac{e}{\hbar} (\vec{\nabla}_k f \cdot \vec{E}) = \left(\frac{\partial f}{\partial t}\right)_{coll} \qquad (1)$$

For simplicity we deal only with an electric field $\vec{E}$.

$\vec{k}$ is the electron wave vector. A familiar assumption for the right hand side of this equation is the existence of a momentum relaxation time τ_m:

$$\left(\frac{\partial f}{\partial t}\right)_{coll} = \frac{f-f_o}{\tau_m} \qquad (2)$$

where $f_o(\vec{k})$ is the zero-field distribution function. For small field intensities $\vec{\nabla}_k f$ may be replaced by $\vec{\nabla}_k f_o$ and the solution of the Boltzmann equation takes the simple form

$$f(\vec{k}) = f_o(\vec{k}) + \frac{e}{\hbar} \tau_m (\vec{\nabla}_k f_o \cdot \vec{E}) \qquad (3)$$

For a nondegenerate electron gas in pure semiconductors f_o is the well-known Maxwell-Boltzmann distribution function

$$f_o(\vec{k}) \propto \exp(-\hbar^2 k^2 / 2mk_B T) \qquad (4)$$

where m is the effective mass of the carriers, and $\vec{\nabla}_k f_o = -\frac{\hbar^2}{mk_B T} f_o \vec{k}$. Hence, the distribution function consists of an isotropic term f_o and an anisotropic term which for the simple case of a scalar effective mass and scalar relaxation time, is spiked in the direction of the applied electric field:

$$f(\vec{k}) = f_o(\vec{k}) \left\{ 1 - \frac{e\hbar}{mk_B T} \tau_m kE \cos(\widehat{\vec{k}\vec{E}}) \right\} \qquad (5)$$

Next we will investigate the integrals [1] of the Boltzmann equation which is given in the form

$$\frac{\partial f}{\partial t} + (\vec{\nabla}_r f \cdot \vec{v}) + \frac{e}{m}(\vec{\nabla}_v f \cdot \vec{E}) + \frac{f-f_o}{\tau_m} = 0 \qquad (6)$$

where for simplicity we have introduced the carrier velocity $\vec{v} = \hbar\vec{k}/m$. For a nonparabolic band structure this has to be replaced by an appropriate expression. Let us multiply this equation by an arbitrary function $Q(\vec{v})$ and integrate over velocity space. The average of Q is defined as

$$<Q> = \frac{1}{n} \cdot \int_{-\infty}^{\infty} Q(\vec{v}) f(\vec{v},\vec{r},t) \, d^3v \qquad (7)$$

where n is the carrier concentration given by

$$n = n(\vec{r},t) = \int_{-\infty}^{\infty} f(\vec{v},\vec{r},t) \, d^3v \qquad (8)$$

The average velocity is the drift velocity $\vec{v}_d$ of the carriers:

$$<\vec{v}> = \vec{v}_d , \qquad (9)$$

Giving Q the value 1 this procedure leads to the continuity equation

$$\partial n/\partial t + \operatorname{div}(n\vec{v}_d) = 0 \qquad (10)$$

which does not contain a collision term since the total density of carriers is not changed by collisions.

Taking for Q the momentum $m\vec{v}$ in the field direction, which may be taken as the direction of the z axis, we obtain the equation of motion which for a negligible dependence on position in the crystal is given by

$$d(mv_d)/dt - eE_z + mv_d/\bar{\tau}_m = 0 \qquad (11)$$

where $\bar{\tau}_m$ is an average over $\tau_m = \tau_m(\vec{v})$ given by

$$1/\bar{\tau}_m = \int_{-\infty}^{\infty} \tau_m^{-1} v_z f d^3v / nv_d \qquad (12)$$

At equilibrium the first term vanishes and the drift velocity is given by the well-known expression

$$v_d = (\frac{e}{m}\bar{\tau}_m) E_z = \mu E_z \qquad (13)$$

The factor μ is denoted as the mobility.

Now we take for Q the energy $\varepsilon = \sum_i \frac{m}{2} v_i^2$. This leads to the "energy balance equation"

$$\frac{d<\varepsilon>}{dt} - eE_z v_d + \frac{<\varepsilon>-\varepsilon_L}{\tau_\varepsilon} = 0 \qquad (14)$$

where we have introduced the thermal energy of a carrier in equilibrium with the crystal lattice

$$\varepsilon_L = \frac{1}{n} \int_{-\infty}^{\infty} \varepsilon f_o(\vec{v}) d^3v \qquad (15)$$

which equals $\frac{3}{2} k_B T$ for a Maxwell-Boltzmann distribution, and the "energy relaxation time" τ_ε given by

$$\frac{<\varepsilon>-\varepsilon_L}{\tau_\varepsilon} = \frac{1}{n} \int_{-\infty}^{\infty} \varepsilon \frac{f-f_o}{\tau_m} d^3v \qquad (16)$$

If τ_m does not depend on the carrier velocity, τ_ε is independent of the average carrier energy and equals $\tau_m = \bar{\tau}_m$. Otherwise after switching off the electric field, the decay of $<\varepsilon>$ with time to its equilibrium value, ε_L, will be non-exponential and τ_ε may be called a "relaxation time" in only a less strict sense of the word.

If the average carrier energy $<\varepsilon>$ is much larger than ε_L we speak of "hot carriers" while for the case, that the difference $<\varepsilon>-\varepsilon_L$ is small compared with ε_L, the term "warm carriers" has been coined. For the latter case the drift velocity v_d can be expanded in a series

$$v_d \propto E_z + \beta E_z^3 + \ldots \qquad (17)$$

which is terminated after the second term; β is a coefficient which depends on the carrier scattering mechanism in the crystal lattice. For acoustic phonon scattering it is simply

$$\beta = -\frac{3\pi}{64}(\mu_o/u_\ell)^2$$

where μ_o is the zero-field mobility and u_ℓ is the longitudinal sound velocity. E.g. in n-type Ge at 77K $\mu_o = 2 \cdot 10^4 \text{cm}^2/\text{Vsec}$, $u_\ell = 5 \cdot 10^5 \text{cm/sec}$, and $\beta \approx -10^{-4} \text{cm}^2/\text{V}^2 = -(1/100 \text{ Vcm}^{-1})^2$.

That is: at about 100 V/cm the deviations from Ohm's law become significant. As shown by the energy balance equation at equilibrium the energy difference $<\varepsilon>-\varepsilon_L$ is to first order proportional to E_z^2. If we replace v_d by $\frac{e}{m}\bar{\tau}_m E_z$ we find that $\bar{\tau}_m$ is a linear function of the energy difference:

$$\bar{\tau}_m \propto 1 + \frac{\beta m}{e^2 \tau_\varepsilon \bar{\tau}_{mo}}(<\varepsilon>-\varepsilon_L) \qquad (18)$$

where $\bar{\tau}_{mo}$ is the zero-field momentum relaxation time.

In this approximation the energy relaxation time τ_ε may still be taken as energy-<u>in</u>dependent for warm carriers although $\bar\tau_m$ does depend on energy.

The deviations from Ohm's law indicated by the nonlinear character of the drift-velocity-vs-field relationship as well as the average carrier energy $<\varepsilon>$ often characterized by a carrier temperature T_e, and the energy relaxation time τ_ε have been determined experimentally by various methods in order to investigate the various scattering mechanisms. Space is too limited to give a complete account of all the work done in this field in recent years at the Ludwig Boltzmann Institute but we will briefly discuss some of the typical results.

For the mobility variation with electric field intensity in n-type Ge let us first show you a diagram published by Ryder [2] many years ago: Fig.1. The lower curve indicates a negative deviation from Ohm's law typical for acoustic and optical phonon scattering at a lattice temperature of 77K while the upper curve at 20K shows in its first part a positive deviation typical for ionized impurity scattering. It may be of interest to plot the coefficient β of warm carriers vs impurity concentration N_I: Fig.2 [3]. The crosses are experimental data by Tschulena of our laboratory, the circles are data by Gunn. The lattice temperature is 77K. In pure samples β is negative while at high values of N_I β is positive and proportional to $1/N_I$ in agreement with theory if predominant ionized impurity scattering is assumed for momentum relaxation.

Experimental determinations of τ_ε in various semiconductors involve microwave and laser techniques. One method is to measure the coefficient β as a function of the frequency of an applied a.c.field. The result of

these measurements in n-type Ge at 100K is shown in
Fig.3 [4]. β is decreased to 1/2 its low-frequency value
at $\omega \simeq 1/\tau_\varepsilon$ which is in the microwave range of frequencies. The full line shows the calculated dependence of β
on $\omega\tau_\varepsilon$ only, while for the dashed line also the dependence on $\omega\tau_m$ has been taken into account. The momentum
relaxation time $\bar{\tau}_m$ is about an order of magnitude smaller
than the energy relaxation time in this case where acoustic phonon scattering is predominant for momentum relaxation and optical phonon scattering for energy relaxation.
Fig.4 shows these relaxation times as a function of temperature in units of the Debye temperature θ. For Ge θ is
430K and T/θ is about 0.2 at the temperature of the experiment. Since the acoustic and optical deformation potentials, ε_{ac} and ε_{opt}, are about equal it is obvious
that $\bar{\tau}_m \ll \tau_\varepsilon$. This would not be the case if optical phonon scattering would account also for momentum relaxation
as indicated by the second full line.

    The actual way of taking the present data of β
has been the application of a microwave field to the
sample parallel to a small d.c.current which serves for
a resistivity determination. The difference in resistivity with and without the microwaves yields β. Another
method is to apply the a.c.field both parallel and perpendicular to the d.c.current and take the ratio of both
resistivities [5]. This method has been practiced at the
far-infrared frequency of a cyanide laser. Since the frequency is about 10^{12}Hz, very short relaxation times can
be measured which occur in III-V compounds where polar
optical scattering is predominant. Fig.5 shows $\mu_o \propto \bar{\tau}_m$
and τ_ε for this case. κ_{opt} and κ are the optical and the
low-frequency dielectric constants, respectively. The relaxation times are multiplied by a factor which e.g. in

n-type GaAs is $6.6 \cdot 10^{-2}$. Hence the times are about 10^{-12} sec at room temperature, or equivalent to the reciprocal of the laser frequency. At low temperatures $\bar{\tau}_m$ and τ_ε are about the same in III-V compounds so pure that ionized impurity scattering can be neglected. Experimental results on n-type GaAs and n-type GaSb are shown in Fig.6 [6]. The insert shows a plot of τ_ε vs the electron temperature T_e for hot carriers in n-type GaSb obtained by a method to be discussed below. The full lines are theoretical. Unfortunately, a detailed discussion is beyond the scope of this report.

A more accurate experimental method for a determination of τ_ε is "harmonic mixing" which is also known as "optical rectification": Two waves with frequencies ω and 2ω are imposed on a homogeneous sample and a d.c.voltage is obtained between the ends of the sample which in magnitude and polarity depends on the phase difference of the waves. In the warm-electron range of field amplitudes it is also proportional to β. The setting ψ of a phase shifter is observed for which no d.c.voltage occurs, as a function of lattice temperature . From the values of ψ the relaxation time τ_ε has been determined. The crossed circles in Fig.7 show for the purest sample these values of τ_ε plotted vs the lattice temperature [7]. The open circles have been obtained by the resistivity-method. The full curve has been calculated with a Maxwell-Boltzmann distribution f_{MB} of the carriers. The dashed curve was obtained by a variational method first applied by Adawi for a calculation of the coefficient β. In the insert is shown the relative deviation from f_{MB} of the variational-method distribution f_v. Acoustic phonon scattering and optical phonon scattering were taken into account with a ratio of deformation

potential constants b = 0.4. The optical deformation
potential constant was adjusted for a best fit at 100K.
The agreement between theory and the harmonic-mixing
experiment is not as good at high temperatures as it
is at low temperatures.

Now let us briefly discuss the variational
method of solving the Boltzmann equation first introduced by Kohler in 1948 [8]. It is based on the assumption that the generation rate of entropy by collision
is a minimum at a constant current density. For numerical solutions the Ritz method is applied. Adawi [9] applied this variational approach to small deviations from
Ohm's law by setting the distribution function equal to
the Maxwell-Boltzmann function times a series of Legendre's polynomials with unknown coefficients ϕ_l for which
a series expansion in powers of the electrical field intensity E is taken:

$$\phi_l = E^l \sum_{n=0}^{\infty} a_{l,l+n}(k) E^{2n} \tag{19}$$

For the new coefficients $a_{l,l+n}$ an infinite system of
equations is obtained by equating the coefficients of
equal powers of E on both sides of the Boltzmann equation.
Considering only deviations from Ohm's law proportional
to E^2, this set of equations has been solved by variational methods and the coefficient of these deviations has
thus been obtained.

Recently Hess of the Ludwig Boltzmann-Institute
succeeded in calculating by this method also the distribution function and the energy relaxation time, which
was possible with the aid of a powerful computer not
available to the previous authors. Acoustic phonon scat-

tering with a small contribution by optical phonons was
considered while electron-electron collisions were neglected.

A disadvantage of the variational method is that
the series expansion in powers of the field intensity
does not converge for the case of hot carriers which is
more interesting for practical purposes than the case
of warm carriers.

We will now consider the experimental determination of energy relaxation times and electron temperatures
in degenerate semiconductors. Due to the high density of
ionized impurities which are required for the electrical
neutrality of the crystal at a high carrier concentration,
deviations from Ohm's law are unmeasurably small. However, another method of determination of the electron temperature is then available; this is the Shubnikov-de Haas
effect. In a strong magnetic field Landau levels are
formed in the allowed energy bands which move on the energy scale as the magnetic field is increased. Each time
such a level passes the Fermi level, a change in magneto-resistance occurs which is more abrupt the lower the carrier temperature. If the magneto-resistance is plotted vs
the magnetic field, oscillations are found which are more
pronounced at lower carrier temperatures. Fig.8a shows
experimental results obtained in n-type InAs at 4.2K and
higher lattice temperatures up to 12.5K [10]. The same
set of curves has been obtained by applying electric field
intensities up to 320 mV/cm and thus heating the carriers
while keeping the lattice at 4.2K (Fig.8b). From these
data a plot of the energy loss rate vs the electron temperature has been obtained which is shown in Fig.9. The
influence of the strong magnetic field on the carrier
temperature is negligible since the magnetic field was

applied parallel to the current direction and the sample is degenerate. The curve shows a kink which indicates the transition from one scattering mechanism to another. At low temperatures the data can be explained by a combination of piezoelectric and acoustic deformation potential scattering while above the kink polar optical scattering is predominant. Observations of these data as a function of time after the application of the electric field using sampling techniques yield the energy relaxation time. For the case of n-type GaSb a plot has been shown in the insert of Fig.6. Data obtained for n-type InAs are similar.

Still another method applicable to degenerate semiconductors involves the determination of the "Burstein shift" as a function of an applied d.c.electric field. Burstein explained the observation that the fundamental optical absorption edge in degenerate material is at higher energies than in non-degenerate material by assuming that the difference is the Fermi energy. When the carriers are heated by an electric field the Fermi energy decreases and eventually becomes negative if the band edge is taken as the energy zero, e.g. the carrier gas becomes nondegenerate, and the Burstein shift disappears. Hence a measurement of the Burstein shift yields the electron temperature if in a second experiment the lattice temperature is raised at zero field and the results of both experiments are compared just.as in the Shubnikov-de Haas experiments. The advantage of this method is that it is not confined to liquid helium temperatures.

Experimental results of T_e vs E in n-type GaSb at 77K are shown in Fig.10 [11]. The curve has been calculated for polar-optical mode scattering assuming a two-band model which unfortunately we cannot discuss here.

If scattering is both <u>inelastic</u> and <u>anisotropic</u>, relaxation times cannot be defined in a strict sense. Furthermore, the resulting electron distribution functions in large electric fields are strongly <u>spiked</u> in the field direction. These are often referred to as "<u>streaming distributions</u>". In these cases the methods discussed so far cannot be applied. Instead two numerical procedures have been developed for the solution of the Boltzmann transport equation which we will review now.

The <u>Monte Carlo procedure</u> [12] is a simulation of the motion of one electron in k-space on a computer. This motion consists of two kinds of contributions. During the times of free flight between consecutive scattering events the electron moves at constant $\vec{k}$-space "velocity":

$$\hbar \dot{\vec{k}} = e \vec{E} \qquad (20)$$

On the other hand, every scattering process causes a discontinuous change of the electron $\vec{k}$-vector, say from $\vec{k}$ to $\vec{k}'$. The probability distribution of $\vec{k}'$ is governed by the scattering rate $S(\vec{k},\vec{k}')$.

The procedure is gratly simplified by the introduction of the concept of <u>self scattering</u>. Let us consider the case of m different scattering processes every one of which is characterized by $S_i(\vec{k},\vec{k}')$ and by a total rate

$$\lambda_i(k) = \int S_i(\vec{k},\vec{k}') \, d^3k' \, . \qquad (21)$$

Then the overall total scattering rate is given by

$$\lambda(k) = \sum_{i=1}^{m} \lambda_i(k) \qquad (22)$$

This will be a complicated function of k which determines the probability distribution of the times of free flight. We now introduce an artificial non-physical scattering process m+1 in such a way that

$$\lambda_{m+1}(k) = \Gamma - \lambda(k) \quad ; \quad S_{m+1}(\vec{k},\vec{k}') = \lambda_{m+1}(k) \, \delta(\vec{k}-\vec{k}') \qquad (23)$$

with Γ being <u>constant</u> all over the $\vec{k}$-space. An electron undergoing this type of scattering is simply not scattered at all but remains at its position $\vec{k}$. The advantage of this artifice is a simple probability distribution P(t) of the times of free flight

$$P(t) = \Gamma \, e^{-\Gamma t} \qquad (24)$$

Random numbers are generated obeying this distribution to simulate the free flight. Furthermore, it is necessary to determine the scattering process which terminates the free flight. This is done by a second set of random numbers. Finally random numbers are required which yield the position $\vec{k}'$ of the electron after the scattering event. The magnitude of $\vec{k}'$ is given by the energy conservation because the energy after the scattering is a unique function of k'. Hence it is sufficient to generate random numbers which simulate the scattering <u>angle</u> according to the probability distribution

$$S(\vec{k},\vec{k}') = S\{k,k',\varkappa(\vec{k},\vec{k}')\} \qquad (25)$$

To obtain the stationary distribution function

from the simulation, the $\vec{k}$-space is divided into discrete cells, and the time is recorded which the electron spends in each cell during the total simulation. Usually some ten thousand scattering processes must be simulated to obtain statistical convergence to the static distribution function. Any macroscopic mean value like drift velocity and mean energy can be easily obtained from this procedure.

Let us now turn to the second type of procedures. The basis of these iterative methods [13] is the calculation of the carriers scattered into the region under consideration from the distribution function obtained by the preceding step of the iteration. This means: If we have arrived at a distribution function $f_n(\vec{k})$ by n iterations we use this to calculate the carriers scattered into any point $\vec{k}$ of the $\vec{k}$-space:

$$g_{n+1}(\vec{k}) = \int f_n(k') S(\vec{k}',\vec{k}) d^3k' \qquad (26)$$

If this function is known the Boltzmann equation reduces to a pure differential equation which is solved to obtain the result f_{n+1} of the iteration step n+1:

$$\frac{e}{\hbar}(\vec{\nabla}_k f_{n+1} \cdot \vec{E}) + \lambda(k) f_{n+1} = g_{n+1}(\vec{k}) \qquad (27)$$

Thus the difficulty arising from the <u>integro</u>-differential **form** of the Boltzmann equation is avoided. In fact, this differential equation can be solved in a straightforward manner. Choosing cylindrical coordinates (k_ρ, k_z) with k_z parallel to the field direction we obtain

$$f_{n+1}(k_\rho,k_z) = -\int_{k_z}^{\infty} dk_z' \frac{\hbar}{zeE} g_{n+1}(k_\rho,k_z') \exp\{\frac{\hbar}{eE} \int_{k_z}^{k_z'} dk_z'' \lambda(k_\rho,k_z'')\} \qquad (28)$$

The constant of integration has been determined by the condition that the distribution function has to vanish at infinite magnitude of $\vec{k}$ in any direction of $\vec{k}$-space. This result is usually transformed by

$$k_z' = k_z - eEt'/\hbar \; ; \; k_z'' = k_z - eEt''/\hbar \qquad (29)$$

Thus we obtain

$$f_{n+1}(\vec{k}) = \int_0^\infty g_{n+1}(\vec{k}-eEt'/\hbar)dt' \cdot \exp\{-\int_0^{t'} dt'' \lambda(\vec{k}-eEt''/\hbar)\} \qquad (30)$$

The pair of formulas for g_{n+1} and f_{n+1} yields the whole iteration procedure. If t', t'' is regarded as time then $k_z' = $ const and $k_z'' = $ const in Eq.(29) describe the motion of a collision-free electron. Therefore these variables are denoted as **path variables** and Budd [13] gave the name "path variable method" to the procedure.

Rees [13] achieved a great simplification of the iteration procedure by introducing the concept of self scattering as discussed above. By substitution of a constant total scattering rate Γ instead of $\lambda(k)$ in Eq.(25) we obtain

$$f_{n+1}(\vec{k}) = \int_0^\infty g_{n+1}(\vec{k} - eEt'/\hbar)dt' \cdot e^{-\Gamma t'} \qquad (31)$$

Here g_{n+1} must be slightly modified in view of Eq.(23):

$$g_{n+1}(\vec{k}) = \int f_n(\vec{k}')S(\vec{k}',\vec{k})d^3k' + \{\Gamma-\lambda(k)\} f_n(\vec{k}) \qquad (32)$$

Evidently Γ must be chosen according to

$$\Gamma \geq \text{Max}\{\lambda(k)\} \qquad (33)$$

because otherwise negative scattering rates would be obtained wherever $\Gamma < \lambda(k)$.

A great advantage of the iterative method is: If Γ is chosen sufficiently large the time dependence of distribution functions is directly obtained. If we start the procedure with $f_o(\vec{k}) = f(\vec{k},t)$ then we obtain $f_1(\vec{k}) = f(\vec{k},t+1/\Gamma)$, ..., $f_n(\vec{k}) = f(\vec{k},t+n/\Gamma)$ and so on.

This property is specially suited to calculate the time response of the electron ensemble to a step in the electric field, say from E_1 to E_2. The stationary distribution function corresponding to E_1 is taken as $f_o(\vec{k})$ of the iteration procedure. The value E_2 is inserted instead of E in Eq.(31) and Eq.(32). Then $f_n(\vec{k})$ is the distribution function at a time n/Γ after the occurrence of the field step. Furthermore, the response to a small field step can be Laplace-transformed to yield the small signal frequency response of the electron ensemble to a small a.c.field superimposed on a large d.c.field. Results of this type can be directly compared to experiments in which the small signal a.c. conductivity of electrons heated by a large d.c.field is measured.

In the remainder of the talk we shall present some results obtained by Zimmerl [15] and Hillbrand [18] using the methods described above.

A convenient test material for hot electron investigations is n-InSb for the following reasons: The high electron mobility implies considerable heating at comparatively low electric fields. At 77K lattice temperature a field range up to 200 V/cm excludes avalanche breakdown and limits the electrons to its minimum of the conduction band in the center of the Brillouin zone. The simple spherical, although non-parabolic, band structure of this "Γ-valley" is well understood and precisely described by the Kane theory [14]. Furthermore, good quality samples are easily available.

The dominant scattering mechanism in the Γ-valley of III-V compounds like InSb is <u>polar optical scattering</u>. However, <u>ionized impurity scattering</u> must be included to account for the behavior of realistic n-InSb samples. The properties of these two relevant scattering mechanisms become apparent from the k-dependence of the total scattering rates shown in Fig.11. Polar absorption of optical phonons λ_{poa} is fairly constant over the range of interest. While it is comparatively low it is still of decisive importance being the only inelastic scattering mechanism at low energy. Polar emission of optical phonons λ_{poe} is only possible at an electron energy above $k_B\theta$ corresponding to a normalized k of 1.05 because of the non-parabolic band structure. Thus there is strong scattering by emission of optical phonons outside the "Debye sphere" while scattering by absorption is weak inside the "Debye sphere". This leads to streaming (spiked) distributions. The spiking is often referred to as <u>cooling effect</u> because the mean <u>transverse</u> energy of the electrons can become less than $k_B T$. Impurity scattering λ_{imp} is seen to exhibit a peak two orders of magnitude in extent of the maximum of λ_{po}. This causes great difficulties because of the tremendous amount of self scattering implied by a choice of $\Gamma \simeq 10^{14} s^{-1}$. Note, however, that impurity scattering being elastic is very inefficient at low electron energies. Thus we have a very high scattering rate with nearly no effect. It is therefore a reasonable approximation to cut off the peak and ne-

glect impurity scattering below a certain energy. Even
with this approximation the impurity scattering rates
remain quite high and the single scattering event is
not effective because of the well known dominance of
small angle scattering.

The properties of polar optical scattering are
demonstrated further by the calculated electron distribution functions [15] along the k_z axis shown in Fig.12.
At zero electric field the well known Maxwell-Boltzmann
distribution appears. At 25 V/cm the maximum of the distribution is higher as a consequence of transverse
"cooling" and a sharp edge appears at the boundary of
the Debye sphere. This edge becomes even more pronounced
at higher fields indicating the increasing influence of
emission processes. It is evident from these curves that
a drifted Maxwellian is a poor approximation to the actual shape of the distribution function. Above 80 Vcm^{-1}
the maximum is at the boundary of the Debye sphere. An
increasing number of electrons escapes to high energies
and finally causes avalanche breakdown. Since electron-hole pair production is not included in the model it is
not reasonable to extend the calculations beyond 200 V/cm.

From the stationary distribution functions the
drift velocity is readily obtained as a function of electric field as shown in Fig.13. Curve (a) corresponds to
a simplified model neglecting both impurity scattering
and non-parabolicity of the band structure. Curve (b)
represents a pure material without impurities. Nonparabolicity is seen to have increasing influence with increasing electric field strength. Finally curve (c) accounts also for impurity scattering, and compares nicely
with the experimental data [16]. The results depend critically on the impurity content. The general shape of the

curves is characterized by a decrease of the differential mobility (slope of curves) with increasing electric field indicating the increasing influence of optical phonon emission.

Next let us discuss the dynamic behavior of the electron ensemble responding to a sudden field step as obtained by the iterative method [15]. Fig.14 shows an example which refers to the simplified model, 65K lattice temperature and a field step from E_1 = 35 V/cm to E_2 = 20 V/cm. The distribution function along the k_z-axis is plotted at intervals of 5 ps after the field step. Within the first 5 ps the distribution is merely shifted to the left which can be easily shown to hold generally for time intervals which are short compared to the reciprocal scattering rates. Let me point out that the population at low energies below the maximum is fed from the population outside the Debye sphere via phonon emission. These low energy electrons are then moving according to the force Eq.(20) and a considerable part of it will arrive at the edge of the Debye sphere because of the weakness of scattering by phonon absorption. Therefore there is a time delay between any change of the low energy distribution function and its "consequence" for the high energy population which is approximately given by the time which the electrons need to travel to the boundary of the Debye sphere. This delay results in <u>damped oscillations</u> of the distribution function shown in Fig.14. These oscillations become clearly apparent in plots of the step response of the normalized drift velocity (mobility) and mean electron energy shown in Fig.15. The time dependence of the mean energy is very similar to that of the electron population at the boundary of the Debye sphere. The oscillations occur only if the maximum of the distribution function is in the <u>interior</u> and not at the boundary of the Debye sphere because only in this case a pronounced difference in the behavior of low and high energy electrons occurs. In the high field case the maximum is at the edge of the Debye sphere and there is only one overshoot of the time response which decays aperiodically.

Fig.16 shows the structure in the frequency domain caused by the damped oscillations. The microwave mobility $\mu(\omega)$ is calculated from the time dependent mobility

$$\mu(t) = \{v(t) - v(0)\}/(E_2-E_1) \qquad (34)$$

simply by a Laplace transformation

$$\mu(\omega) = \{s \int_0^\infty \mu(t)e^{-st}dt\}_{s=j\omega} \qquad (35)$$

The real part of $\mu(\omega)$ is plotted vs the frequency with the d.c.electric field strength as a parameter in Fig.16. The damped oscillations cause the minima in Fig.16 at $E = 22.5 Vcm^{-1}$, $f_{min} = 48$ GHz, and $E = 32.5$ Vcm^{-1}, $f_{min} = 74$ GHz. Note that f_{min} is proportional to the d.c.field intensity and $1/f_{min}$ is roughly the period of the damped oscillations. This is consistent with the explanation of the oscillations by the motion of the electrons from the low energy range to the edge of the Debye sphere.

The same results are plotted vs electric field strength in Fig.17. In this form they can be obtained experimentally by measuring the small signal a.c.conductivity vs the heating d.c.electric field at constant frequency. Actually by measurements at 134 GHz Löschner [17] has found a structure of the type shown in Fig.17 and the calculations [15] presented here were carried out to exhibit the physical origin of this structure. Simplified approaches based on drifted Maxwellians or on momentum and energy balance equations and relaxation times were found insufficient to explain Löschner's results.

Unfortunately space is too limited to present the results concerning the Gunn effect in GaAs and InP obtained by the Monte Carlo method [18]. For those who are familiar with the Gunn effect we show an example in Fig.18. It shows various calculated drift velocity vs. field characteristics (curves 4-8) of InP together with experimental results [19] obtained by microwave methods (dashed curves). The material constants used in these calculation are given in Table I. These characteristics should be helpful to determine the mechanism of electron transfer to higher conduction band minima in InP. Curves of the type 6, 7, 8 result if there is only one kind of upper minima involved while 4 and 5 result from a so-called "three level model". In this latter case there is a lower set of minima weakly coupled to the Γ-valley and an upper set of strongly coupled valleys. The calculations - although confined to the case of a parabolic band structure - seem to support the two-level rather than the three-level mechanism. However, doubt has been raised by the Hilsum group [20] concerning these experiments and new experimental material is announced to become available in the near future [21] which might lead to the solution of this interesting problem of semiconductor physics.

REFERENCES

[1] L. Spitzer, "Physics of Fully Ionized Gases", Interscience Publ., New York 1967, Appendix.

[2] E.J. Ryder, Phys. Rev. $\underline{90}$ (1953) 766.

[3] K. Seeger, Zeitschr.f.Physik $\underline{244}$ (1971) 439;

[4] K. Seeger and K.F. Hess, Zeitschr.f.Physik $\underline{237}$ (1970) 252;
K. Seeger, Zeitschr.f.Physik $\underline{172}$ (1963) 68.

[5] K. Seeger, F. Kuchar, and A. Philipp, unpublished

[6] n-GaAs: K. Hess and H. Kahlert, J.Phys.Chem.Solids $\underline{32}$ (1971) 2262;
n-GaSb: for T>160K: H. Heinrich, K. Hess, W. Jantsch, and W. Pfeiler, J.Phys.Chem.Solids $\underline{33}$ (1972) 425;
n-GaSb for T = 4.2K: H. Kahlert and G. Bauer, phys.stat.sol.(b) $\underline{46}$ (1971) 535.

[7] K. Hess and K. Seeger, Zeitschr.f.Physik $\underline{218}$ (1969) 431;
K. Seeger, G. Bauer, F. Kuchar, and H. Kuzmany, Acta Phys.Austr. $\underline{35}$ (1972) 195.

[8] M. Kohler, Zeitschr.f.Physik $\underline{124}$ (1948) 772; $\underline{125}$ (1949) 679.

[9] I. Adawi, Phys.Rev. $\underline{115}$ (1959) 1152; $\underline{120}$ (1960) 118.

[10] G. Bauer and H. Kahlert, Phys.Rev. B$\underline{5}$ (1972) 566.

[11] H. Heinrich and W. Jantsch, Phys.Rev. $\underline{B}$4 (1971) 2504

[12] T. Kurosawa, Proc.Int.Conf.Phys.Semicond.Kyoto (1966) 424; W. Fawcett, A.D. Boardman, and S. Swain, J.Phys.Chem.Solids $\underline{31}$ (1970) 1963

[13] H. Budd, Proc.Int.Conf.Phys.Semicond.Kyoto (1966) 420;
H.D. Rees, J.Phys.Chem.Solids 30 (1969) 643;
H.D. Rees, IBM J.Res.Develop.13 (1969) 537.

[14] E.O. Kane, J.Phys.Chem.Solids 1 (1957) 249.

[15] O. Zimmerl, Thesis, Techn. Hochschule Wien (1972);
Proc. Int. Conf. Phys. Semicond. Warsaw (1972).

[16] E.Bonek, J.Appl.Phys.43 (1972) Nov. issue.

[17] H. Löschner, J.Appl.Phys. 43 (1972), (in print);
Proc. Int. Conf. Phys. Semicond. Warsaw (1972).

[18] H. Hillbrand, Thesis, Techn. Hochschule Wien (1972)
J. Phys. C, to be published

[19] P.M. Boers, Electronics Lett. 7 (1971) 625;
G.H. Glover, Appl.Phys.Lett.20 (1972) 224;
H.'T Lam and G.A. Acket, Electronics Lett.7(1971)722;
L.D. Nielsen, Solid State Comm.10 (1972) 169; Phys. Lett.38A (1972) 221.

[20] C. Hilsum, Proc. Int. Conf. Phys. Semicond. Warsaw (1972).

[21] R. Kaul, H.L. Grubin, G.O. Ladd Jr., and J.M. Berak, Trans.IEE-ED, in print;
B.A. Prew, Proc. ESDERC Lancaster (Sept. 1972).

		effective masses			energy separation		dielectric constants		Debye temperatures		acoustic deformation potential			intervalley coupling constants				
		$\dfrac{m^*}{m_0}$	$\dfrac{m_L^*}{m_0}$	$\dfrac{m_X^*}{m_0}$	$\Delta E_{\Gamma L}$	$\Delta E_{\Gamma X}$	ε_0	ε_∞	intra	inter	$E_{l\Gamma}$	E_{lL}	E_{lX}	$\Xi_{\Gamma L}$	$\Xi_{\Gamma X}$	Ξ_{LX}	Ξ_{LL}	Ξ_{XX}
		--			eV		--		K		eV			10^9 eV/cm				
InP 3-1	4,5	.08	.4	.4	.6	.8	12.35	9.52	500	350	7	12	7	.1 / .3	1	.5	.18	1
InP 2-1(X)	6,7	.08	--	.4	--	.9	12.35	9.52	500	350	7	--	7	--	1	--	--	1
InP 2-1(L)	8	.08	.4	--	.6	--	12.35	9.52	500	350	7	12	--	.5	--	--	.18	--

Table I : Material constants used in the calculations for InP.

CAPTIONS

Fig.1 : Field dependent mobility in n-type germanium at 20K and 77K.

Fig.2 : Absolute values of β vs. ionized impurity concentration in n-Ge. Experimental data by Tschulena in <100>-direction (crosses) and Gunn (circles).
Calculations: for combined acoustical phonon scattering and ionized impurity scattering (dashed); by Adawi using a variational method (dash - dotted); by Adawi using a Maxwell - Boltzmann distribution function (dotted).

Fig.3 : Frequency - dependence of β calculated for energy relaxation (full curve) and for combined momentum and energy relaxation (dashed).

Fig.4 : Energy relaxation time and momentum relaxation time averaged over a Maxwell - Boltzmann distribution, for optical deformation potential scattering, and momentum relaxation time for acoustic deformation potential scattering.

Fig.5 : Energy relaxation time and mobility as a function of temperature for polar optical scattering; for the calculation of the mobility, data valid for n-type gallium arsenide have been applied.

Fig.6 : Warm - carrier energy relaxation time observed by harmonic mixing in n-type GaAs and n-type GaSb as a function of temperature. The insert shows the energy relaxation time observed by the Shubnikov de Haas effect in n-type GaSb as a function of the electron temperature at a lattice temperature of 4.2K.

Fig.7 : Observed and calculated energy relaxation time as a function of temperature for n-type Ge. The full curve has been calculated with a Maxwell - Boltzmann distribution, f_{MB}, with an optical deformation potential constant of $D = 4.8 \times 10^8$ eV/cm. The dashed curve was obtained for $D = 8 \times 10^8$ eV/cm by a variational method first applied by Adawi for a calculation of β; the relative deviation of the variational-method distribution f_v from f_{MB}, devided by $y = 3\pi\, \mu_{ac}^2\, E^2/16 u_\ell^2$, where μ_{ac} is the "acoustic" mobility, E is the electric field strength, and u_ℓ is the longitudinal sound velocity, is shown in the insert; acoustic phonon scattering was taken into account with a ratio of deformation potential constants, b = 0.4; acoustic deformation potential constants: Ξ_u = 19.3 eV; Ξ_d = 9eV.

Fig.8 : Oscillatory component of the longitudinal magnetoresistance; left-hand side: measurement under ohmic conditions and different lattice temperatures between 4.2 and 12.5K; right-hand side: measurements at a constant lattice temperature of 4.2K and different electric fields between 2 and 320 mV/cm.

Fig.9 : Dependence of the energy-loss rate P on the electron temperature; dashed curve: energy-loss rate due to polar-optical scattering; dash-dotted curve: energy-loss rate due to both types of screened acoustic scattering; full curve: total energy loss rate; open circles: data obtained from SdH measurements; full circles: data obtained from mobility measurements.

Fig.10 : Experimental (full and open circles) and calculated results (solid line) of the electron temperature as a function of the electric field strength in n-type GaSb from Burstein-shift experiments.

Fig.11 : Total scattering rates for InSb accounting for the non-parabolic band structure vs. the normalized wave number $\hbar k_z/\sqrt{2m^* k_B \theta}$
$m^* = 0.012\, m_o$, $\quad n = 10^{14}\,\text{cm}^{-3}$
energy gap 225 meV, $N_I = 4 \times 10^{14}\,\text{cm}^{-3}$
The onset of polar-optical emission is indicated by the vertical line at 1.05.

Fig.12 : Distribution functions at different values of the electric field strength plotted along the axis of cylindrical symmetry as a function of $\hbar k_z/\sqrt{2m^* k_B \theta}$

Fig.13 : Drift velocity vs. electric field strength.
a) $m^* = 0.014\, m_o$, parabolic band structure, no impurity scattering
b) $m^* = 0.012\, m_o$, non-parabolic band structure, no impurity scattering
c) $m^* = 0.012\, m_o$, non-parabolic band structure, $n = 10^{14}\,\text{cm}^{-3}$, $N_I = 4 \cdot 10^{14}\,\text{cm}^{-3}$
full circles: Experimental data by Bonek [16].

Fig.14 : Development of the distribution function with time plotted at intervals of 5 ps after a field step from 35 V/cm to 20 V/cm; parabolic band structure; no impurity scattering; 65K lattice temperature.

Fig.15 : Step response of the time-dependent mobility and the mean electron energy showing damped oscillations.

Fig.16 : Real part of the complex small signal a.c. mobility vs. angular frequency at different values of the electric field strength as calculated from the step responses of the time dependent mobility.

Fig.17 : Real part of the complex small signal a.c. mobility vs. the strength of the heating d.c. electric field at different values of the frequency. The data are identical with those presented in Fig.16 but plotted in a different way.

Fig.18 : Static velocity-field characteristics of InP calculated by Hillbrand [18]. Numbers refer to different sets of material constants indicated by table I. Broken curves: Experimental results [19].

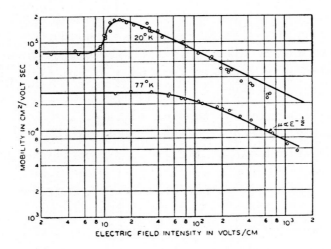

Fig. 1

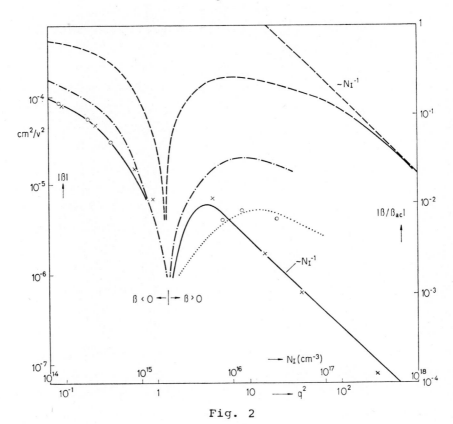

Fig. 2

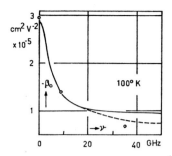

Fig. 3

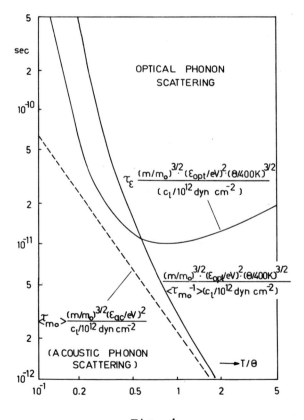

Fig. 4

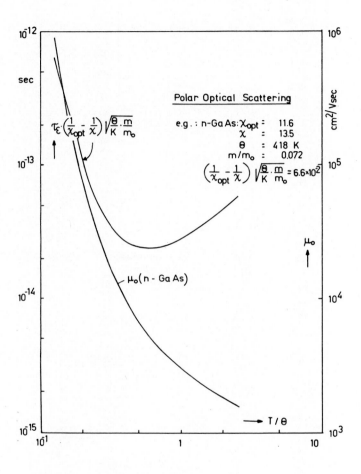

Fig. 5

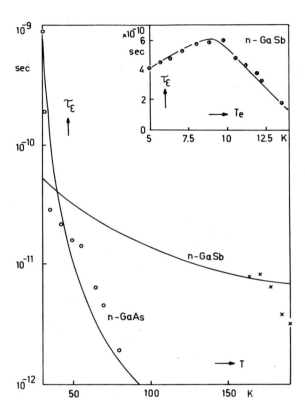

Fig. 6

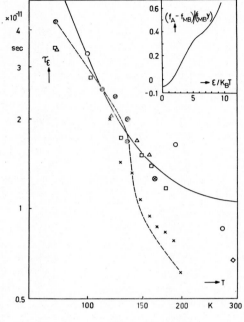

Fig. 7

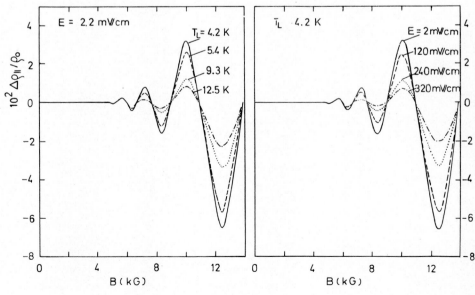

Fig. 8

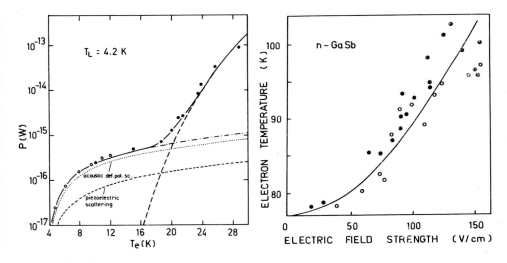

Fig. 9   Fig. 10

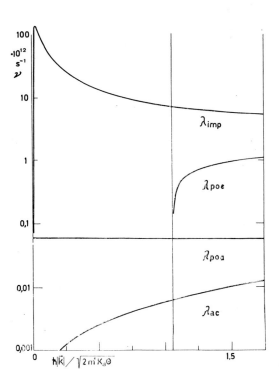

Fig. 11

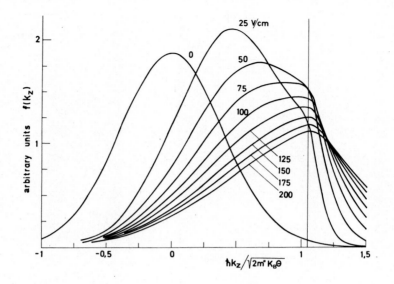

Fig. 12

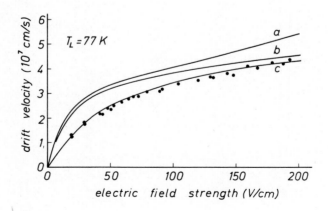

Fig. 13

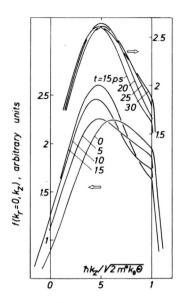

Fig. 14

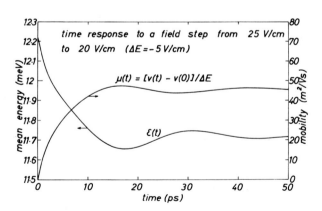

Fig. 15

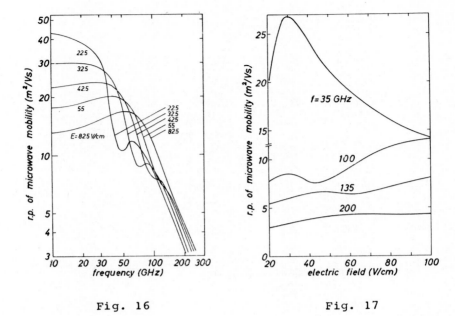

Fig. 16

Fig. 17

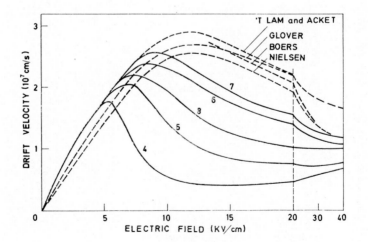

Fig. 18

SOME PROBABILISTIC ASPECTS OF THE
BOLTZMANN-EQUATION

MARK KAC

The Rockefeller University, New York

ABSTRACT

Discussion of the probabilistic nature of the Boltzmann equation centered around the Master Equation approach. Some purely mathematical implications as well as some recent work on adding fluctuating terms to the Boltzmann equation will also be reviewed briefly.

1º In my discussion of probabilistic aspects of the Boltzmann equation, I am forced by the vastness of the subject and by limitations of my knowledge to address myself only to two rather general questions:

    (a) In what sense is the Boltzmann equation a "probabilistic" equation?

and

    (b) How can one justify adding fluctuating terms to the Boltzmann equation?

These questions are not wholly unrelated, and the second is even more than of academic interest.

2º Before he was forced by logical difficulties to adapt a statistical viewpoint, Boltzmann considered

$$Nf(\vec{r},\vec{v},;t)\Delta\vec{r}\Delta\vec{v}$$

to be the <u>actual</u> number of particles having at time t velocity $\vec{v}$ within $\Delta\vec{v}$ and position $\vec{r}$ within $\Delta\vec{r}$.

    The averages which are needed to derive e.g. the equations of hydrodynamics were to Boltzmann in (1872!) ordinary averages, the kind a Bureau of Census statistician would use in calculating, say, the average height of male inhabitants of New York in the age group of 20-25 years.

    Thus e.g. the average velocity at $\vec{r}$ (within $\Delta\vec{r}$) is

$$\frac{\vec{v}_1+\ldots+\vec{v}_n}{n}$$

where $\vec{v}_1,\ldots,\vec{v}_n$ are the <u>actual</u> velocities of the n particles at $\vec{r}$ within $\Delta\vec{r}$.

It seemed natural and absolutely straightforward to replace the above average by

$$\frac{\int \vec{v} f(\vec{r},\vec{v};t) d\vec{v}}{\int f(\vec{r},\vec{v};t) d\vec{v}}$$

and recognize that

$$\nu(\vec{r};t) = \int f(\vec{r},\vec{v};t) d\vec{v}$$

is the number density at $\vec{r}$ at time t. Given then that $f(\vec{r},\vec{v};t)$ satisfies the famous equation whose centenary we are celebrating, we have the necessary machinery to go on.

$3°$ It soon became clear (and I do not have to review again the fascinating history) that a much trickier statistical element was sneaked in through the innocent sounding "Stosszahlansatz".

There was now no way out but to interpret $Nf(\vec{r},\vec{v};t)$ which satisfied the Boltzmann equation as the average density of particles, keeping the rest (i.e. definitions of average velocity, stress tensor, heat flow, etc.) unchanged.

But if Nf had to be an average, then what kind of an average? It was no longer possible to think of it in the primitive terms used in descriptive statistics. One had to go back to Γ-space, postulate an initial probability density ρ_o, allow it to evolve according to Liouville's equation, and finally average with respect to ρ_t -- the evolved density at time t. This is, of course, how one goes about deriving the Boltzmann equation via the BBGKY hierarchy, and if one could only do it rigorously or at least if one could clearly understand the nature of the approximate procedure, then an important advance would be made.

4⁰ We have moved however quite a ways from the simple intuitively appealing 1872 derivation of Boltzmann based on considering the number of particles in the collision cylinder.

Boltzmann's original derivation can however be couched in probabilistic terms by the simple device of treating binary collisions as random events whose probabilities are calculated in accordance with the Stosszahlansatz. One is led in this way to what is called the Master Equation.[1]

Rather than to review the derivation of the realistic equation, I shall present an extremely simplified artificial model invented by McKean [1] which has many (but as we shall see, far from all) of the relevant features. McKean considers n "particles" each capable of having only the "velocities" +1 and -1. When a "particle" with "velocity" e_1 (= ±1) "collides" with a "particle" with "velocity" e_2 (= ±1), they emerge with velocities e_1^{**} and e_2^{**} respectively, where the probability $\frac{1}{2}$

$$e_1^{**} = e_1 e_2$$

$$e_2^{**} = e_2$$

---

[1] This is actually intermediate between Boltzmann's original approach and the approach based on the Liouville equation. In fact the Master Equation is like the Liouville equation insofar as it concerns the probability distribution of all N particles but unlike it because it introduces a stochastic element through the treatment of collisions as random events.

and with probability $\frac{1}{2}$

$$e_1^{::} = e_1$$
$$e_2^{::} = e_1 e_2 \; .$$

During a time interval dt a "collision" between the i-th and j-th "particle" takes place with probability

$$\frac{2}{n} dt \; .$$

With probability dt/n the (e_i, e_j) "collision" results in $(e_i e_j, e_j)$ and with the same probability (i.e. dt/n) in $(e_i, e_i e_j)$. The probability that nothing happens during dt is

$$1 - \binom{n}{2}\frac{2}{n}dt = 1 - (n-1)dt \; .$$

Denoting by

$$f(e_1, \ldots, e_n; t)$$

the probability that at time t the "velocities" are $e_1, e_2, \ldots, e_n$, we obtain easily the equation

$$\frac{\partial f}{\partial t} = \frac{2}{n} \sum_{1 \leq i < j \leq n} [\frac{1}{2} f(e_1, \ldots, e_i e_j, \ldots, e_j, \ldots, e_n; t) +$$

$$+ \frac{1}{2} f(e_1, \ldots, e_i, \ldots, e_i e_j, \ldots, e_n; t) - f(e_1, e_2, \ldots, e_n; t)] \quad (4.1)$$

which describes the evolution of the probability distribution f and is the Master Equation for the McKean model.

5° It is perhaps worth recalling that the Master Equation for a gas of hard spheres is of the form

$$\frac{\partial f}{\partial t} = \frac{\delta^2}{\nu} \frac{1}{n} \sum_{1 \le i < j \le n} \int d\vec{\ell} \{f(\vec{v}_1,\ldots,\vec{v}_i^*(\vec{\ell}),\ldots,\vec{v}_j^*(\vec{\ell}),\ldots,\vec{v}_n;t)$$

$$- f(\vec{v}_1,\ldots,\vec{v}_n;t)\}((\vec{v}_i-\vec{v}_j)\cdot\vec{\ell})^+ \quad (5.1)$$

where $\vec{\ell}$ is a unit vector

$$\vec{v}_i^*(\vec{\ell}) = \vec{v}_i + (\vec{v}_j-\vec{v}_i)\cdot\vec{\ell}\vec{\ell}$$

$$\vec{v}_j^*(\vec{\ell}) = \vec{v}_j - (\vec{v}_j-\vec{v}_i)\cdot\vec{\ell}\vec{\ell}$$

and, in general,

$$a^+ = \begin{cases} a, & a > 0 \\ 0, & a < 0 \end{cases}.$$

For Maxwellian molecules $((\vec{v}_i-\vec{v}_j)\cdot\vec{\ell})^+$ is replaced by an appropriate function of the <u>angle</u>, between $\vec{v}_i-\vec{v}_j$ and $\vec{\ell}$, <u>only</u>.

The McKean model is thus of the "Maxwell type" with the added simplification that $\vec{\ell}$ can have only two values $(+1,-1)$.

6° The Master Equation (4.1) as well as the more realistic (5.1) is <u>linear</u>. To obtain the corresponding (nonlinear!) Boltzmann equation, one needs an assumption and a theorem.

The assumption is that at $t = 0$ the particles are independent and identically distributed i.e.

$$f(e_1,e_2,\ldots,e_n;0) = f(e_1;0)\, f(e_2;0)\ldots f(e_n;0) \ . \qquad (6.1)$$

The theorem is that in an asymptotic sense property (6.1) remains valid for all times ("propagation of chaos").

To be precise, what one proves is that for k fixed (but arbitrary)

$$\lim_{n\to\infty} \sum_{e_{k+1},\ldots,e_n} f(e_1,e_2,\ldots,e_n;t) = f(e_1;t)\ldots f(e_k;t) \ , \qquad (6.2)$$

and (as a consequence) $f(e;t)$ fulfills the Boltzmann equation

$$\frac{df(e;t)}{dt} = \sum_{\varepsilon=\pm 1} f(e\varepsilon;t)[f(\varepsilon;t) + f(e;t)] - 2f(e;t)$$

$$= f(e;t)f(1;t) + f(-e;t)f(-1;t) - f(e;t) \ . \qquad (6.3)$$

(Use has been made of the fact that $f(e;t)+f(-e;t) = 1$).

The great virtue of the McKean model is that (6.3) can be solved explicitly obtaining

$$f(e;t) = \frac{1}{2} + \frac{e\Delta}{2} \frac{e^{-t}}{1-(1-e^{-t})\Delta} \qquad (6.4)$$

where

$$\Delta = f(1;0) - f(-1;0) \ . \qquad (6.5)$$

7° The master equation approach suffers from a major deficiency. It is limited to the spatially homogeneous case. It seems impossible to bring in streaming terms while at

the same time treating collisions as random events. The
explanation of this I believe lies in the fact that in
a gas streaming and collisions come from the same source
i.e. the Hamiltonian of the system. It thus appears that
the full Boltzmann equation (i.e. with streaming terms)
can be interpreted as a probabilistic equation only by
going back to the Γ-space and postulating an initial probability density ρ_0.

There are other drawbacks e.g. that inspite of
many efforts propagation of chaos has not yet been proved
for a single realistic case.

But there are also some advantages, mainly of a
mathematical nature. The idea of "propagation of chaos"
proved applicable in a variety of situations, some having
little or even nothing to do with kinetic theory that
gave it birth. There is unfortunately no time to even
briefly review the ever growing literature on this subject, but the interested reader is referred to McKean [2]
for an exposition of some of the current work. Of special
interest is § 5 in which a most interesting conjecture
concerning Burger's equation

$$\frac{\partial u}{\partial t} + u\frac{\partial u}{\partial x} = \nu \frac{\partial^2 u}{\partial x^2}$$

as arising from an appropriate master equation is discussed.

8° I now turn to the question of adding fluctuating terms
to the Boltzmann equation.

This is a rather recent concern which owes its origin to an observation of Landau and Lifshitz [3] that equations of hydrodynamics should include fluctuating terms. Guided by irreversible thermodynamics, they proposed specific equations, and in his Rockefeller University Thesis R.F. Fox [4] (see also R.F. Fox and G.E. Uhlenbeck [5]) has shown that the Landau-Lifshitz equations when linearized imply the Langevin equation of the theory of Brownian motion.

Since ordinary hydrodynamical equations are derivable from the Boltzmann equation, the question naturally arose as to how should one modify the Boltzmann equation so as to obtain the Landau-Lifshitz equations.

Fox in his Thesis l.c. has shown how to do it for the <u>linearized</u> Boltzmann equation (i.e. near equilibrium) but the resulting equation is largely ad hoc, and its meaning remains dark.

Let me say first of all that it seems rather clear what the origin of fluctuating terms in hydrodynamical equations is.

In deriving equations of hydrodynamics, we must interpret $Nf(\vec{r},\vec{v};t)\Delta\vec{r}\Delta\vec{v}$ as the <u>actual</u> number of particles, while in the classical Boltzmann equation $Nf(\vec{r},\vec{v};t)\Delta\vec{r}\Delta\vec{v}$ <u>must</u> be interpreted as the <u>average</u> number of particles. There is therefore a confusion which results from identifying a random variable with its expectation (average). The confusion is completely harmless if one can neglect the fluctuations, but if fluctuations is what we are interested in, then, of course, the classical Boltzmann equation is insufficient.

9° I shall now illustrate in some detail what is involved on the McKean model. Let

$$\delta_+(e) = \begin{cases} 1, & e = +1 \\ 0, & e = -1 \end{cases}$$

and similarly

$$\delta_-(e) = \begin{cases} 0, & e = +1 \\ 1, & e = -1 \end{cases} .$$

At $t = 0$ the "velocities" are subject to the initial (factorized) distribution

$$f(e_1,\ldots,e_n;0) = f(e_1;0)\ldots f(e_n;0) ,$$

and at time t the velocities are random variables $e_1(t),\ldots,e_n(t)$ whose disjoint distribution is

$$P\{e_1(t) = e_1,\ldots,e_n(t) = e_n\} = f(e_1,e_2,\ldots,e_n;t) ,$$

where $f(e_1,e_2,\ldots,e_n;t)$ evolved from $f(e_1,\ldots,e_n;0)$ by the Master Equation (4.1).

Consider now

$$F_n(1;t) = \frac{1}{n} \sum_{k=1}^{n} \delta_+(e_k(t))$$

i.e. the actual (random) proportion of particles having "velocity" +1.

We have

$$F_n(\pm 1;t) = \frac{1}{n} \sum_{k=1}^{n} <\delta_\pm(e_k(t))> + \frac{1}{n} \sum_{k=1}^{n} (\delta_\pm(e_k(t)) - <\delta_\pm(e_k(t))>) .$$
(9.1)

For $n \to \infty$

$$\frac{1}{n} \sum_{k=1}^{n} <\delta_+(e_k(t))> \to f(1;t) \quad,$$

and we can write

$$F_n(1;t) \sim f(1;t) + \frac{1}{\sqrt{n}} \frac{1}{\sqrt{n}} \sum_{k=1}^{n} (\delta_+(e_k(t)) - <\delta_+(e_k(t))>) \quad (9.2a)$$

$$F_n(-1;t) \sim f(-1;t) + \frac{1}{\sqrt{n}} \frac{1}{\sqrt{n}} \sum_{k=1}^{n} (\delta_-(e_k(t)) - <\delta_-(e_k(t))>) \quad . \quad (9.2b)$$

Let now

$$X_n^+(t) = \frac{1}{\sqrt{n}} \sum_{k=1}^{n} (\delta_+(e_k(t)) - <\delta_+(e_k(t))>) \quad , \quad (9.3)$$

and rewrite (9.2a) and (9.2b) in the form

$$F_n(e;t) \sim f(e;t) + \frac{1}{\sqrt{n}} X(e;t) \qquad (e = +1,-1) \quad (9.4)$$

where

$$X(1;t) = X^+(t)$$

$$X(-1;t) = X^-(t)$$

and I have dropped the subscript n in $X(e;t)$.

Differentiating (9.4) with respect to t, we obtain (formally!)

$$\frac{dF_n(e;t)}{dt} \sim \frac{df(e;t)}{dt} + \frac{1}{\sqrt{n}} \frac{dX(e;t)}{dt} \quad , \quad (9.5)$$

and for large t

$$\frac{df(e;t)}{dt} = -Kf(e;t) \quad , \qquad (9.6)$$

where $-K$ is the <u>linearized</u> Boltzmann operator.

Since by (9.1)

$$f(e;t) = F_n(e;t) - \frac{1}{\sqrt{n}} X(e;t) \quad ,$$

we can rewrite (9.5) in the form

$$\frac{dF_n(e;t)}{dt} \sim -KF_n(e;t) + \frac{1}{\sqrt{n}} \{\frac{dX(e;t)}{dt} + KX(e;t)\} \quad . \qquad (9.7)$$

It is easily seen that

$$-Kf = -\frac{1}{2} f(e;t) + \frac{1}{2} f(-e;t) \qquad (9.8)$$

and the Fox-Uhlenbeck "Ansatz" is that

$$\frac{dX(e;t)}{dt} + KX(e;t) = \tilde{c}(e;t) \quad , \qquad (9.9)$$

where

$$\langle \tilde{c}(e;t_1) \tilde{c}(e';t_2) \rangle = 2K(e,e') \delta(t_2-t_1) \qquad (9.10)$$

and $K(e,e')$ is the matrix defining K i.e.

$$K = \begin{pmatrix} \frac{1}{2} & -\frac{1}{2} \\ -\frac{1}{2} & \frac{1}{2} \end{pmatrix} \quad . \qquad (9.11)$$

It is now a matter of common knowledge that $X^+(e;t)$ is a stationary, Gaussian, Markoffian process with covariance

$$\langle X^+(t_1) \, X^+(t_2) \rangle = \frac{1}{4} e^{-|t_2-t_1|} . \qquad (9.12)$$

10° The process $X^+(t)$ (as well as $X^-(t) = -X^+(t)$) is however <u>explicitly</u> defined as being the limiting version (in the limit as $n \to \infty$) of the process

$$X_n^+(t) = \frac{1}{\sqrt{n}} \sum_{k=1}^{n} (\delta_+(e_k(t)) - \langle \delta_+(e_k(t)) \rangle) , \qquad (10.1)$$

and therefore (9.12) should be derivable from the probabilistic description of the process.

We are dealing with a stationary, Markoffian process

$$(e_1(t), \ldots, e_n(t)) \qquad (10.2)$$

(stationary because the system is in equilibrium, Markoffian because of the <u>postulated</u> collision mechanism which governs transitions from one state to another), and we are concerned with the statistical properties of the process $X_n^+(t)$. Since $X_n^+(t)$ provides only a <u>contracted description</u> of the underlying process (10.2), it need not be Markoffian, and the fact that it actually is requires proof. Similarly, while it follows from the most elementary version of the central limit theorem that for each t $X_n^+(t)$ is (approximately) Gaussian, the fact that the <u>joint</u> distribution of

$$X_n^+(t_1), \ldots, X_n^+(t_r) , \qquad r = 2, 3, \ldots$$

is also Gaussian (in the limit $n \to \infty$) again requires proof.

The subtlety of these problems can be appreciated by considering $\langle X_n^+(t_1) X_n^+(t_2) \rangle$ which is equal to

$$<\delta_+(e(t_1))\delta_+(e(t_2))> - <\delta_+(e(t_1))><\delta_+(e(t_2))>$$

$$+ (n-1)[<\delta_+(e_1(t_1)\delta_+(e_2(t_2))> - <\delta_+(e(t_1))><\delta_+(e(t_2))>]$$

and noting that the (n-1) cross terms may yield a non-vanishing contribution. One must therefore calculate cross-correlations

$$<\delta_+(e_1(t_1))\ \delta_+(e_2(t_2))> \qquad (10.3)$$

to order $1/n$ i.e. to the next higher order than is necessary in the classical Boltzmann theory.

For the McKean model it appears (though a rigorous proof is lacking) that even to order $1/n$ the correlations (10.3) are all equal to $1/4$ and that therefore

$$\lim_{n\to\infty} <x_n^+(t_1)x_n^+(t_2)> = -\frac{1}{4}\lim_{n\to\infty} <\delta_+(e(t_1))\delta_+(e(t_2))>\ .\quad(10.4)$$

It is easy to calculate

$$\lim_{n\to\infty} <\delta_+(e(t_1))\ \delta_+(e(t_2))>$$

for setting

$$P(e_o|e;t) = \lim_{n\to\infty} \text{Prob}\{e_1(0) = e_o|e_1(t) = e\}\quad,$$

we can derive in a few lines the equation

$$\frac{dP(e_o|e;t)}{dt} = -\frac{1}{2}P(e_o|e;t) + \frac{1}{2}P(e_o|e;t)\quad,\qquad(10.5)$$

and hence

$$P(1|1;t) = \tfrac{1}{2} + \tfrac{1}{2} e^{-t} \ . \tag{10.6}$$

Finally,

$$\lim_{n\to\infty} \text{Prob}\{e_1(0) = 1, e_1(t) = 1\} =$$

$$\tfrac{1}{2} P(1|1;t) = \tfrac{1}{4} + \tfrac{1}{4} e^{-t} \tag{10.7}$$

which is in agreement with (9.12).

11° The situation is much more complicated for a realistic model (e.g. gas of hard spheres) and even already for my caricature of the Maxwell gas (Kac [6]).

In the case of my model we are again dealing with a stationary Markoffian process

$$(x_1(t), x_2(t), \ldots, x_n(t)) \tag{11.1}$$

subject to the constraint

$$\sum_{k=1}^{n} x_k^2(t) = n \ , \tag{11.2}$$

the stationary distribution being the uniform distribution on the sphere (11.2).

During time dt with probability

$$\frac{\nu}{n} \frac{d\theta}{2\pi} dt$$

and "(i,j) collision" takes place changing $x_i(t)$ into $x_i(t)\cos\theta + x_j(t)\sin\theta$, $x_j(t)$ into $-x_i(t)\sin\theta + x_j(t)\cos\theta$ and leaving the other x's unchanged. With probability

$$1 - \nu(n-1) \, dt$$

nothing happens during dt.

The problem now is to describe the statistical properties of the process

$$Y_n(t) = \frac{\sum_{k=1}^{n} (g(x_k(t)) - <g(x_k(t)>)}{\sqrt{n}} \qquad (11.3)$$

for an "arbitrary" function g.

The Fox-Uhlenbeck Ansatz would imply that if g is an eigenfunction of the <u>linearized</u> Boltzmann equation for my model i.e.

$$\nu \int_{-\infty}^{\infty} dy \frac{e^{-y^2/2}}{\sqrt{2\pi}} \frac{1}{2\pi} \int_{-\pi}^{\pi} \{g(x \cos\theta + y \sin\theta) + \qquad (11.4)$$

$$g(-x \sin\theta + y \cos\theta) - g(x) - g(y)\} d\theta = \lambda g(x)$$

then the process $Y_n(t)$ becomes in the limit $n \to \infty$ an O-U process with covariance

$$(\frac{1}{\sqrt{2\pi}} \int_{-\infty}^{\infty} g^2(y) \, dy) e^{\lambda t} \quad , \quad \lambda \leq 0 \quad . \qquad (11.5)$$

To prove this, we need the cross-correlations

$$<g(x_k(t)) \, g(x_\ell(t))> \quad , \quad k \neq \ell \quad , \qquad (11.6)$$

to order 1/n and there is no doubt that in this case the 1/n corrections are absolutely essential.

It is easy to calculate (to order 1) the correlation

$$<g(x(t_1) \, g(x(t_2))> \quad ,$$

and one finds (again if g is an eigenfunction of (11.4))
that it decays as exp(μt) where μ is now the eigenvalue
of the equation

$$\nu \int_{-\infty}^{\infty} dy \frac{e^{-y^2/2}}{\sqrt{2\pi}} \frac{1}{2\pi} \int_{-\pi}^{\pi} \{g(x\cos\theta + y\sin\theta) - g(x)\} dx = \mu g(x) .$$
(11.7)

(It should be noted that g is also an eigenfunction of
(11.7)).

In general $\mu \neq \lambda$, and hence the (n-1) terms (11.6)
must conspire to cancel the wrong exp(μt) terms and to
introduce the right (exp(λt)) one.

Calculations are now in progress to check this,
and while I have no doubt of what the outcome will be, I
found the verification trickier and more laborious than
I had expected and hoped for.

It took nearly one hundred years to modify the
Boltzmann equation by addition of fluctuating terms. I
hope it will be soon that one will be able to report that
even these terms "sit" as it were in Boltzmann's original
framework and that not only zeroth law and entropy but
even the Onsager-Machlup basic postulate of Irreversible
Thermodynamics is part of his grand design.

≠) At the meeting in Vienna Professor I.M. Khalatnikov informed me that a Boltzmann equation with a fluctuating
term added was already considered in a paper by Abrikosov
and Khalatnikov in 1958.
"See A.A. Abrikosov and I.M. Khalatnikov, Scattering of
Light in a Fermi Liquid, Soviet Physics JETP **34** (7) Nr.1
July 1958, pp. 135-138.

Appendix I

It may be amusing to note as was done by Dr. D.B. Abraham[1] that the master operator L of the McKean model as defined in (4.1) can be written in terms of Pauli matrices as follows:

$$L = \frac{n-1}{2n} \sum_{j=1}^{n} (\sigma_j^z + \sigma_j^x - 1) - \frac{1}{2n} \sum_{1 \le i < j \le n} (\sigma_i^z \sigma_j^x + \sigma_i^x \sigma_j^z) \quad . \tag{AI.1}$$

It is most convenient to use the representation

$$\sigma^z = \begin{pmatrix} 1 & 0 \\ 0 & -1 \end{pmatrix} \qquad \sigma^x = \begin{pmatrix} 0 & 1 \\ 1 & 0 \end{pmatrix} \quad .$$

L is thus a rather bizarre Hamiltonian of a spin system. Its eigenvalues are, of course, the inverses of the time constants with which an arbitrary initial distribution decays to the equilibrium distribution. Since factorized distributions decay according to the (non-linear) Boltzmann equation (6.3) whose solution (6.4) is a linear combination of

$$e^{-kt} \quad ; \qquad k = 0, 1, 2, \ldots \tag{AI.2}$$

it follows that zero and negative integers are (approximate for large n) eigenvalues of L. I rather suspect that (except for degeneracies) these are *all* the eigenvalues. Be this as it may, it could be of interest to relate more realistic Hamiltonians of spin systems to non-equilibrium problems of artificial models and try to gain information about spectra in this somewhat unorthodox way.

---

1. Private communication.

Appendix II

It may be recalled that in their celebrated Encyclopedia article the Ehrenfests have used the "wind-tree" model to illustrate the origin of the reversibility paradox in the Stosszahlansatz.

Using a simplified version of the "wind-tree" model which I introduced some years ago, one can show a little more, namely how the Boltzmann equation should be modified by an addition of a fluctuating term.

On a circle consider n equidistant points, some of which are marked with probability μ to form a (random) set S of "scatterers".

On each of the n points there is a ball which can be either black (b) or white (w), and during time intervals of unit duration each ball moves one step counterclockwise. Balls upon leaving S change color and for the sake of simplicity let us assume that initially all balls were black. Let $N_b(t)$, $N_w(t)$ be the numbers of black and white balls respectively at time t and $N_b(S;t)$, $N_w(S;t)$ the corresponding numbers of black and white balls in the set S.

Clearly

$$N_b(t+1) = N_b(t) + N_w(S;t) - N_b(S;t)$$

$$N_w(t+1) = N_w(t) + N_b(S;t) - N_w(S;t)$$

(AII.1)

and introducing the Stosszahlansatz

$$N_b(S;t) = \mu N_b(t)$$

$$N_w(S;t) = \mu N_w(t) \quad,$$

we obtain the Boltzmann equation for the model in the form

$$N_b(t+1) - N_w(t+1) = (1-2\mu)(N_b(t) - N_w(t))$$

or better yet setting

$$\Delta(t) = N_b(t) - N_w(t)$$

$$\Delta(t+1) - \Delta(t) = -2\mu\Delta(t) \ .$$

Let

$$\varepsilon_p = \begin{cases} -1, & p \in S, \\ 1, & p \notin S, \end{cases}$$

and

$$n_p(t) = \begin{cases} 1 & \text{if at time } t \text{ the ball at } p \text{ is black} \\ -1 & \text{otherwise} \end{cases} \ .$$

We have

$$n_p(t) = n_{p-t}(0)\varepsilon_{p-1}\varepsilon_{p-2}\cdots\varepsilon_{p-t} = \varepsilon_{p-1}\varepsilon_{p-2}\cdots\varepsilon_{p-t} \ ,$$

and hence

$$\Delta(t) = \sum_p n_p(t) = \sum_p \varepsilon_{p-1}\varepsilon_{p-2}\cdots\varepsilon_{p-t} \ .$$

Clearly

$$N_b(S;t) - N_w(S;t) = \sum_p \frac{1-\varepsilon_p}{2} \varepsilon_{p-1}\varepsilon_{p-2}\cdots\varepsilon_{p-t}$$

and the average $\langle N_b(S;t) - N_w(S;t) \rangle$ turns out to be $\mu(1-2\mu)^t$ i.e.

$$\langle N_b(S;t) - N_w(S;t)\rangle = \mu \langle N_b(t) - N_w(t)\rangle \quad .$$

Thus the Stosszahlansatz is a correct statement only about <u>averages</u>.

To introduce fluctuations, we rewrite equations (AII.1) in the form

$$\Delta(t+1) = N_b(t+1) - N_w(t+1) = \Delta(t) - 2(N_b(S;t) - N_w(S;t))$$

$$= \Delta(t) - \sum_p (1-\epsilon_p)\epsilon_{p-1}\epsilon_{p-2}\cdots\epsilon_{p-t} =$$

$$(1-2\mu)\Delta(t) + \sum_p (1-\epsilon_p-2\mu)\epsilon_{p-1}\epsilon_{p-2}\cdots\epsilon_{p-t} \quad .$$

Although the summands in the sum above are not independent, the dependence extends only over a distance $t \ll n$. Because of this the central limit theorem is still applicable, and one can show that

$$Z_n(t) = \frac{1}{\sqrt{n}} \sum_p (1-\epsilon_p-2\mu)\epsilon_{p-1}\cdots\epsilon_{p-t}$$

is (in the limit $n \to \infty$) a Gaussian process with covariance

$$4\mu(1-\mu)(1-2\mu)^t \quad .$$

The fluctuating term in the equation

$$\Delta(t+1) - \Delta(t) = -2\mu\Delta(t) + \sqrt{n}\, Z_n(t) \qquad (AII.2)$$

is directly traceable to the fluctuation in the number of balls in the set S which is the analogue of the collision cylinder.

The "classical" Boltzmann equation

$$\Delta(t+1) - \Delta(t) = -2\mu\Delta(t) \qquad (AII.3)$$

is seen to be an equation governing only the average.

BIBLIOGRAPHY

[1] H.P. McKean, Jr., An Exponential Formula for Solving Boltzmann's Equation for a Maxwellian Gas, J. of Comb. Theory, 2 (1967) pp. 358-382

[2] H.P. McKean, Jr., Propagation of Chaos for a Class of Non-linear Parabolic Equations, in Lecture Series in Differential Equations, Vol. II, Ed. A.K. Aziz, Van Nostrand Reinhold Co., 1969.

[3] L.D. Landau and E.M. Lifshitz, Fluid Mechanics, Pergamon Press, 1959. See especially Chapter XVII.

[4] R.F. Fox, Contributions to the Theory of Non-Equilibrium Thermodynamics, Thesis, Rockefeller University, 1969.

[5] R.F. Fox and G.E. Uhlenbeck, Contributions to the Theory of Non-Equilibrium Thermodynamics I and II, Physics of Fluids, 13 (1970) pp.1893-1902 and 2881-2890.

[6] M. Kac, Foundations of Kinetic Theory, Proc. Third Berkeley Symp. on Math. Stat. and Prob., Vol. III (1956) pp. 171-197.

THE STATISTICAL INTERPRETATION OF
NON-EQUILIBRIUM ENTROPY

I. PRIGOGINE

Faculté des Sciences
Université Libre de Bruxelles
and
Center for Statistical Mechanics
University of Texas, Austin

ABSTRACT

Boltzmann's original scheme leading to the statistical interpretation of non-equilibrium entropy may be summarized as follows: Dynamics → Stochastic Process (kinetic equation) → Entropy. Recent computer experiments as well as spin echo experiments in dipolar coupled systems illustrate clearly the difficulties in Boltzmann's derivation. Indeed, they display situations for which a Boltzmann type of a kinetic equation is not valid. The main purpose of this communication will be to show that we can now construct a more general microscopic model of entropy which shows the expected monotoneous approach to equilibrium even in non-Boltzmannian situations such as experiments involving "negative time evolution".

First dynamic and thermodynamic descriptions of time evolution will be compared. The time inversion symmetry present in the dynamic equations is broken in the thermodynamic description (such as the Fourier equations). The relation between this symmetry breaking and causality will be discussed.

A brief summary of non-equilibrium statistical mechanics leading to the master equation will be given. While this master equation is rigorous it is not well suited for the discussion of the statistical interpretation of entropy mainly because of its non-local character in time. However it leads to a discussion of the dissipativity condition. Briefly this condition means, that the collision operator as defined in this theory is non-vanishing and has a part which is <u>even</u> in the Liouville-von Neumann operator L. As the result the time inversion symmetry of the dynamic equations is broken. Examples of simple systems for which the dissipativity condition can be rigorously verified (in an asymptotic sense when the system becomes large) will be given.

A formulation of dynamics in which the even part of L are explicitly displayed will be indicated. This formulation may be called the "causal" or "obviously causal" formulation of dynamics as causality is now incorporated into the <u>differential equations</u> (and not only as in the usual formulation in the integral representation of the solutions). The transformation from the initial representation to the causal representation conserves averages of all observables. It leads, therefore to equivalent (but <u>not</u> unitary equivalent) representations of dynamics. Examples will be given. In the causal formulation of dynamics there appears a Liapounoff function which is positive and can only decrease in time. This leads directly to a statistical model for non-equilibrium entropy. One of the important features of this new model is that it contains all non-equilibrium correlations which may be introduced through initial conditions. As an application, experiments involving "negative time evolution" will be discussed. It is shown that Loschmidt's paradox is now solved as during each time interval in such experiments, the entropy production is now positive.

It is concluded that the second law of thermodynamics is valid for all initial value problems when formulated for mechanical systems which satisfy the dissipativity condition indicated above.

I. Boltzmann's H-Theorem

It is one hundred years ago that Boltzmann's paper entitled "Further Studies on Thermal Equilibrium between Gas Molecules" [1] was published, having made its appearance the 8th of February 1872. It was the sixth paper on the second law published by Boltzmann. This explains the word "further" studies in the title of this paper.

Since the beginning of his scientific career, Boltzmann was fascinated with the second law. Already his second paper bore the title: "Mechanical Explanation of the Second Principle" [2]. Why was Boltzmann so fascinated with the second law? What attracted him to a problem to an extent that he devoted his whole life to its discussion? We have a hint of this in an article in his "Populäre Schriften" [3]. There he writes, "if one would ask me which name we should give to this century, I would answer without hesitation that this is Darwin's century". This is a most interesting statement. Boltzmann was obviously fascinated by the idea of evolution and specially by the "mechanical interpretation" of the process of evolution. It is also interesting that he dedicated his "Populäre Schriften" to the memory of Schiller and Beethoven, the two romantic heroes of Germany. From all this, we may infer that Boltzmann's spirit was deeply immersed in the romantic mood of his days. To the classic and static view of the world he opposed the romantic concept of historicism put forward by some of the most distinguished philosophers and writers of his time such as Hegel.

A few years earlier Clausius had already clearly recognized the unique feature of the second law. Indeed,

the quantity which appears in the second law was named by
him the "entropy", which in Greek means <u>evolution</u>. It was,
therefore, natural that Boltzmann became deeply involved
in the problem of trying to understand what the entropy
meant in terms of more basic "mechanical" concepts. He
hoped in this way to obtain an interpretation of thermodynamic evolution somewhat analogous to the interpretation
of biological evolution given by Darwin. This fascination
with entropy is common to many great thinkers of the last
one hundred years. In his "Evolution Créative" Bergson
called it the <u>most metaphysical of all laws of physics</u> [4]
<u>Spengler</u> dates the decline of occidental science explicit-
ly to the formulation of the second law of thermodynamics
[5] . This list could be continued.

Boltzmann's first attempts [2] in this direction
were in the line of other investigators. He gave a mechanical meaning to the heat received by the system, and obtained some interesting results. But the really characteristic and unique feature of entropy, the positive entropy
production, was not considered in these studies. He addressed
himself to this problem in 1872 [1] and then pursued it
throughout his life. His starting point was to deal with
<u>large</u> systems. Here again, he made analogies with social
and biological situations. He noticed that we cannot predict the individual fate of an individual, but that we can
describe accurately the average behaviour of groups. Therefore he believed that the first step towards an understanding of the second law should be the <u>replacement</u> of
dynamics by some form of probability calculus. This idea
was not new: Kronig in 1856 [7], Clausius in 1857 [8],
Maxwell in 1859 [9], Boltzmann himself in 1868 [10] had
already introduced probability concepts in the framework

of kinetic theory. But now [1], Boltzmann went much further. Using both probabilities and mechanical concepts, he established his famous kinetic equation for the velocity distribution function

$$\frac{\partial f}{\partial t} = \int d\Omega \int d\underline{v}_1 \, \sigma(\Omega) |\underline{v}_1 - \underline{v}| (f'f'_1 - ff_1) \, . \qquad (I.1)$$

He then introduced the H_B quantity

$$H_B = \int d\underline{v} \, f \log f \qquad (I.2)$$

and established the inequality

$$\frac{dH_B}{dt} \leq 0 \, . \qquad (I.3)$$

This basic inequality led him to the identification of H_B with the thermodynamic entropy S,

$$S = -k \, H_B \, . \qquad (I.4)$$

In modern language H_B provides us with a "Liapounoff function" to which Boltzmann ascribed a basic thermodynamic meaning [11].

Therefore, we may summarize Boltzmann's scheme in the following way:

<u>Dynamics → Stochastic Process (Kinetic Equation) → Entropy</u>

We see why Boltzmann's investigations are so central. They link various levels of description which had been introduced independently in the history of science:

<u>the dynamical description expressed through the laws of mechanics, the description in terms of probabilities and the thermodynamical description</u>.

Boltzmann's H-theorem has led to a very strange situation, perhaps unique in the history of science. On the one hand, Boltzmann's kinetic equation has been applied successfully to a very large range of physical phenomena, such as transport processes in dilute gases, in plasmas, shock waves, hydrodynamics and chemical reactions. (A good account may be found, for instance, in the textbook by Hirschfelder, et al. [12]). Molecular experiments, such as those performed by Alder and Wainwright [13], and more recently by Bellemans and Orban [14], completely verify Boltzmann's predictions. The quantity H_B indeed decreases in a fluctuating fashion and the system reaches thermodynamic equilibrium after a time corresponding to a few collisions per particle. The fluctuations decrease when the number of particles is increased. There is little doubt that in the thermodynamic limit (volume $V \to \infty$, number of particles $N \to \infty$, N/V = constant) we would observe for a dilute gas the monotonous behaviour approach to equilibrium predicted by Boltzmann. (See figure I.1).

On the other hand, Boltzmann's ideas met with violent objections coming from theoretical physicists and mathematicians [15]. In fact, these discussions never stopped. In which sense Boltzmann's microscopic model of entropy "explains" irreversibility, what is its domain of validity? The classical objections are often formulated in the form of paradoxes. The two most famous are:

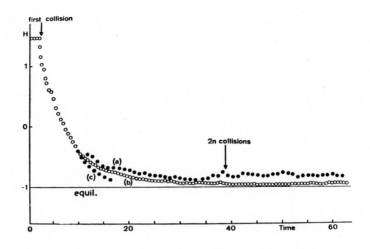

Fig. I.1 Evolution of H_B with time [12]
(a) Total number of disks (n) : 100
(b) (n) = 484
(c) (n) = 1225

1. <u>Zermelo's Recurrence Paradox</u> [16]: This paradox is based on Poincaré's recurrence theorem which states: "For almost all initial states an arbitrary function of phase space will infinitely often assume its initial value within an arbitrary error, provided the system remains in a finite part of the phase space". (Poincaré's cycle). As a result, it seems that irreversibility is incompatible with the validity of this theorem.

2. <u>Loschmidt's Reversibility Paradox</u> [17]: Since the laws of mechanics are symmetrical with respect to the inversion of time,

$$t \rightarrow -t$$

to each process there belongs a corresponding time reversed process. This again seems to be in contradiction with the existence of irreversible processes.

The answer to Zermelo's paradox has been given by Boltzmann [18]. He pointed out that the recurrence time increases beyond imagination with the number of degrees of freedom of the system.

On the other hand, the answer to Loschmidt's paradox is not so straightforward. We shall in fact devote much of this report to its discussion. Let us first ask: Is Loschmidt's paradox at all justified? It is easy to make a computer experiment to test it. Bellemans and Orban have calculated Boltzmann's quantity for two-dimensional hard spheres (hard disks) [14]. They start with disks on lattice sites with an isotropic velocity distribution. The results are shown in Figure I.2.

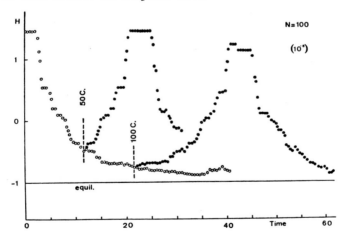

Fig. I.2 Evolution of H_B with time for a system of 100 disks when velocities are inverted after o, 50 collisions, 100 collisions.

Now, if after 50 or 100 collisions, we invert the velocities we obtain a new ensemble. We may then again follow the evolution of the corresponding quantity H_B in time. The results are given in Fig. I.2.

We see that indeed the entropy (given by $-H_B$) first decreases after the velocity inversion. The system deviates from equilibrium over a period of 50 to 60 collisions (corresponding to about 10^{-6} seconds in a dilute gas).

A similar situation exists also in spin echo experiments or plasma-echo experiments [19-22]. Of special interest are the spin echo experiments in dipolar coupled spin systems as recently described by Rhim, Pines and Waugh [21], because these experiments involve an N-body situation quite similar to a classical gas. There also, over limited periods of time "anti-Boltzmannian" behaviour corresponding to an increase of H_B may be observed. Rhim, Pines and Waugh [21] introduce even a "Loschmidt demon" to deal with this situation.

Boltzmann's point of view was that his derivation of the kinetic equation depended in addition to the laws of mechanics on the initial conditions which were assumed [18,23]. He recognized quite clearly that as first emphasized by Burbury, the molecules have to be uncorrelated before the collision [18,24]. This is an important remark to which I shall come back. He considered that this was a "natural" assumption that the probability of a situation in which this would not be so could be neglected [18,23].

Now both the computer experiments and the spin experiments show that we have to be careful: conditions in which correlations would exist between molecules before a collision (or a "recollision") occurs between them can be realized at least temporarily. The Ehrenfests recognized quite clearly that Boltzmann's equation cannot be valid both before and after velocity inversion [15]. What is the consequence of this situation? Does the fact that Boltzmann's equation is not valid imply a set-back of Boltzmann's interpretation of entropy or of the second law itself?

Let us remind the very definition of an irreversible process as introduced in thermodynamics [25]. According to Planck's definition it is a process which once performed leaves the world in an altered state. Planck insists: <u>By no experimental device whatever the ingenuity of the experimenter, should it be possible to restore the initial state</u>. Now if we consider the velocity inversion experiment the positive entropy produced in the period $0 - t_0$ would be compensated by a <u>negative</u> entropy production, in contradiction with the very definition of an irreversible process.

Therefore, I consider that it is very fortunate indeed that as we shall show, we can now construct a more general microscopic model of entropy which remains valid even for "negative time evolution" experiments. We shall be even able to make explicit the thermodynamic price of an experiment involving time inversion.

Before doing so, I would like to stress that independently of the Loschmidt-paradox, the task of constructing a general statistical non-equilibrium thermodynamics is of special interest today.

Non-equilibrium thermodynamics is going today through a period of rapid growth. It has been applied with success to many new problems. Some of them correspond to

near equilibrium conditions ("linear" non-equilibrium
thermodynamics [26,27]) others, perhaps the most interesting, correspond to systems in far from equilibrium
conditions [11]. New and unexpected phenomena such as the
formation of dissipative structures have been predicted
theoretically and discovered experimentally. But till now,
it has been possible to approach such problems only in the
case of "local equilibrium". That corresponds to situations
in which the variables used are the same as would appear
in equilibrium theory [11]. Clearly many situations such as
treated in lasers [28] or in plasma physics [29] do not
belong to this class. In cases when one had to go beyond
"local-equilibrium" rather specific ad-hoc assumptions were
generally introduced. Some clarification is therefore
needed and this can only come from statistical mechanics.

II. Dynamic and Thermodynamic Descriptions

Time-inversion symmetry

Let us start from classical or quantum dynamics as expressed by Hamilton's equations of motion

$$\frac{dq}{dt} = \frac{\partial H}{\partial p}, \quad \frac{dp}{dt} = -\frac{\partial H}{\partial q} \quad (\text{II}.1)$$

or Schrödinger's equation

$$i \frac{\partial \psi}{\partial t} = H\psi . \quad (\text{II}.2)$$

Both descriptions may be unified through the Liouville-von Neumann equation for the distribution function (or density matrix [30])

$$i \frac{\partial \rho}{\partial t} = L \qquad (II.3)$$

with

$$L\rho = \begin{cases} -i\{H,\rho\} & \text{Poisson bracket} \\ [H,\rho]_- & \text{commutator} \end{cases} \qquad (II.4)$$

In both cases L is a hermitian operator (better super-operator, see [45]). Therefore

$$L = L^\dagger \quad . \qquad (II.5)$$

A basic feature of equation (II.3) is its "<u>Lt-invariance</u>". If we perform the operations

$$L \to -L$$
$$t \to -t \qquad (II.6)$$

the equation (II.3) remains invariant. Note that <u>if L is a possible Liouville operator so is -L</u>. On the other hand macroscopic equations involving thermodynamic quantities do not present such an invariance. Let us consider the heat equation

$$\frac{\partial T}{\partial t} = K \frac{\partial^2 T}{\partial x^2} \quad . \qquad (II.7)$$

As the heat conductivity K is a positive quantity it has no meaning to reverse its sign. Therefore the time inversion

$$t \to -t \qquad (II.8)$$

now leads to the different equation

$$\frac{\partial T}{\partial t} = - K \frac{\partial^2 T}{\partial x^2} \qquad (II.9)$$

we may call the "anti-Fourier" equation. Both equations have a physical meaning. Equation (II.8) corresponds to the situation in which we have an initial value problem and we want to calculate the temperature distribution in the future. The solutions of (II.7) are "retarded solutions". On the contrary in the case of (II.9) we have a "final value" problem as may arise as the result of a fluctuation in a system which was isolated for a long time. The uniform distribution corresponds then to the far distant past ($t \to -\infty$). The solution of (II.9) is an "advanced" solution.

We have now a pair of equations in each of which the direction of time play a different role. The time symmetry present in the dynamical description (II.3) has been broken.

There is obviously a relation between this "symmetry breaking" and causality. This has been emphasized by many people [30,34,35]. But the symmetry breaking goes beyond causality as it leads to a description of natural processes in terms of differential equations in which the direction of time is specified. On the contrary causality is a more general concept which may be applied to the solution of a differential equation in which both directions of time play a symmetrical role.

We may say that the second law of thermodynamics

$$\frac{dS}{dt} \geq 0 \qquad (II.10)$$

summarizes in a single inequality all laws such as Fourier's law, friction phenomena... which are precisely described by phenomenological equations involving a privileged direction of time. This is the situation we associate with thermodynamic irreversibility.

What are the supplementary conditions we have to require in addition to causality to generate thermodynamic behaviour? What is the class of systems which will manifest this behaviour? These are the questions we have to investigate now.

III. Non-Equilibrium Statistical Mechanics - The Master Equation

Let us start with the Liouville equation (II.3). Its formal solution is

$$\rho(t) = e^{-iLt} \rho(0) = \frac{1}{2\pi i} \int_C dz \, e^{-izt} \frac{1}{L-z} \rho(0) \quad . \tag{III.1}$$

In the case of the initial value problem the contour has to be traced in the upper half plane corresponding to the complex variables z [29,30] (that is for Im z > 0, see fig. III.1a).

With this choice of the contour we calculate the retarded solution of the Liouville equation

$$\rho^r(t) = \frac{1}{2\pi i} \int_{C^+} dz \, e^{-izt} \frac{1}{L-z} \rho(0) \quad . \tag{III.2}$$

Similarly the "advanced" solution would be given by

$$\rho^a(t) = \frac{1}{2\pi i} \int_{C^-} dz \, e^{-izt} \frac{1}{L-z} \rho(0) \quad . \tag{III.3}$$

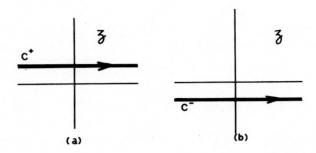

Fig. III.1 (a) retarded solution
(b) advanced solution

Both solutions satisfy the same differential equation. However causality leads to the two different integral representations (III.2), (III.3). If the evaluation of the integrals (III.2) and (III.3) leads to a difference we may expect "thermodynamic behaviour".

To proceed further it is convenient to introduce orthogonal hermitian projection operators P, Q, such that

$$P + Q = 1 \ . \tag{III.4}$$

It is then easy to verify the identity for the resolvent $R(z) \equiv (L-z)^{-1}$ [36,37,38]:

$$R(z) = \{P + C(z)\} \frac{1}{PLP+\psi(z)-z} \{P + \mathcal{D}(z)\} - \mathcal{P}(z) \tag{III.5}$$

with

$$\psi(z) = -PLQ \frac{1}{QLQ-z} QLP \qquad (III.6)$$

$$\mathcal{D}(z) = -PLQ \frac{1}{QLQ-z} \qquad (III.7)$$

$$C(z) = -\frac{1}{QLQ-z} QLP \qquad (III.8)$$

$$P(z) = -\frac{1}{QLQ-z} \quad . \qquad (III.9)$$

We have here the basic operators of our formulation of non-equilibrium statistical mechanics; the choice of the projection operators fixes the "language" in which we formulate our results. Generally P is chosen as projecting onto the diagonal elements in the representation in which some model-Hamiltonian H_o is diagonal:

$$<m|P\rho|n> = <m|\rho|m> \delta_{mn} \quad . \qquad (III.10)$$

For this reason $P\rho$ is called the "vacuum of correlations". Of special importance is the collision operator $\psi(z)$: reading (III.6) from right to left it corresponds to a transition from the vacuum of correlations followed by a dynamical evolution in the correlation space and finally followed by a return to the vacuum of correlations.

When we introduce the formal decomposition (III.5) into (III.2) we derive directly the "master equation" for the density matrix ρ [30]. For the diagonal elements ρ_o of ρ we obtain

$$i \frac{\partial \rho_o(t)}{\partial t} = \int_o^t d\tau \, G(t-\tau) \, \rho_o(\tau) + F(t;Q\rho(0)) \qquad (III.11)$$

where

$$G(t) = \frac{1}{2\pi i} \int_C dz \, e^{-izt} \, \psi(z) \qquad (III.12)$$

and

$$F(t; Q\rho(0)) = \frac{1}{2\pi i} \int_C dz \, e^{-izt} \, \mathcal{D}(z) \, Q\rho(0) \, . \qquad (III.13)$$

It should be emphasized that the master equation (III.11) is exact. No approximations have been introduced in its derivation from the Liouville equation.

The denomination "master equation" (for III.11) may lead to confusion. Indeed the master equation as introduced by Kac and Uhlenbeck [39] refers to a model equation incorporating features of probability theory. There are certainly supplementary assumptions to be made to permit such an interpretation of equation (III.11). Indeed let us compare it to Boltzmann's kinetic equation (I.1). We see that there are two main differences: One is the non-local character of the collision operator in (III.11), as the time change at t depends on the previous history of the system ("non-Markoffian" equation). The second one is the occurrence of the memory term F depending on the initial correlations. The master equation (III.11) has been applied successfully to many interesting problems [40]. It has also been shown that it describes correctly the evolution of a system both before and after velocity inversion [41,42]. Suppose that we start at $t = 0$ with no correlations. During the evolution from 0 till t_0 the diagonal elements evolve according to (III.11) (with $F = 0$); correlations appear progressively as the result of the interactions. However when we inverse the velocities at $t = t_0$ we form a new ensemble. In the subsequent evolution from t_0 till $2t_0$ the memory term F plays an essential role as inversion of velocities leads to long range correlations. At time $2t_0$ we recover the original state at $t = 0$ with reversed velocities.

It is certainly very interesting that the master equation permits to describe both "Boltzmann" and "anti-Boltzmannian" behaviour. However the non-local form of (III.11) makes it difficult to discuss such basic questions as the time inversion symmetry and the validity of the second law as expressed by the "local inequality" (II.10). We are reminded here of a most interesting <u>controversy between Einstein and Ritz [43].</u> Ritz considered that it would be essential to include explicitly causality in the laws of physics such as electromagnetism. Einstein's reaction was that this would lead to a "non-local" formulation of dynamics and would therefore destroy the possibility to express simply basic principles of physics such as the conservation laws.

Non-equilibrium statistical mechanics incorporates causality in the formulation of dynamics but the price we had to pay was precisely the non-locality. However as I shall show in the later part of this paper it is now possible to formulate non-equilibrium statistical mechanics in a new form which combines causality and locality.

Before we do so let us consider the limiting case in which (III.11) reduces to a local description. We proceed formally in the following way: we neglect the memory term F and assume that

$$\int_0^t d\tau\, G(\tau)\, \rho_0(t-\tau) \simeq \int_0^t d\tau\, G(\tau)\, \rho_0(t) \simeq [\int_0^\infty d\tau\, G(\tau)]\rho_0(t) \ .$$
(III.14)

Then (III.11) takes the Boltzmannian "local" form.

$$i \frac{\partial \rho_o(t)}{\partial t} = \psi(+i0)\, \rho_o(t) \qquad \text{(III.15)}$$

with (see (III.6))

$$\psi(+i0) = \lim_{z \to +i\epsilon} \psi(z) \qquad \text{(III.16)}$$

$$= \lim_{z \to +i\epsilon} -PLQ \frac{1}{QLQ-z} QLP \; . \qquad \text{(III.17)}$$

The transformation (III.14) has only be made to indicate the manipulations we have to make to go from the non-local form (III.11) to the local form (III.15) and to introduce in this way the operator $\psi(0)$[+]. This operator plays a central role in our whole approach and we would like to discuss it now in more detail.

[+] In order to avoid misunderstandings let us stress that no claim of equivalency is made between (III.11) and (III.15).

IV. Dissipativity Condition

In the limit of large systems, the sum over the intermediate states Q ("the correlations") involves an integration. Therefore a formal representation of (III.17) is given by

$$-i\psi(0) = -\pi PLQ\delta(QLQ)QLP + iPLQ \frac{1}{QLQ} QLP \qquad (IV.1)$$

where the second term is understood as a principal part. Suppose first that this expression is meaningful and that the first term does not vanish. Then we see that

a) $-i\psi(0)$ may be split into an even and an odd contribution in respect to the transformation $L \to -L$ (see (II.6))

b) the even part of $-i\psi(0)$ is a hermitian operator (see (II.5)), the odd is antihermitian

c) the even part of $-i\psi(0)$ is a negative operator.

We may consider the class of dynamic systems for which the Hamiltonian is of the form

$$H = H_o + \lambda V . \qquad (IV.2)$$

For example we may consider a N-body system such as described by

$$H = \sum_k \varepsilon_k a_k^\dagger a_k + \lambda \sum_{k\ell pq} v(k,\ell,p,q) a_k^\dagger a_\ell^\dagger a_p a_q \qquad (IV.3)$$

The explicit form of the collision operator ψ is then obtained using perturbational techniques [30,29] based on formal expansion in powers of λ or C (the concentration) and on partial resummation. In the lowest order in the relevant expansion parameter one derives in this way the standard forms of the collision operator such as the Fokker-Planck operator or the Boltzmann operator. In all these cases only the even part in (IV.1) is not vanishing. The δ-function expresses simply conservation of the unperturbed energy in the collision process.

But the basic question is the existence of $i\psi(0)$ <u>independently of any perturbational approach</u>. This question can now be answered in simple cases. Let us consider two examples. The first corresponds to a Hamiltonian of the form

$$H = \sum_n a_n^\dagger a_n + \lambda v \sum_k \{a_k^\dagger a_{k+1} + a_{k+1}^\dagger a_k\} . \qquad (IV.4)$$

Obviously the number of particles is conserved. We may therefore consider separately the 1-particle sector. It is easy to obtain from (IV.4) the Liouville operator and then using (III.5) to obtain the expression of $i\psi(0)$. Of special interest is the infinite volume limit when the spectrum of H becomes continuous. The result obtained by Stey and Grécos for the leading terms of the matrix elements of the asymptotic expression of $i\psi(0)$ is [38]

$$-i<m|\psi(0)|n> = -4|\lambda v| \frac{1}{2\pi} \int_{-\pi}^{\pi} d\phi \, e^{i(m-n)\phi} \, |\sin \tfrac{\phi}{2}| + O(\tfrac{1}{N}) .$$
$$(IV.5)$$

It is easy to verify that the general properties we have stated are satisfied (moreover the odd part of $i\psi(0)$ vanishes in this example).

Similar results may be obtained for the Friedrichs model [45-47] (which corresponds to the one particle sector of the Lee model). More precisely we consider in this case a discrete level (in the absence of interactions) which is coupled to a set of quantum states. Again the limiting process to large volume may be performed. In the infinite volume limit we recover the Friedrichs model as originally formulated. The leading terms of the matrix elements of the asymptotic expression for $\psi(0)$ are given by [46]

$$\psi(0) = \begin{pmatrix} a^{-1} & -\frac{2\pi}{L} a^{-1} \frac{\lambda^2 |v(\omega')|^2}{|\eta(\omega')|^2} \\ -\frac{2\pi}{L} a^{-1} \frac{\lambda^2 |v(\omega)|^2}{|\eta(\omega)|^2} & (\frac{2\pi}{L})^2 a^{-1} \frac{\lambda^2 |v(\omega)|^2}{|\eta(\omega)|^2} \frac{\lambda^2 |v(\omega')|^2}{|\eta(\omega')|^2} \end{pmatrix} \quad (IV.6)$$

Here $\eta(z)$ is defined as (ω_0 the unperturbed energy of the discrete state, $\bar{v}_m = L^{-1/2} v(\omega_m)$)

$$\eta(z) = \omega_0 - z - \lambda^2 \sum_m \frac{\bar{v}_m \bar{v}_m^*}{\omega_m - z} \quad . \quad (IV.7)$$

In the limit of an infinite volume $\eta(z)$ has a cut along the real axis. We have then to consider the function $\eta^+(z)$ (or $\eta^-(z)$) which is analytic in the upper half plane and may be continued analytically into the lower (or upper) half plane

$$\eta^+(z) = \omega_0 - z - \lambda^2 \int_{C^+} d\omega \frac{|v(\omega)|^2}{\omega - z} \quad . \quad (IV.8)$$

Also in (IV.6) a represents the integral

$$a = -\frac{1}{2\pi i} \int_{-\infty}^{+\infty} d\omega \, |\eta(\omega)|^2 \quad . \quad (IV.9)$$

There are two cases: either the dispersion equation admits

a <u>real</u> solution

$$\eta^+(z) = 0 \qquad (IV.10)$$

then as we have shown [45] $i\psi(0)$ vanishes. Either $\eta^+(z)$ admits no real solution then $\psi(0)$ is given by (IV.6). We may verify that $i\psi(0)$ as given by (IV.6) has again the properties we have enumerated. Here also the odd part vanishes.

We may therefore conclude that the collision operator exists at least in an asymptotic sense in the limit of large systems. We may therefore use this operator to classify dynamic systems (or better the limits to which dynamical systems tend when the volume and eventually the number of particles is increased):

  a) the even part of $i\psi(0)$ vanishes;
  b) the even part of $i\psi(0)$ is different from zero;
  c) the even part of $i\psi(0)$ becomes undefined.

An example of class c) is the case of particles interacting through long range forces such as gravitational forces [48]. Of foremost interest will be for us the <u>second class</u> of systems. We expect that for such systems a thermodynamic description will be possible. Indeed as $i\psi(0)$ has an even part the kinetic equation (III.15) has lost its Lt-invariance. If we reverse both time and L the equation is no more invariant and we go over to a different "antikinetic equation". Because of this basic property we have called the condition

$$i\overset{e}{\psi}(0) \neq 0 \qquad (IV.11)$$

the <u>condition of dissipativity</u>. This condition when satisfied leads to the breaking of the time inversed symmetry. There are many interesting questions which arise in this connexion such as the relation between dissipativity and the nature of the invariants as well as the relation between dissipativity and ergodicity and phase mixing. These questions are considered elsewhere [45,46,49].

It is also interesting to note that Uhlenbeck (see [39] Appendix I by Uhlenbeck) had clearly recognized the necessity to formulate a kind of dynamic version of ergodic theory. The dissipativity condition (IV.11) plays precisely this role.

In the preceding paragraph the kinetic equation (III.15) was obtained from the exact non-local master equation (III.11) through drastic simplifications (see III.14). We want therefore now to indicate a new formulation of non-equilibrium statistical mechanics which makes explicit the appearance of even terms in L leading to Lt symmetry breaking without going through the non-local master equation (III.11)$^{+)}$.

+) A fuller account may be found in papers prepared for publications by our group, see specially [33,32].

V. Causal Dynamics

Let us go back to the integral representation (III.2) for the "retarded" solution of the Liouville equation. By a careful analysis of the singularities it can be shown that this formula may be written as [33,32]

$$\rho^r(t) = \Lambda(L)\, \rho^{p,r}(t) \qquad (V.1)$$

where the time independent operator $\Lambda(L)$ acts on the "physical" distribution function $\rho^{p,r}(t)$ taken at the same time t. Causality is now incorporated in the form of $\Lambda(L)$. We may consider (V.1) as the transformation from our representation of the density matrix to another. The essential point is that for the advanced solution we have the _different_ transformation formula

$$\rho^a(t) = \Lambda(-L)\, \rho^{p,a}(t) \quad . \qquad (V.2)$$

Indeed because of the Lt-invariance time inversion is equivalent to L inversion.

There is a condition which this transformation has to satisfy: we may perform similar transformations on observables A. Now in the Heisenberg representation A satisfies the Liouville equation (II.3) with L replaced by -L. Therefore for A we obtain instead of (V.1)

$$A^r(t) = \Lambda(-L)\, A^{p,r}(t) \quad . \qquad (V.3)$$

If we want to consider the equations (V.1), (V.3) as transformations to a new representation we have to require that all averages remain invariant. Therefore

$$\langle A \rangle = \mathrm{tr}\, \rho^\dagger A = \mathrm{tr}(\rho^p)^\dagger A^p \quad . \tag{V.4}$$

Using (V.1) and (V.3) we see that this implies

$$\Lambda(L)\, \Lambda^\dagger(-L) = \Lambda^\dagger(-L)\, \Lambda(L) = 1 \quad . \tag{V.5}$$

We have called such transformations "<u>star-unitary transformations</u>" as we have introduced the notion star to denote the <u>combined operations of taking the hermitian conjugate and reversing L</u> [33,31]

$$\Lambda^*(L) = \Lambda^\dagger(-L) \quad . \tag{V.6}$$

We have verified that the operator $\Lambda(L)$ we derived satisfies the basic conditions. Except in the case where $\Lambda(L)$ does not depend on the sign of L, star unitary transformations are <u>non</u>-unitary transformations. Still they preserve the average values of observables. They correspond therefore to equivalent (but not unitary equivalent) representations of dynamics.

Let us consider the equations of motion in the causal representation. Using (II.3), (V.1), (V.5) and (V.6) we obtain immediately for the retarded solution in the physical representation

$$i\, \frac{\partial \rho^{p,r}}{\partial t} = \phi\, \rho^{p,r} \quad \text{with} \quad \phi = \Lambda^* L \Lambda \quad . \tag{V.7}$$

Similarly we have for the advanced solution

$$i\, \frac{\partial \rho^{p,a}}{\partial t} = -\phi'\, \rho^{p,a} \quad \text{with} \quad \phi' = -\Lambda^{*\prime} L \Lambda' = -\Lambda^\dagger L \Lambda' \tag{V.8}$$

where the prime means L inversion

$$\Lambda'(L) \equiv \Lambda(-L) \; . \qquad (V.9)$$

As we did for the collision operator in § 4, we may always decompose ϕ into an even part and an odd part with

$$\overset{e}{\phi}(L) = \overset{e}{\phi}(-L) \; , \qquad \overset{o}{\phi}(L) = - \overset{o}{\phi}(-L) \; . \qquad (V.10)$$

Then (V.7), (V.8) become

$$i \frac{\partial \rho^{p,r}}{\partial t} = (\overset{e}{\phi} + \overset{o}{\phi}) \, \rho^{p,r} \qquad (V.11)$$

$$i \frac{\partial \rho^{p,a}}{\partial t} = (-\overset{e}{\phi} + \overset{o}{\phi}) \, \rho^{p,a} \; . \qquad (V.12)$$

The L_t symmetry of the Liouville equation is broken if and only if the even part $\overset{e}{\phi}$ may be defined and does not vanish.

We can make some general statements about the structure of ϕ. The hermiticity of L (II.5) together with the star unitarity of Λ (see (V.5)) implies that ϕ has a basic invariance property:

$$[i\phi(-L)]^{\dagger} = i\phi(L) \; . \qquad (V.12')$$

Introducing again the "star" notation (V.6) we may say that $i\phi$ is a <u>star-hermitian operator</u>. Star-hermiticity may be realized in two ways:

a) hermiticity together with positive parity in respect to L inversion;

b) antihermiticity together with negative parity in respect to L inversion.

In this terminology the collision operator $i\psi(0)$ as given in (IV.1) is a star-hermitian operator. Each of its two parts corresponds to one of the two possible realizations of star-hermiticity.

There is one more basic property of $-i\psi(0)$ we mentioned in § 4, it is the fact that its even part is a negative operator. In all cases where we have succeeded to construct the transformation operator either through perturbation expansion or exactly, this is also true for $-i\overset{e}{\phi}$. This is due ultimately to the analytical continuation introduced in the integral representation (III.2). It is also a necessary condition to give a meaning to the distinction between the retarded and the advanced solutions in (V.11), (V.12). Moreover the non-vanishing of $\overset{e}{\phi}$ is connected to the basic dissipativity condition (IV.11). For this reason we shall call <u>dynamic dissipative systems</u> systems for which $\overset{e}{\phi}$ can be explicitly constructed and has the forementioned basic properties.

The evolution of dynamic dissipative systems may be described in terms of a Liapounoff function. Let us consider the quadratic functional <u>in the causal representation</u> (we drop the superscript p,r)

$$\Omega = \text{tr}\, \rho^{\dagger} \rho > 0 \ . \qquad (V.14)$$

It is easy to show using (V.11) that

$$\frac{1}{2} \frac{d\Omega}{dt} = - \text{tr}\, \rho^{\dagger} (i\overset{e}{\phi}) \rho \leq 0 \ . \qquad (V.15)$$

Similarly for the advanced solution Ω can only increase. It is essential to use the causal representation. In the initial representation satisfying the Liouville equation Ω would be a constant. It is the fact that the two representations are linked through a <u>non</u>-unitary transformation which makes the introduction of the Liapounoff function possible.

The existence of the Liapounoff function for dissipative systems is of course of upmost importance for the thermodynamic interpretation of the dynamical evolution. We shall discuss in § 8 its link with entropy.

It is certainly surprising that so different descriptions as given by the initial Liouville equation on one side, the causal equations (V.11), (V.12) on the other may coexist and may even be related through the transformation Λ which guarantees the equivalence of the two descriptions (see (V.4)). The causal representation makes simply <u>explicit</u> properties which appear otherwise in the <u>solution</u> of the mechanical equations of motion. Still this possibility is very unexpected. Let us therefore display the two representations explicitly in a very simple example due to Grécos.

VI. The Grécos Model [50]

Let us consider a model system whose time evolution is described by a Liouville equation (II.3) with

$$L = \nu \delta(\nu-\nu') \quad , \quad -\infty < \nu < +\infty \quad . \tag{VI.1}$$

Similarly the observables A are described by a Heisenberg type equation identical to (II.3) with (VI.1) but with L replaced by $-L$. To L inversion corresponds here the operation

$$\nu \to -\nu \quad . \tag{VI.2}$$

Average values are represented as in (V.4) through

$$\langle A \rangle = \mathrm{tr}\, \rho^{\dagger} A = \int_{-\infty}^{+\infty} d\omega\, \rho^{\dagger}(\omega)\, A(\omega) \quad . \tag{VI.3}$$

In the Schrödinger representation in which ρ is considered as time dependent we have

$$\langle A(t) \rangle = \int_{-\infty}^{+\infty} d\omega\, e^{i\omega t}\, \rho^{\dagger}(\omega, t=0)\, A(\omega) \quad . \tag{VI.4}$$

The main simplifying feature of this model is that L is here an operator (and <u>not</u> a superoperator [33,32,31]). Obviously this model is Lt-invariant, exactly as the original Liouville equation.

Note also that under well known conditions the Riemann-Lebesgue theorem leads to

$$\lim_{t \to \pm\infty} \langle A(t) \rangle = \lim_{t \to \pm\infty} \int_{-\infty}^{+\infty} d\omega\, e^{i\omega t}\, \rho^{\dagger}(\omega, 0)\, A(\omega) \to 0 \quad . \tag{VI.5}$$

There is an approach to "equilibrium" independently of the direction of time.

Suppose that there exists a <u>second</u> description corresponding to (V.11) (we omit the superscript r)

$$i\, \frac{\partial \rho^P}{\partial t} = \phi\, \rho^P \tag{VI.6}$$

with

$$\phi = \omega \delta(\omega - \omega') - i v(\omega)\, v^{*}(\omega') \quad . \tag{VI.7}$$

We impose on $v(\omega)$ the parity condition

$$v(\omega) = \pm v(-\omega) \quad . \tag{VI.8}$$

The operator $-i\phi$ has then all the properties we discussed in §§ 4-5:

1) it is the sum of an even and an odd contribution in respect to the transformation $L \to -L$ (see (VI.2)).

2) it is "star-hermitian" in the sense of (V.13); the even part is represented by a hermitian operator, the odd by an anti-hermitian operator.

3) the even part is a negative operator as for an arbitrary vector

$$(u, -i\overset{e}{\phi}u) = -[\int d\omega |v(\omega)\, u(\omega)|]^2 < 0 \quad . \tag{VI.9}$$

Of course, equation (VI.6) is no more L_t-invariant. The L inversion of (VI.6) would lead to the equation (V.12) satisfied by the advanced solution.

The Liapounoff function Ω as given in (V.14)

$$\Omega = \int d\omega \, |\rho|^2 \geq 0 \tag{VI.10}$$

has indeed the requested properties. It satisfies the inequality

$$\frac{1}{2}\frac{d\Omega}{dt} = -|\int d\omega \, v(\omega)\, \rho(\omega)|^2 \leq 0 \quad . \tag{VI.11}$$

The retarded solution tends to "equilibrium" in the future as

$$\lim_{t \to +\infty} \int d\omega \, |\rho|^2 \to 0 \quad . \tag{VI.12}$$

Similarly the advanced solution tends to equilibrium in the past.

Clearly then the second description (VI.6) is a thermodynamic one in contrast with the first one (VI.1). Now in this model the transformation operator $\Lambda(L)$ may be explicitly calculated [50]. The result is

$$<\nu|\Lambda|\omega> = c(\nu)\{\alpha(\nu)\delta(\omega-\nu) + i\frac{v^*(\omega)v(\nu)}{\omega - \nu}\} \qquad (VI.13)$$

with

$$\alpha(\nu) = 1 - i \int\!\!\!\!\!\!\!\!\!\diagup d\omega \frac{|v(\omega)|^2}{\omega - \nu} \qquad (VI.14)$$

and

$$c(\nu) = \{\alpha^2(\nu) - \pi^2|v(\nu)|^4\}^{1/2} \qquad (VI.15)$$

where the second term in (VI.13) and the integral in (VI.14) are understood as principal parts.

On these expressions one may easily verify that the transformation satisfies (V.5) and is therefore star-unitary. As the result the two descriptions while deeply different from the physical point of view are "equivalent" as average values $<A(t)>$ are preserved. In a similar way we may calculate $\Lambda(-L)$ which would lead to the "advanced solution".

VII. The Friedrichs Model - Dynamics and Stochastic Processes

We have already discussed briefly the Friedrichs model in § 4. Recently F. Henin and M. De Haan [47] have given the explicit form of the dynamic operator ϕ in the causal representation. We want to make here a few remarks concerning the evolution of the diagonal elements of the density matrix as this leads to interesting conclusions concerning the relation between the causal representation of dynamics and stochastic theory.

A remarkable feature of the causal representation is that the diagonal elements of the density matrix satisfy a <u>closed equation</u>. This is expressed by the concept of subdynamics we have introduced recently [51,31]. As has been shown by George the elements of the density matrix may be classified according "the type of correlation" they describe; the elements corresponding to a given type satisfy a separate equation, see for more details [33]. **The** part of $i\phi$ which corresponds to the evolution of the diagonal elements may be written in the matrix form

$$-i\phi = \begin{pmatrix} W_{oo} & W_{ok} \\ W_{ko} & W_{kk'} \end{pmatrix} \qquad (VII.1)$$

where o is the discrete state (in absence of interactions). Exactly as in the case of the Grecos' model the collision operator $i\phi$ is even and real. Therefore according to the general properties enumerated in § 5, $-i\phi$ is then a symmetric, negative operator. We have

$$W_{ok} = W_{ko} , \quad W_{kk'} = W_{k'k} . \qquad (VII.2)$$

The matrix (VII.1) is therefore "doubly stochastic" in the terminology used in the theory of Markoff chains. Moreover the trace of ρ is preserved in time. We have also in agreement with the fact that $-i\phi$ is a negative operator

$$W_{oo} < 0 , \quad W_{kk} < 0 . \qquad (VII.3)$$

If we now limit ourselves to weakly coupled systems that is if we neglect in the W_{ij} contributions of higher order than two in the coupling constant λ (see (IV.6), (IV.7)) we have in addition

$$W_{ok} = W_{ko} > 0 , \quad W_{kk'} = W_{k'k} > 0 . \qquad (VII.4)$$

Then causal dynamics (for the diagonal elements of the density matrix) reduces to a Markoff process. The same is true for N-body situations which we can of course only explore through perturbation methods. For example, it is well known that the quantum mechanical Pauli equation valid for weakly coupled systems corresponds to a Markoff process [29,30]. There is therefore a remarkable link between dynamics and the theory of stochastic processes.

However this link is in general limited to the <u>weakly coupled case</u>. Exact expressions have been derived for the matrix elements W_{ok}, $W_{kk'}$ [47]. Their sign depends on the details of the solution of the dispersion equation (IV.10) as a function of the coupling constant. Similar conclusions have been reached in the case of the study of N-body systems [32].

Therefore the analogy stressed so forcefully by Boltzmann between the dynamical behaviour of large systems and stochastic processes breaks down once one goes beyond the lowest approximation. It is interesting that Boltzmann anticipated somewhat this conclusion as in § 91 of his "Lectures in Gas Theory" he mentions that the application of probability calculations in molecular physics rests <u>both</u> on the great number of molecules and the <u>length</u> of their free path [18]. This second condition corresponds to a requirement of effective "weak coupling".

VIII. Approach to Equilibrium – Statistical Model of Entropy

Let us come back to the Liapounoff function introduced in (V.14). Using the notation

$$\rho_{ij} \equiv \rho_{i-j}\left(\frac{i+j}{2}\right) \equiv \rho_\nu(N) \qquad (VIII.1)$$

we may write Ω more explicitly

$$\Omega = \sum_{ij} \rho^\dagger_{ij} \rho_{ji} = \sum_N \rho_o^2(N) + \sum_{\nu N} |\rho_\nu(N)|^2 \ . \qquad (VIII.2)$$

Ω therefore includes both the diagonal elements and the off-diagonal elements $\rho_\nu(N)$ (the "correlations"). The distribution function is assumed to be normalized

$$\text{tr } \rho = \sum_N \rho_o(N) = 1 \ . \qquad (VIII.3)$$

It is obvious that the minimum of (VIII.1) subject to the constraint (VIII.3) is given by

$$\rho_0(N) = \text{constant independent of } N$$

$$\rho_\nu(N) = 0 \; . \tag{VIII.4}$$

We see that statistical equilibrium has a very simple meaning in the causal formulation of dynamics: all quantum states have the same probability and random phases. Situations which can be described in terms of the diagonal elements $\rho_0(N)$ <u>alone</u> correspond to what George, Rosenfeld and I have called the "macroscopic level of description" in a recent publication [31]. As mentioned above the diagonal elements in the causal representation satisfy a closed equation. As the consequence in such situations we have in addition to the dynamic description a second "reduced" description. This applies to all equilibrium laws as well as to the linear range of non-equilibrium processes [11]. As we mentioned in § 1 the corresponding thermodynamic description is the so-called "local equilibrium description".

We want now to go further and introduce a statistical definition of entropy which would remain valid in the whole range of the causal formulation of dynamics. The expression for the H quantity

$$H - H_{eq} = \frac{1}{2} \log \frac{\operatorname{tr} \rho^\dagger \rho}{(\operatorname{tr} \rho^\dagger \rho)_{eq}} \tag{VIII.5}$$

together with

$$S = -k\, H \tag{VIII.6}$$

satisfies all conditions known to us. Indeed

$$H - H_{eq} \geq 0$$

$$\frac{dH}{dt} \leq 0 \qquad (VIII.7)$$

H is an additive function for independent systems.

Moreover the factor 1/2 in (VIII.5) insures that this expression gives in the neighborhood of equilibrium identical results as the Boltzmann expression [32]. The term $(tr\rho^{+}\rho)_{eq}$ has a simple meaning as at equilibrium in virtue of (VIII.4)

$$(tr\ \rho^{+}\ \rho)_{eq} = \frac{1}{\nu} \qquad (VIII.8)$$

where ν is the common value of $\rho_o(N)$. Because of the normalization (VIII.3), ν is equal to the number of accessible quantum states.

It is of course essential that we express ρ in the causal representation, if not (VIII.5) would be a constant in time. It is also noteworthy that even in this representation the Gibbs formula

$$H = tr\ (\rho\ \ln\rho) \qquad (VIII.9)$$

would not satisfy the second inequality (VIII.7). As a consequence, the property [53] derived from expression (VIII.9)

$$H(AB) = H(A) + H_A(B) \qquad (VIII.10)$$

for two interacting subsystems A, B is not valid for (VIII.5) except in the case where A and B are independent.

While property (VIII.10) is important in the context of information theory [53] we do not know any thermodynamic property which would be violated as the consequence of the non-validity of (VIII.10).

The most important property of H as given in (VIII.5) is that correlations are included as seen from (VIII.2). Let us emphasize that in agreement with (VIII.4) there are no <u>equilibrium</u> correlations in the causal formulation. The equilibrium correlations which would appear in other representations are <u>included</u> in the diagonal elements. The correlations which are part of the entropy (VIII.5) are therefore <u>non-equilibrium</u> correlations which ultimately die out. This separation between "natural" correlations included in the diagonal elements and transient correlations introduced by initial conditions is of course very important.

Let us now illustrate these conclusions by showing that with our new statistical definition of entropy the Loschmidt paradox disappears.

IX. Applications to Negative Time Evolution - The Resolution of Loschmidt's Paradox

Let us now come back to the velocity inversion experiment (see fig. I.) and let us now represent H defined through (VIII.5), as a function of time. Suppose first we start at t = 0 with a state without correlations (ρ_ν = 0) and isotropic velocity distribution. During time 0 till t_o, the diagonal elements tend to their equilibrium distribution. This gives rise to a decrease of H. At time t_o we reverse the velocities. This introduces long range non-thermodynamical correlations $\rho_\nu \neq 0$ (see also the discussion in § 3) [54]. As a consequence (see (VIII.2)) of this external action on the system H increases. The system is ordered. During the time t_o till $2t_o$ the abnormal correlations die progressively out. The details of the calculations are given elsewhere [32,54]. The time variation of H is represented schematically in fig. IX.1. At time $2t_o$ the system is in the same state as at $t = 0$ but with reversed velocities. As at $t = 0$ the system had an isotropic velocity distribution, this is the same state.

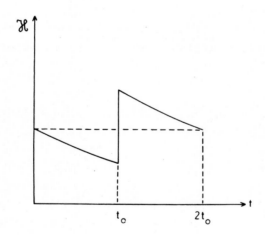

Fig. IX.1 Time variation of H as defined in (VIII.5).

We have thermodynamic behaviour even when Boltzmann's model of entropy would predict "antithermodynamic" behaviour. We can now consider the entropy balance for the thermodynamic cycle over the time interval 0 till $2t_o$

$$\oint dS = \oint d_i S + \oint d_e S = 0 \qquad (IX.1)$$

where $d_i S$ is the entropy production and $d_e S$ the entropy flow. As the entropy production is now positive everywhere in agreement with the second law, the entropy flow has to be negative. This corresponds to a flow of "information" received by the system.

The sign of the entropy flow can be directly calculated from (VIII.5) and we may verify that it is negative [32]. These conclusions remain true even if we start with an arbitrary initial state (no more isotropic in the velocities and which may include initial correlations). The cycle may then for example look as represented on fig. IX.2.

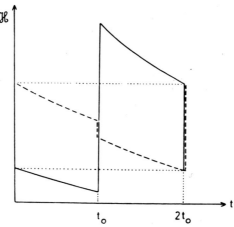

Fig. IX.2

Velocity inversion at time t_0 brings the system <u>nearer</u> to thermodynamic equilibrium (by destroying partially the abnormal correlations which existed at time $t = 0$). We have to introduce a second velocity inversion at time $2t_0$ to bring back the system to the initial state. Whatever the initial conditions the total entropy flow for the complete cycle is again negative [32].

We can therefore calculate the thermodynamic prize for cycles involving "negative time evolution". Of course the "prize of rejuvenation" is a functional of the initial state.

X. Conclusions

When Boltzmann one hundred years ago proposed his kinetic equation, he had obviously in mind some scheme of successive approximations. In the case of N-body systems, we can still only proceed by approximations, but we can now study simpler dynamic systems, of the type I mentioned in this report. We can show in this way that the physical concepts introduced by Boltzmann such as the collision operator, are amenable to an <u>exact</u> treatment. The basic problem is to reformulate the laws of dynamics in a way which displays explicitly in the differential equations the terms which are even in L and are responsible for the loss of the Lt-invariance. Once this is done, a Liapounoff function can easily be identified and its relation with entropy discussed.

Therefore, instead of Boltzmann's scheme mentioned in § 1,

Dynamics → Stochastic Theory → Entropy,

we have the scheme:

Dynamics of dissipative systems → Liapounoff function
→ Entropy

In special cases, the dynamics of dissipative systems leads to a Markoff process (see § 7). There are then two statistical models for non-equilibrium entropy. The use of one or the other is a question of commodity. No contradictions can arise as there may exist more than one Liapounoff function.

It is precisely one of the most interesting aspects of our approach that it eliminates any appeal to stochastic processes. The fact that we deal with ensembles is unavoidable in the quantum mechanics and is also natural in classical dynamics as the initial conditions are never exactly known. This has not to be confused with a probabilistic description.

We therefore hope that our approach brings us nearer to the goal formulated by Boltzmann, to give the "mechanical interpretation" of the second law.

We have described in some detail the Loschmidt paradox. It gives us a striking example of the role "non-thermodynamic" correlations play. The beautiful discussion of Loschmidt's paradox as given by Kac [39] and based on the Ehrenfest model [15], had already solved the logical problem involved. But it could of course not show how the reconciliation of time reversibility and irreversible behaviour was possible in classical or quantum mechanics.

We hope to apply our statistical model of entropy to other situations in which "non thermodynamic correlations" are important. This includes lasers, stability, problems in plasmas and so on.

Let us close with a few remarks on the relation between thermodynamic irreversibility and causality. The Liapounoff function we have introduced exists only if the causal representation leads to <u>two</u> formulations of dynamics (see equation (V.11)), which are Lt conjugates. This is an unavoidable consequence of the Lt-invariance of the initial formulation. There is however no ambiguity introduced in this way. On the contrary only the retarded solution predicting thermodynamic equilibrium in the future has to be used in conjunction with initial value problems (see § 2) and it is for these situations that we derive for dissipative systems an increase of entropy with increasing time. Let us repeat a statement made forcefully by Rosenfeld: "<u>The inclusion of a specification of the conditions of observation into the account of the phenomena is ... an indispensable part of their objective description</u>". It is this account which introduces the distinction between retarded and advanced solutions. As we suppose that the observer may make this distinction it belongs to the type of primitive concepts which we have to introduce in our theoretical scheme at the start (Rosenfeld [31], see also [45]), and which permits us to select the retarded form of causal dynamics. But then using in addition the laws of mechanics and the conditions of dissipativity we may go much further: we may now measure the irreversibility content of a time interval, through the corresponding entropy production.

In the first stage of dynamics, time was associated with motion, with displacement of wave packets. Causal dynamics permits to reach the next level in the description of time, associated now with irreversibility. Time's arrow, the entropy, according to Eddington's famous expression [55], is now explicitly displayed.

The concept of time has always been the central problem of natural philosophy. These new developments therefore lead to the possibility of a new approach to the relation between science and philosophy. This question is discussed more in detail in a recent publication of the author [56].

As well known, Boltzmann was deeply interested in problems of philosophy and epistemology. I am therefore sure that he would have been pleased to see that the ideas which he originated may indeed bring closer together human activities so different at first as science and philosophy but which both try to give us an account of the world in which he could recognize us ourselves.

Acknowledgments

The ideas summarized in this report are the outcome of years of work by the members of our group in Brussels and Austin. The specific aspects considered here have been developed and discussed in papers to appear shortly in collaboration with Professors C. George, A. Grécos, F. Henin and L. Rosenfeld.

The final version of this report was prepared during a stay of the author at General Motor's Technical Center. I want to thank Dr. R. Herman for stimulating discussions.

This research is being supported by the Fonds de la Recherche Collective (Belgium) and the R. Welch Foundation (Houston, Texas, U.S.A.).

References

[1] L. Boltzmann, Weitere Studien über das Wärmegleichgewicht unter Gasmolekülen, Wien. Ber. <u>66</u> 275 (1872). see Wissenschaftliche Abhandlungen, Vol. 1., Verlag von Johann Ambrosius Barth, Leipzig, 1909.

[2] L. Boltzmann, Mechanische Bedeutung des Zweiten Hauptsatzes, Wien. Ber. <u>53</u>, 199 (1866).

[3] L. Boltzmann, Der Zweite Hauptsatz der Mechanischen Wärmetheorie in Populare Schriften, Verlag von Johann Ambrosius Barth, Leipzig, 1919, p. 25.

[4] H. Bergson, "Evolution Créatrice", Eds. du Centenaire, Presses Universitaires de France, Paris 1963.

[5] O. Spengler, Der Untergang des Abendlandes,

[6] For a modern survey, see L. Brillouin, Tensors, Dover New York, 1946.

[7] A. Kronig, Ann. Physik, <u>99</u>, 315 (1856).

[8] R. Clausius, Ann. Physik, <u>100</u>, 353 (1857).

[9] J.C. Maxwell, Phil. Mag., <u>19</u>, 19 (1860).

[10] L. Boltzmann, Wien. Ber., <u>58</u>, 517 (1869).

[11] see P. Glansdorff & I. Prigogine, Stability, Structure and Fluctuations, Wiley-Interscience, 1971, french edition, Masson, Paris, 1971.

[12] J.O. Hirschfelder, C.F. Curtiss & R.B. Bird, The Molecular Theory of Gases and Liquids, Wiley, New York (1959).

[13] B.J. Alder & T.E. Wainwright, J. Chem. Phys., <u>33</u>, 1434 (1960).

[14] A. Bellemans and J. Orban, Phys. Letters <u>24A</u>, 620 (1967).

[15] see P.T. Ehrenfest, Begriffliche Grundlagen der Statistischen Auffassung der Mechanik, Encycl. Math. Wiss. 4, 4 (1911).
[16] E. Zermelo, Ann. Physik, 57, 485 (1896); 59, 793 (1896).
[17] J. Loschmidt, Wien. Ber., 73, 139 (1876).
[18] see specially L. Boltzmann, Lectures on Gas Theory, §§ 87-91, An English translation by St. G. Brush is available, University of California Press, 1964.
[19] J.C. Powless and P. Mansfield, Phys. Letters, 2, 58 (1962).
[20] P. Mansfield, Phys. Rev., 137, 1961 (1965).
[21] W.K. Rhim, A. Pines and J.S. Waugh, Phys. Rev. B3, 684 (1971).
[22] D. Walgraef and P. Brockmans, Physica 59, 37 (1972).
[23] L. Boltzmann, Wien. Ber., 75, 62 (1877).
[24] S.H. Burbury, Nature 51, 78 (1894).
[25] M. Planck, Thermodynamik, Berlin & Leipzig, De Gruyter, 1930, p.83.
[26] S.R. de Groot and P. Mazur, Non Equilibrium Thermodynamics, North Holland Publ. Co., Amsterdam, 1962.
[27] I. Prigogine, Thermodynamics of Irreversible Processes, 3rd. edition, Wiley-Interscience, New York, 1967.
[28] R. Graham and H. Haken, Phys. Lett. 29A, 530 (1969).
[29] R. Balescu, Statistical Mechanics of Charged Particles, Wiley-Interscience, New York, 1963.
[30] I. Prigogine, Non Equilibrium Statistical Mechanics, Wiley-Interscience, New York, 1962.
[31] Cl. George, I. Prigogine and L. Rosenfeld, Koningl. Dansk. Vid. Mat-Phys. Medd. 38, 12, (1972).
[32] I. Prigogine, Cl. George, F. Henin and L. Rosenfeld, to appear Proc. Roy. Swedish Acad., Stockholm, 1972.

[33] Cl. George, Physica, to appear 1972.
[34] M. Reichenbach, The Direction of Time, Univ. California Press, Berkeley and Los Angeles, 1956.
[35] O. Costa de Beauregard, Information and Irreversibility Problems, in Time in Science and Philosophy, ed. by J. Zeman, Czechoslovak Academy of Sciences, Prague 1971, p. 11.
[36] M. Baus, Acad. Roy. Belg. Bull. Cl. Sci., 53, 1291, 1332, 1352, (1967).
[37] L. Lanz and L.A. Lugiato, Physica 44, 532 (1969).
[38] A. Grécos, Physica 51, 50 (1970).
[39] see M. Kac, Probability and related topics in physical sciences, Interscience, New York, 1959.
[40] see specially the work of P. Résibois, M. De Leener and others in Physica and Phys. Rev. during the years 1966, 1969 and 1971.
[41] I. Prigogine and P. Résibois, Atti di Simposia Lagrangiano, Accademia delle Science, Torino, 1964.
[42] R. Balescu, Physica 36, 433 (1967).
[43] A. Einstein and W. Ritz, Phys. Zs., 10, 323 (1909).
[44] G. Stey, to appear Physica 1972.
[45] A. Grécos and I. Prigogine, Physica 59, 77 (1972).
[46] A. Grécos and I. Prigogine, P.N.A.S. 69, 1629 (1972).
[47] F. Henin and M. De Haan, papers to appear in Physica and Acad. Roy. Belg., Bull. Cl. Sc. 1972.
[48] I. Prigogine and G. Severne, Physica 32, 1376 (1966), G. Severne, Physica 61, 307 (1972).
[49] I. Prigogine and A. Grécos, Volume in honour of H. Fröhlich, ed. by H. Haken, to appear 1973.
[50] A. Grécos, private communication
[51] I. Prigogine, Cl. George and F. Henin, Physica 45, 418 (1969).

[52] R. Balescu and J. Wallenborn, Physica 54, 477 (1971).
[53] A.I. Khinchine, Mathematical Foundations of Informations Theory, Dover Publ. Inc., New York, 1957.
[54] Duk In Choi, Acad. Roy. Belg. Bull. Cl. Sci. 47, 1054 (1971)
[55] A.S. Eddington, The Nature of the Physical World, Cambridge University Press, 1929.
[56] I. Prigogine, La Naissance du temps, Communication at the "Académie Internationale de Philosophie des Sciences", Drongen-Gand, September 1972 and also Acad. Roy. Belg. Bull. Cl. Sci. to appear 1972.

A REVIEW OF COMPUTER STUDIES IN THE KINETIC THEORY OF FLUIDS[*]

W.W. WOOD

Los Alamos Scientific Laboratory, University of California
Los Alamos, New Mexico 87544

ABSTRACT

The role of computer studies in the kinetic theory of fluids is briefly reviewed, with emphasis upon the results for hard-core systems. After a short discussion of the advantages and limitations of computer methods, we first consider computer studies directly related to the Boltzmann equation, e.g., calculations of the one-particle distribution function $f(\underline{r},\underline{v},t)$, by the methods of molecular dynamics and by numerical solution of the Boltzmann equation itself. There then follows a review of the molecular dynamics studies of time-correlation functions and the associated transport coefficients.

[*]) Work performed under the auspices of the U.S. Atomic Energy Commission.

I. Introduction

Since the development of the Monte Carlo method for equilibrium classical statistical mechanics by Metropolis, Rosenbluth, Rosenbluth, Teller, and Teller in 1953 and the molecular dynamics method by ALDER and WAINWRIGHT [1958] in 1956, computer methods have played an important role in the development of both equilibrium and nonequilibrium statistical mechanics. This review is limited to the nonequilibrium results relating to the kinetic theory of classical fluids, and therefore to a greater or lesser degree to the Boltzmann equation, whose centenary is the occasion for this meeting. Accordingly we will be concerned for the most part with the molecular dynamics method, although we shall have some occasion to mention Monte Carlo methods also.

In essence the molecular dynamics method is simply the application of a suitable numerical integration method to the Newtonian differential equations of motion of the N-body system, starting from given initial positions and velocities. Equilibrium properties are calculated as time averages of mechanical quantities after transients associated with the initial state have decayed. And indeed, most "nonequilibrium" properties are actually calculated in the form of equilibrium time-correlation functions, although we shall also describe several examples of results relating to the approach to equilibrium. Details regarding the method for hard-core systems are given by ALDER and WAINWRIGHT [1959], and for systems with continuous pair interactions by RAHMAN [1964] and VERLET [1967].

There are several reasons for the popularity of data from computer studies for use in comparison with theoretical results. Subject to the limitations to be discussed below, the method is exact for the assumed interactions between the particles. Thus one avoids the uncertainties associated with the interaction law for real molecules, and the resultant possible confusion between the effects of statistical mechanical approximations and of assumptions regarding the intermolecular forces. In addition one can specify some simple form of intermolecular forces, e.g., hard spheres, for which the theoretical analysis may be more easily carried out. Finally, but very importantly, the computer studies often yield information at a microscopic level, which it would be difficult or impossible to obtain in a physical experiment.

The principal limitation of the computer methods is the small size of the systems which can be studied, due primarily to the speed of even the fastest existing computers, and secondarily to their storage limitations. Typical values of the largest feasible number N of particles are presently 1000 to 2000, although systems as large as 8000 or more have been used in some of the work on Lorentz models. Periodic boundary conditions are usually used to minimize the effects of the small size. The limit on N in turn implies a lower limit on the densities which can be studied, since one clearly should have the mean free path much less than the period of the system. This restriction is not severe when several hundred or more particles are used. But in the calculation of time correlation functions, which is one of the most active current applications of molecular dynamics to kinetic theory problems, the finite size of the system seriously limits the time interval over which the functions can be calculated. At times somewhat less than the time required for traversal of a sound wave across the period of the system, irregular oscillations in the time correlation function appear and slowly increase in magnitude; an example is given below. For these reasons, and others, the question naturally arises as to how closely systems of a few hundred or a few thousand particles can approximate macroscopic behavior. For most properties of interest the correction can be assumed to be of $O(N^{-1})$, but usually the coefficient is unknown. Thus the usual procedure is to verify that the computer results are not sensitive to the value of N used.

The accuracy with which the equations of motion of the finite system are integrated depends upon the finite-difference approximation to the differential equations of motion, and upon the accumulation of round-off errors, i.e., the number of significant digits carried in the machine. For continuous intermolecular forces the first of these sources of error may dominate, while for hard core systems only the last is present. Indications of the accuracy of a given machine code can be obtained by monitoring the calculated values of dynamical invariants such as the energy, and by checking the dynamical reversibility of the machine code by reversing the particle velocities at some point along the trajectory. A few examples will be discussed later in conjuction with specific investigations. As has been emphasized by BERNE and FORSTER [1971], there is often insufficient evidence of control of these sources of error in the molecular dynamics literature.

Under some circumstances the molecular dynamics method, as well as the Monte Carlo method for equilibrium properties, may fail to converge satisfactorily toward equilibrium behavior. It is believed that in the work under review here, such errors are insignificant.

Finally, the calculated results contain statistical errors associated with averages taken over a finite number of dynamical trajectories of the N-body system, each of which yields time averages extending over a finite interval of time. Simple statistical techniques yield estimates of the reliability of any particular datum, but there may remain residual effects such as severe serial correlation (in the sense of mathematical statistics) of the errors in estimating a time-correlation function as a function of time.

Due to limitations of time, space, the availability of detailed results from modern kinetic theory, my personal familiarity and experience, and the relevance of the work to the theory of gases as distinct from liquids, this review will be limited to computer studies on hard-core systems. Regretfully, this results in only occasional mention of the excellent work of Rahman, of Verlet, of Berne, and their respective collaborators, on systems with continuous forces. Fortunately this work is well covered in the useful review by BERNE and FORSTER [1971]. Also omitted is any discussion of the now voluminous literature on computer studies of plasma systems. Within these limitations, the emphasis will be on the discussion of computer results viewed as a body of "experimental" data on systems with known force laws, and on comparisons with theoretical results where available.

Most of the computer studies have been directed toward the determination of the dependence of various properties, e.g., transport coefficients, on the density, and accordingly have been carried out at relatively high densities where departures from strictly Boltzmann behavior become important. For this reason it is natural, particularly for hard-core systems, to compare the results in the first instance with the predictions of Enskog's modification of the Boltzmann equation. In particular we will often use as a reduced time variable the number of collisions per particle (Alder and Wainwright use the term "mean collisions")

$$s = t/t_o'$$

where $1/t_o'$ is the actual collision rate, related to the low-density "Boltzmann" collision rate $1/t_o$ through Enskog's factor χ: $1/t_o' = \chi/t_o$. Throughout the paper V_o will denote the close-packed volume (area, for two dimensional systems) occupied by the N-body system. Thus $V_o = \frac{1}{2}\sqrt{3}\, N\sigma^2$ for hard disks, $\frac{1}{2}\sqrt{2}\, N\sigma^3$ for hard spheres, and $N\sigma^2$ for parallel squares, with σ being the hard disk or sphere diameter, or the edge of a square.

II. Computer Studies Directly Related to the Boltzmann Equation

A. Molecular Dynamics Calculations of $f(\underline{r},\underline{v},t)$.

In view of the occasion of this conference it seems fitting to make some mention of computer studies which have been carried out at the level of description of the Boltzmann equation, i.e., the one-particle distribution function $f(\underline{r},\underline{v},t)$, even though they are not extensive and are mostly qualitative in character.

Consider first molecular dynamics calculations of $f(\underline{r},\underline{v},t)$ for a given initial distribution $f(\underline{r},\underline{v},0)$ which might in principle be compared with the corresponding solution of the Boltzmann equation to investigate the latter's

(generally undoubted) correctness at low densities, and
its deviation at higher densities. So far as I know, only
one rather qualitative study of this kind has been carried
out, by ALDER and WAINWRIGHT [1958]. The Boltzmann equation was solved (numerically) for the initial distribution
$f(\underline{r},\underline{v},0) = F(v)$, i.e., a spatially homogeneous distribution,
uniform also in the direction of the velocity. F(v) was
taken to be somewhat more peaked than the Maxwell distribution with the same mean square velocity. The molecular
dynamics velocity distribution[*] was computed from a trajec-

---

[*] The function $f(\underline{r},\underline{v},t)$ at any instant of time along a
single trajectory of a finite system is a delta function
in positions and velocities. For the calculation described,
it was converted into a histogram velocity distribution
by counting the velocity magnitudes falling into specified
velocity intervals, ignoring the positions. In the subsequent calculation of the Boltzmann function H(t), spatial
homogeneity was <u>assumed</u>, and the velocity integral replaced
by a sum. The work described was evidently based on interpreting $f(\underline{r},\underline{v},t)$ as the actual number density of particles
in a single experiment, for which the Boltzmann equation
is expected to apply only in the limit of a large number
of particles. For small systems it might be preferable to
adopt the alternative view of $f(\underline{r},\underline{v},t)$ as the expected
number density averaged over a large number of experiments,
in each of which the initial phase is sampled from an N-
particle distribution function chosen so that integration
over the phases of all but one of the particles gives
$N\sigma^{d-1}$ = constant and V constant; d denotes the number of
dimensions) might also be worthwhile. See GRAD [1958] for
a discussion of these points.

tory started from a lattice initial configuration and velocities of equal magnitudes and random directions. These two distribution functions were compared by shifting the molecular dynamics time scale so as to match the height of the peak of the molecular dynamics distribution function with that of the initial peak of the distribution function obtained from the Boltzmann equation, see Fig. 1. Thus only a qualitative agreement between the Enskog theory and the machine results was to be expected. They also demonstrated a rough agreement between their "equilibrium" velocity distribution function and the Maxwell distribution. The work is mentioned primarily to point out that it could now be improved upon in a number of ways, if problems of this kind were to become theoretically interesting.

Having obtained $f(\underline{r},\underline{v},t)$ one can of course compute the Boltzmann H-function

$$H(t) = \int d\underline{r} \int d\underline{v}\, f(\underline{r},\underline{v},t)\, \ln f(\underline{r},\underline{v},t)\ .$$

ALDER and WAINWRIGHT [1958] did so, observing in a qualitative way at all densities the expected decay toward equilibrium. Similar calculations have been done several times since≠, among which is an interesting one by ORBAN and

---

≠ A calculation similar to the one discussed in the text was done by A. AHARONY [1971], by perturbing the hard disk collision law at each collision in such a way that the equations of motion were not reversible in time. ORBAN and BELLEMANS [1969] have reported another calculation of the Boltzmann function H(t) in conjunction with an investigation of the pair contribution to a cluster expansion of $H(t) - N^{-1} \int d\underline{r}^N$. $\int d\underline{v}^N f_N \ln f_N$ where N is the N-particle distribution function.

BELLEMANS [1967]. They studied a system of 100 hard disks at reduced area $V/V_0 = 25$ and the initial condition

$$f(\underline{r},\underline{v},0) = \delta(v-v_0) \sum_{i=1}^{N} \delta(\underline{r}-\underline{R}_i)$$

corresponding to all particles on the sites $\underline{R}_i$ of a square lattice and randomly oriented velocities all of the same magnitude, although again only a single trajectory was calculated - i.e., no averaging over more than one set of initial values was done. The results are shown in Fig. 2, and are of interest particularly in connection with the effects of accumulated round-off error on the dynamical reversibility of the calculation. In Fig. 2a we see that when the particle velocities are reversed after 50 total collisions, H(t) retraces its steps quite closely; but when the inversion is postponed to 100 total collisions, deviations become detectable on the scale of the figure, and H(t) does not quite recover its original value before starting to decrease. Reversibility does not, of course, imply accuracy, but it seems reasonable to assume that irreversibility implies inaccuracy; i.e., it seems likely that the calculated phase corresponding to the open circle at 200 total collisions is beginning to deviate significantly from the correct one. The smaller maximum values of H(t) attained after velocity inversion with introduction of additional random errors are asserted to lend some credence to a conjecture by BALESCU [1967] that most initial distributions in the neighborhood of a "bad" one [i.e., one for which H(t) increases appreciably with t] are "less bad".

B. Computer Solutions of the Boltzmann Equation.

It seems desirable to make some mention of computer methods for solution of the non-linear Boltzmann equation, although it is not an area in which I have any experience. As a single example we consider the solution by HICKS, YEN, and REILLY [1972] for steady shock-wave boundary conditions in a hard-sphere gas, using a Monte Carlo integration technique to evaluate the 5-fold collision integrals. This is the same problem which Professor Foch discussed, from the standpoint of the successive approximations of the Chapman-Enskog development. Their results for shock thickness as a function of Mach number are shown in Fig. 3. The slow convergence toward Navier-Stokes behavior with decreasing Mach number, Fig. 3a, is particularly interesting. Even at quite low Mach numbers they find considerable departures from linear relationships between fluxes and gradients, as shown in Fig. 4 for the "effective reduced viscosity" μ_{rel}, defined as the ratio of the stress to the velocity gradient in units of the Navier-Stokes viscosity coefficient at the local temperature. The deviation from Navier-Stokes behavior at the downstream (hot) boundary, $\hat{n} = 1$ in the figure, is quite surprising.

In the strong shock region, Fig. 3b, the Monte Carlo results for shock thickness are in approximate statistical agreement with the Mott-Smith approximation, but there are significant differences in the details of the shock structure, including the "effective reduced viscosity" mentioned previously.

The investigation seems to be carefully done with respect to the details of the Monte Carlo calculation and consideration of various possible sources of error. I would have appreciated more discussion of the difference approximation (9 to 17 points within the shock) to the differential Boltzmann equation, especially in the vicinity of the boundary points, where the collision term vanishes as equilibrium is approached.

I am indebted to Professor Uhlenbeck for the remark that existence and uniqueness of the solution of the Boltzmann equation for shock-wave boundary conditions do not seem to have been proven.

Mention should also be made of the "direct simulation" method of BIRD [1970], whose sampling approximations appear to be related to those in the elementary derivation of the Boltzmann equation.

It is interesting to note that such numerical solutions of the non-linear Boltzmann equation are beginning to be used as a standard of comparison in evaluating approximate treatments of shock structure; see, for example, SEGAL and FERZIGER [1972].

III. Computer Studies of Time-Correlation Functions and Transport Coefficients

The relationship between time-correlation functions[*]

---

[*] The term generically denotes quantities of the form $<f(t')g(t'+\tau)>$, in which $f(t)$ and $g(t)$ are functions of the phase at time t of the system of interest. The angular brackets denote an equilibrium ensemble average, so that the quantity is actually independent of t', which is therefore often set equal to zero.

and transport coefficients is discussed in the reviews by ZWANZIG [1965] and by BERNE and FORSTER [1971], and this connection, along with that between time-correlation functions and neutron scattering and light scattering [BERNE and FORSTER, 1971] results in these functions being the principal focus of attention in recent molecular dynamics calculations. As mentioned in the Introduction, we have chosen to concentrate here on the results for hard-core systems, and so omit much interesting work [see BERNE and FORSTER, 1971].

Much theoretical attention [ALDER and WAINWRIGHT, 1970; WAINWRIGHT, ALDER, and GASS, 1971; DORFMAN and COHEN, 1970, 1972; ERNST, HAUGE, and van LEEUWEN, 1970, 1971a, b; ZWANZIG and BIXON, 1970; KAWASAKI, 1970, 1971; DUFTY, 1972] has recently been given to the long-time behavior of the time-correlation functions associated with the transport coefficients, in connection with Alder and Wainwright's see below for specific references observation that at least some of them apparently decay like $t^{-d/2}$, where d denotes the number of dimension and that in any case they show deviations from exponential decay at relatively long times. This aspect of the subject will accordingly be the focus of most of our attention.

A. Velocity Autocorrelation Function and Self-Diffusion Coefficient.

The normalized velocity autocorrelation function

$$\rho_D(t) = (N\beta md)^{-1} \sum_{i=1}^{N} <\underline{v}_i(0)\cdot\underline{v}_i(t)> \qquad (1)$$

and the self-diffusion coefficient D are related by

$$D = \beta m \int_0^\infty dt\, \rho_D(t) \qquad (2)$$

where $\underline{v}_i(t)$ is the velocity of particle i at time t, $\beta = 1/k_B T$, m is the particle mass, and the angular brackets denote an equilibrium average. Eq.(2) can be shown [ERNST, 1967] to be the limit of the time-correlation function expression for the binary mutual diffusion coefficient as the two species become mechanically identical. The Einstein mean-square displacement law

$$\Delta_D(t) = N^{-1} \sum_{i=1}^{N} \langle \Delta \underline{r}_i(t)^2 \rangle \sim 2dDt, \qquad (3)$$

with $\Delta \underline{r}_i(t) = \underline{r}_i(t) - \underline{r}_i(0)$ and $\underline{r}_i(t)$ denoting the position of particle i at time t, can be obtained using Eq.(2), providing the integral is convergent.

As already mentioned, we will concentrate mainly on the long-time behavior of $\rho_D(t)$ and its implications for D, but it may be worth mentioning that for hard-core systems the Boltzmann-Enskog theory gives the correct value of the initial slope $[\partial \rho_D(t)/\partial t]_{t=0}$ of the velocity autocorrelation function.

1. Hard Disks and Spheres

To the accuracy which was practicable at the time, ALDER and WAINWRIGHT's [1958] early calculations of $\rho_D(t)$ for hard spheres suggested an exponential decay at low densities. At higher densities, beginning at about $V/V_0 = 1.767$, they found that $\rho_D(t)$ became negative after several collisions per particle, and attributed this to "backscattering" due to correlated collisions involving more than two particles[*].

Definite data for hard spheres showing positive deviations from exponential behavior in $\rho_D(t)$ at low densities[##] ($V/V_o \gtreqqless 2$) and at times long (ten to thirty collisions per particle) compared to the time constant (1.5 collisions per particle) of the initial exponential decay, were not reported until almost 10 years later [ALDER and WAINWRIGHT, 1967]. No explicit form was given for the time dependence of the decay. The long time scale of the positive structure suggested a hydrodynamic analogy with the flow of a small fluid element initially set into motion relative to the surrounding fluid. This in turn led ALDER and WAINWRIGHT [1969] to examine the correlation

---

[#] One of these early calculations ($V/V_o = 1.767$) seemed to suggest that $\rho_D(t)$ returned to positive values in the interval from 8 to 12 collisions per particle, and then either decayed toward zero through positive values, or oscillated around zero with diminishing amplitude. Subsequent more accurate data (references given in the text) show no late-time positive phase at this density, and in fact there are no reliable published molecular dynamics data showing a second positive phase in $\rho_D(t)$ at any density for either hard spheres or disks. However, ALDER and WAINWRIGHT [1970, 1972] have observed such behavior in unpublished data for hard disks at $V/V_o = 1.4$, and in other cases it might be obscured by statistical fluctuations or appear only at times longer than can presently be calculated.

[##] At higher densities the curves show the negative backscattering behavior from the crossover (5 to 8 collisions per particle, depending on the density) out to the maximum time calculated (roughly 30 collisions per particle).

between the velocity of a typical hard-disk particle and
the velocities of its neighbors. They found the double-
vortex pattern shown in Fig. 5. Note the suggestion that
a fraction of the momentum initially transferred from the
moving test particle to the fluid ahead of it is returned
to it from behind after a number of intervening collisions.
This highly correlated motion extending over 25 or more
particles, according to Fig. 5, suggested a serious attempt
at a hydrodynamic description. Using a machine program to
solve the Navier-Stokes equations for the motion of such
a small fluid element, as well as an analytical treatment
of the hydrodynamic flow at long times, ALDER and WAINWRIGHT
[1970] obtained flow patterns in close agreement with Fig. 5,
and the asymptotic decay law

$$\rho_D(t) \sim \alpha_D \, s^{-d/2} \quad . \tag{4}$$

The decay law given by Eq.(4) is also obtained by
the other previously mentioned theoretical treatments, in
particular by the kinetic theory analysis of DORFMAN and
COHEN [1970, 1972] with reservations as to whether it represents the true long-time behavior or an intermediate
region, and by the hydrodynamical theory of ERNST, HAUGE,
and van LEEUWEN [1970, 1971a, 1971b] as the presumed long-
time limit.

The agreement of the molecular dynamics results
for hard disks with this decay law is shown in Fig. 6. The
data for the larger system ($N = 984$) do indeed appear to
be approximately linear in s^{-1}, from roughly 10 collisions
per particle out to the largest times calculated, $s \approx 22$
for $V/V_0 = 5$ and $s \approx 29$ for $V/V_0 = 2$. In addition, the
values of the coefficient α_D obtained by empirically fit-

ting the molecular dynamics results are in good agreement with the theoretical predictions; see the figure in Professor Dorfman's paper.

The only comparison of Eq.(4) with molecular dynamics results for hard spheres so far published is shown in Fig. 7. The apparent agreement at the longest times (maximum s = 25) between the molecular dynamics data and the hydrodynamic points shown in the figure is slightly misleading, in that the hydrodynamic points do not include [ALDER and WAINWRIGHT, 1972] the correction factor

$$F = \left[\frac{\eta/\eta_o}{\eta/\eta_o + 6D_E/5D_o}\right]^{3/2}$$

(in the notation of ALDER and WAINWRIGHT [1970] and ALDER, GASS, and WAINWRIGHT [1970]) for diffusive motion of the test particle. ALDER, GASS, and WAINWRIGHT [1970] give this factor for hard spheres at $V/V_o = 3$ as $F = 0.761$. This correction factor is needed in order to produce agreement of the Alder and Wainwright hydrodynamic calculation of α_D with the other previously cited theories. Thus the actual theoretically predicted asymptotic behavior is given by a line parallel to the one shown in the figure, but displaced downward from it by an increment equal to $|\log F|$. This displacement is fairly small, but significant, on the scale of the figure. On the basis of the observed agreement between theory and molecular dynamics in two dimensions, one is perhaps inclined to believe the theory in three dimensions, and accordingly to attribute the observed discrepancy to times of the order of s = 10 to 25 being insufficient for attainment of asymptotic behavior in three dimensions at this density. In any case there is for hard spheres no "experimental" confirmation of Eq.(4) comparable in precision to that which has been obtained for hard disks. Assuming the validity of the preceding interpretation of the existing results, the verification of Eq.(4) for hard spheres appears to be doubly difficult; (1) calculations are needed at longer times than for hard disks; and (2) for a fixed value of the time, one presumably needs to use larger systems in the three-dimensional case, in order to minimize finite system effects.

If Eq.(4) correctly represents the asymptotic behavior of $\rho_D(t)$, then in two dimensions the self-diffusion coefficient obtained from the autocorrelation formula, Eq.(2), is infinite. The same $t^{-d/2}$ behavior is found theoretically for the time-correlation functions relevant to the coefficients of viscosity and thermal conduction, with some support from molecular dynamics results for hard disks, as will be seen below, so that these coefficients may also be infinite in two dimensions. The theoretical treatments suggest that the same behavior may apply to more general force laws. If such is the case, then the conventional Navier-Stokes hydrodynamics description in two dimensions, and the Burnett corrections in three dimensions, are invalid. See the paper of Professor Dorfman at this conference as well as the already-cited papers for further discussion of these and related points[#].

Recently we [ERPENBECK and WOOD, 1972] have independently calculated $\rho_D(t)$ for hard disks at $V/V_o = 2$,

---

[#] As discussed by WAINWRIGHT, ALDER, and GASS [1971], the "self-consistent" decay law $\rho_D(t) \sim \alpha_D(s\sqrt{\ln s})^{-1}$, proposed by them and by several others (see ERNST, HAUGE, and van LEEUWEN, [1971b] for references), is not distinguishable from Eq.(4) as far as the existing molecular dynamics data are concerned.

using a combination of the Monte Carlo method (for averaging over the initial phase) and molecular dynamics (for the dynamical part of the calculation, as well as for time averaging), obtaining results that confirm those of Alder and Wainwright, as shown in Fig. 8. The largest system studied by ALDER and WAINWRIGHT [1970] was 986 disks, for which their data (see Fig. 6) extend out to $s \approx 29$. Our results display small oscillations which become noticeable at somewhat earlier times, and appear to slowly increase in magnitude with increasing s, with no obviously unique choice of a cut-off time. We did not calculate beyond the maximum time ($s \approx 53$) shown in the figure. Results for smaller systems suggest that the disturbances become relatively large at a time approximately equal to the time required for a sound wave to traverse a period of the system ($s \approx 61$) for the 1512 particle system. We interpret our data to suggest strongly that $\rho_D(t)$ remains linear in t^{-1} at least out to 50 collisions per particle.

For a normal diffusion process, the diffusion coefficient is of course equal to the ratio of the diffusion current to the negative concentration gradient. It has been questioned whether the velocity autocorrelation formula, Eq.(2), gives the correct result for this ratio. Does the long tail contribute? We owe to Professor E.G.D. Cohen the suggestion that it should be possible to study this question in a computer calculation. It lead us to consider the following "experiment" [ERPENBECK and WOOD, 1972]. Construct the initial state by first sampling a phase $(\underline{r}^N, \underline{v}^N)$ from the canonical probability density for a periodic system of N particles in the volume V. Then tag the particles according to their x-coordinates x_i and a probability $p(x_i) = n^{-1} n(x_i, 0)$, each particle being tagged

with this probability independently of the others; here
n = N/V. The tagging process produces an initial phase
in which the expectation value of the number density of
tagged particles at position x is n(x,0). The expectation
value of the number density of tagged particles at position x and time t in the above "experiment" is given by

$$n(x,t) = \int_{-L/2}^{L/2} d\xi \, n(\xi,0) \sum_{\nu=-\infty}^{\infty} \phi(x + \nu L - \xi, t) \qquad (5)$$

where $L = V^{1/d}$ and

$$\phi(\Delta x, t) = N^{-1} \sum_{i=1}^{N} <\delta[\Delta x_i(t) - x]> \qquad (6)$$

is the probability density of x-displacements at time t
(essentially it is the van Hove function). Considering
for convenience an initial state with a single Fourier
component of wavelength L/k, $n(x,0) = n_o + n_k \cos(2\pi k x/L)$
and calculating from (5) the diffusion coefficient $D_k(t)$,
defined as the ratio of the current of tagged particles
to the negative gradient of n(x,t), we obtain

$$D_k(t) = \frac{L}{2\pi k} \frac{<v_{1x}(t) \sin[2\pi k \Delta x_1(t)/L]>}{<\cos[2\pi k \Delta x_1(t)/L]>} \qquad (7)$$

where $v_{1x}(t)$ is the x-component of the velocity of particle 1 at time t. Note that $D_k(t)$ is independent of x
and of n_k, the amplitude of the initial disturbance, and
therefore it is also independent of the magnitude of the
initial gradients. The denominator of Eq.(7) is essentially the intermediate scattering function, and the numerator
is the Fourier transform of a current-displacement time-correlation function.

In the long wavelength limit Eq.(7) reduces to

$$D_o(t) = \int_0^t dt'\, \rho_D(t') \qquad (8)$$

and therefore in the long wavelength, long-time limit we recover the self-diffusion coefficient given by the usual velocity autocorrelation function, as expected. One can pose a number of interesting questions, among them the following: For fixed wavelength, what is the behavior of $D_k(t)$ for large t? For fixed t large enough for the long-time structure in $\rho_D(t)$ to be included in $D_o(t)$, over what range of wavelengths is the difference $D_o(t) - D_k(t)$ small compared to the contribution of the tail to $D_o(t)$? Some preliminary molecular dynamics results[*] for $D_k^*(t) = t_o^{-1} \langle v^2 \rangle^{-1} D_k(t)$ for hard disks are shown in Fig. 9, for the longest and shortest wavelengths so far investigated. The k = 1 wavelength is long enough that, for the times considered here, the corresponding $D_1^*(t)$ shown in the figure can be taken as a close approximation to $D_o^*(t)$. Note that it is quite linear in ln t beyond 10 collisions per particle, as expected from Fig. 8. The short wavelength (about 22 mean free paths) points begin to scatter beyond 30 or 35 collisions per particle, and so can best be said only to be consistent with a ln t divergence. At about 29 collisions per particle, i.e., near the upper time limit of the Alder and Wainwright work, where the precision in $D_{10}^*(t)$ is still quite good, we find a marginally significant value for $D_1^* - D_{10}^*$, which is small compared to the contribution to D_o^* from the tail of $\rho_D(t)$ between 10 and 29

---

[*] Here t_o again denotes the Boltzmann mean free time and $\langle v^2 \rangle$ is the mean square velocity.

collisions per particle. Hence we provisionally infer that the long-time tail contributes strongly to $D_k(t)$ at such times for wavelengths as small as 22 mean free paths, or about 5.1 disk diameters at this density.

We will of course analyze these calculations more closely, and we hope to extend them to shorter wavelengths, other densities, possibly to larger systems, and of course to hard spheres. LEVESQUE and VERLET [1970] make a passing reference to a formula related to our Eq.(7); see also BERNE [1971].

For hard spheres, where the computer studies and the theoretical work both indicate that the self-diffusion coefficient is finite, ALDER, GASS, and WAINWRIGHT [1970] have tabulated the best available values of D over the fluid range of densities, using the $t^{-3/2}$ decay law to extrapolate their molecular dynamics results to infinite time, and extrapolating the finite system results to infinite N[*]. Previously, using their small system data without these corrections, ALDER and WAINWRIGHT [1967] made the rough estimate $D_1 = -1.5$ for the coefficient of the first density correction in the theoretical expansion

$$\frac{D}{D_o} = 1 + D_1 \frac{V_o}{V} + D_2' (\frac{V_o}{V})^2 \ln(\frac{V_o}{V}) + D_2 (\frac{V_o}{V})^2 + \ldots$$

discussed by Professor Cohen in his talk; D_o is the low-density Boltzmann coefficient. Professor Sengers, in his talk, presented calculations of the collision integrals for 3 hard spheres which give the value $D_1 = -1.69$. The

---

[*] HERMANN and ALDER [1972] report velocity autocorrelation functions and diffusion coefficients for one hard sphere in a system of N-1 other hard spheres of the same diameter but different mass.

Alder and Wainwright data at low densities is not extensive enough to make their estimate very precise, and they did not quote any estimate of their probable error. It would be of some interest to improve the molecular dynamics estimate, inasmuch as it is plausibly assumed, but not as far as I know proven, that the resummations required to regularize the expansion leave this coefficient unchanged. An estimate of the coefficient D_2' would be even more interesting, but that presents a much more difficult problem.

2. Parallel Hard Squares

CARLIER and FRISCH [1972] report molecular dynamics calculations on systems of parallel hard squares indicating the presence of a positive long-time t^{-1} decay in $\rho_D(t)$. At $V/V_o = 2$ it is observed over the interval 12 to 25 collisions per particle, and appears after a negative minimum. At $V/V_o = 4.5$ they obtain a positive t^{-1} decay with no intermediate negative phase. The calculations appear to be carried to a rather smaller number of total collisions in the time averages than used by Alder and Wainwright, and the t^{-1} decay may be less firmly established than in the case of the hard disk system.

3. Lorentz Models

In view of their relative simplicity one might have supposed that one or another of the Lorentz models (independent point particles moving in a system of infinitely massive scatterers) would have been a subject of an early molecular dynamics investigation. But in fact, so far as I am aware, none was undertaken until after the theoretical investigations by HAUGE and COHEN [1967, 1969] on Ehrenfest's wind-tree model (in which the scatterers

are overlapping or nonoverlapping parallel squares) and
by van LEEUWEN and WEIJLAND [1967] with overlapping and
nonoverlapping disks and spheres as scatterers.

At the suggestion of Professor Cohen, we [WOOD and
LADO, 1971] carried out a molecular dynamics-Monte Carlo
investigation of diffusion in the wind-tree system. Fig.
10 shows the velocity autocorrelation function as a function of $t^* = t/t_o$, where t_o is again the Boltzmann mean
free time. In the inset figure a line drawn with a slope
corresponding to the Enskog correction would be in close
agreement with the points for nonoverlapping squares.
Fig. 11 shows the reduced mean-square displacemnt function
$S = <\Delta r^2>/\ell_o^2 t^*$ versus t^*, ℓ_o being the Boltzmann m.f.p.
The apparently constant asymptotic behavior of S in the
NOV (nonoverlapping) case confirms Hauge and Cohen's
prediction of normal diffusion, while the apparent power
law decay in the OV case agreed qualitatively with their
prediction of abnormal diffusion, i.e., a <u>vanishing</u> diffusion constant (the exact opposite of the abnormal hard
disk behavior!). Van BEYEREN and HAUGE [1972] have since
been able to account for the power-law behavior by a further consideration of the "retracing events" described by
Hauge and Cohen.

Van LEEUWEN and BRUIN [1972] have undertaken molecular dynamics calculations for the Lorentz model with
overlapping disks, in an attempt to verify the presence
of the logarithmic terms in van LEEUWEN and WEIJLAND's [1967]
result for the density dependence of the diffusion constant,

$$v\sigma D^{-1} \sim c_1\rho + c_2'\rho^2 \ln\rho + c_2\rho^2 + c_3''\rho^3(\ln\rho)^2$$

in which numerical values are known for the four coefficients and $\rho = n\sigma^2$ is the reduced density. The results so far obtained show promise of providing convincing evidence for the presence of the c_2' term. This represents, so far as I know, the first "experimental" evidence for the presence of non-analytic terms in the transport properties of fluid systems.

B. The Coefficients of Viscosity and Thermal Conduction, and Their Related Time-Correlation Functions.

Molecular dynamics calculations of the coefficients of viscosity and heat conduction, and their related time-correlation functions, have been reported only for hard disks and spheres. Omitting a presentation of the rather complicated expressions for the time-correlation functions, we simply recall that they are conventionally separated into three terms called the kinetic, potential, and cross terms, denoted by $\rho^k(t)$, $\rho^p(t)$ and $\rho^c(t)$. To these symbols are appended subscripts η and λ, for the functions appropriate to the viscosity and thermal conductivity respectively. As a framework for the discussion of the molecular dynamics results it will be helpful to first summarize the present state of the theoretical results.

a. The kinetic (Dorfman and Cohen) and hydrodynamic (Alder and Wainwright; Ernst, Hauge, and van Leeuwen) theories all predict that ρ^k_η and ρ^k_λ have $t^{-d/2}$ tails.

b. The same hydrodynamic theories predict the absence of $t^{-d/2}$ tails in ρ^c_η and ρ^p_η; the situation seems to be unclear for ρ^c_λ and ρ^p_λ. The kinetic theory analysis has not been carried out for the cross and potential contributions.

For hard disks at $V/V_0 = 2$, WAINWRIGHT, ALDER, and GASS [1971] report the molecular dynamics results for the long-time behavior of ρ_η^k and ρ_λ^k shown in Fig. 12. The machine results can be said to be consistent with the t^{-1} decay predicted by the theory and perhaps even to lend it some support, although the scatter is rather large and the data extend only out to about 20 collisions per particle. However, they state that a similar graph of the machine results for the cross and potential terms would also suggest a t^{-1} decay, in contradiction to the hydrodynamic theories for the viscosity, at least. The speculate that the time scale of 10 to 25 collisions per particle, while marginally long enough for the kinetic terms to reach their asymptotic behavior, is not long enough for the potential and cross terms. One must conclude that the asymptotic bevavior of these time-correlation functions remains uncertain.

For hard spheres, ALDER, GASS, and WAINWRIGHT [1970] present results for the time-correlation functions over the fluid range of densities, but the data are too inaccurate to permit a useful comparison with the theoretically predicted long-time $t^{-3/2}$ behavior. Values of the coefficients of viscosity and thermal conductivity are tabulated, relative to the Enskog approximation, as obtained for the finite systems studied without attempting to extrapolate the time-correlation functions to infinite time.

IV. Conclusion

I have attempted in this limited review to give an impression of the stimulating interaction which it is possible to achieve between kinetic theory and computer studies. One may speculate that Boltzmann, and certainly Ehrenfest, would have found it a fascinating subject.

I am pleased to express my gratitude to Professor E.G.D. Cohen to whom I am deeply indebted for many insights and suggestions, as well as for a careful reading of the manuscript. I am likewise in debt to my colleague, J.J. Erpenbeck for many discussions and collaborations, for reading the manuscript, and for agreeing to publish some of the preliminary results of our current work. Finally, I thank J.R. Dorfman, J.D. Foch, Jr., T.E. Waunwright and J.W. Dufty for useful comments and encouragement; also H.L. Frisch and C. Carlier, and J.M.J. van Leeuwen and C. Bruin, for permission to discuss their work prior to publication.

References

AHARONY, A., 1971, Physics Letters 37A, 45 (1971).
ALDER, B.J., GASS, D.M., and WAINWRIGHT, T.E., 1970, J. Chem. Phys. 53, 3813.
ALDER, B.J., and WAINWRIGHT, T.E., 1958, International Symposium on Statistical Mechanical Theory of Transport Processes, Brussels, 1956, I. Prigogine, editor (Interscience, 1958), p. 97.
ALDER, B.J., and WAINWRIGHT, T.E., 1959, J. Chem. Phys. 31, 459.
ALDER, B.J., and WAINWRIGHT, T.E., 1967, Phys. Rev. Letters 18, 968.
ALDER, B.J., and WAINWRIGHT, T.E., 1969, J. Phys. Soc. Japan 26, Supplement, p. 267 (Proceedings of the International Conference on Statistical Mechanics, Kyoto, 1968).
ALDER, B.J., and WAINWRIGHT, T.E., 1970, Phys. Rev. A1, 18.
BALESCU, R., 1967, Physica 36, 433 (1967).
BERNE, B.J., 1971, in Physical Chemistry, An Advanced Treatise (H. Eyring, D. Henderson, and W. Jost, editors; Academic Press) Vol VIII B, Liquid State, page 539.
BERNE, B.J., and FORSTER, D., 1971, Ann. Rev. Phys. Chem. 22, 563.
BIRD, G.A., 1970, Phys. of Fluids 13, 2676; see also earlier work mentioned therein.
CARLIER, C., and FRISCH, H.L., 1972, Phys. Rev. A6, 1153 (1972).
DORFMAN, J.R., and COHEN, E.G.D., 1970, Phys. Rev. Letters 25, 1257.

DORFMAN, J.R., and COHEN, E.G.D., 1972, Phys. Rev. $\underline{A6}$, 776.
DUFTY, J.W., 1972, Phys. Rev. $\underline{A5}$, 2247.
ERNST, M.H., 1967, "Transport Coefficients from Time Correlation Functions" in "Lectures in Theoretical Physics", Vol. IX C, "Kinetic Theory", W.E. Brittin, editor (Gordon and Breach, Inc. N.Y. 1967), p. 441.
ERNST, M.H., HAUGE, E.H., and van LEEUWEN, J.M.J., 1970, Phys. Rev. Letters, $\underline{25}$, 1254.
ERNST, M.H., HAUGE, E.H., and van LEEUWEN, J.M.J., 1971a, Physics Letters $\underline{34A}$, 419.
ERNST, M.H., HAUGE, E.H., and van LEEUWEN, J.M.J., 1971b, Phys. Rev. $\underline{A4}$, 2055.
ERPENBECK, J.J., and WOOD, W.W., 1972, preliminary unpublished results.
GRAD, H., 1958, in Encyclopedia of Physics (S. Flügge, ed., Springer, Berlin), Vol. XII, page 205.
HAUGE, E.H., and COHEN, E.G.D., 1967, Phys. Letters $\underline{25A}$, 78.
HAUGE, E.H., and COHEN, E.G.D., 1969, J. Math. Phys. $\underline{10}$, 397.
HERMAN, P.T., and ALDER, B.J., 1972, J. Chem. Phys. $\underline{56}$, 987.
HICKS, B.L., YEN, S.M., and REILLY, B.J., 1972, J. Fluid Mech. $\underline{53}$, 85.
KAWASAKI, K., 1970, Physics Letters, $\underline{32A}$, 379.
KAWASAKI, K., 1971, Prog. Theoretical Physics (Kyoto) $\underline{45}$, 1691.
LEVESQUE, D., and VERLET, L., 1970, Phys. Rev. A $\underline{2}$, 2514.
ORBAN, J., and BELLEMANS, A., 1967, Physics Letters $\underline{24A}$, 620.
ORBAN, J., and BELLEMANS, A., 1969, J. of Statistical Physics, $\underline{1}$, 467.
RAHMAN, A., 1964, Phys. Rev. $\underline{136}$, A405.
SEGAL, B.H., and FERZIGER, J.H., 1972, Phys. of Fluids $\underline{15}$, 1233.

van BEYEREN, H., and HAUGE, E.H., 1972, Physics Letters 39A, 397.
van LEEUWEN, J.M.J., and BRUIN, C., 1972, Private communication.
van LEEUWEN, J.M.J., and WEIJLAND, A., 1967, Physica 36, 457.
VERLET, L., 1967, Phys. Rev. 159, 98.
WAINWRIGHT, T.E., ALDER, B.J., and GASS, D.M., 1971, Phys. Rev. A4, 233.
WOOD, W.W., and LADO, F., 1971, Journal of Computational Physics 7, 528.
ZWANZIG, R., 1965, Ann. Rev. Phys. Chem. 16, 67.
ZWANZIG, R. and BIXON, M., 1970, Phys. Rev. A2, 2005.

Added in proof:

ALDER, B.J., and WAINWRIGHT, T.E., 1972, private communication.
METROPOLIS, N.A., ROSENBLUTH, A.W., ROSENBLUTH, M.N., TELLER, A.H., and TELLER, E., 1953, J. Chem. Phys. 21, 1087.

Figure Captions

Fig. 1.   Qualitative comparison of the decay of the peaks of velocity distribution functions obtained by molecular dynamics (N = 100) and from the Boltzmann equation, for initial distributions more peaked than the equilibrium Maxwell distribution; (a) V/V_0 = 32; (b) V/V_0 = 1.767. The upper abscissa scale is the time scale (arbitrary units) for the solution of the Boltzmann equation (solid curve) and of the Enskog modification for dense hard spheres (dashed curve). The lower abscissa scale gives the total number of collisions in the molecular dynamics calculation (points connected by straight lines); note that it does not begin at zero, due to the matching described in the text. The ordinate is the number of particles lying in the velocity interval (width not given) containing the peak of the velocity distribution function. [ALDER and WAINWRIGHT, 1958].

Fig. 2.   The Boltzmann function H(t) versus t, in arbitrary units; the vertical dotted lines denote the times corresponding to 50 and 100 total collisions, i.e., 1 and 2 collisions per particle, for a system of 100 hard disks, V/V_0 = 25, and the initial conditions discussed in the text. The open circles show the behavior as the trajectory is developed normally using 27 binary digit, floating-point arithmetic. The solid circles show the behavior after velocity inversion after 1 and 2 collisions per particle, in (a) with no perturbation of the velocities at the time of inversion, and in (b) and (c) with random perturbations in the range $\pm 10^{-5}$ and $\pm 10^{-2}$, respectively, in the velocity of each particle at inversion. [ORBAN and BELLEMANS, 1967].

Fig. 3. The maximum-slope reciprocal shock-thickness, in units of the upstream (cold) mean-free path, versus the Mach number M_1 for steady shock waves in a dilute hard-sphere gas. The points, with error bars drawn for the 50% confidence interval, are the Monte Carlo solution of the Boltzmann equation. The weak shock regime is shown in (a), the solid line giving the Navier-Stokes solution. In the strong shock regime (b), the curve is the Mott-Smith solution. [HICKS, YEN, and REILLY, 1972].

Fig. 4. The variation of the "effective reduced viscosity" μ_{rel} (see text) through a Mach 1.2 shock in a hard sphere gas. The abscissa $\hat{n}$ is $(n-n_1)/(n_2-n_1)$, where n is the local number density at the given position within the shock, and n_1 and n_2 are the values at the cold and hot boundaries, respectively. In the Navier-Stokes approximation μ_{rel} would be equal to unity throughout the shock. [HICKS, YEN, and REILLY, 1972].

Fig. 5. The velocity correlation between a central particle and its neighborhood at a density corresponding to 1/2 of close-packing for 224 hard disks at 4.48 (dashed arrow) and 8.96 (heavy arrows) collisions per particle. Because of symmetry only the upper half of the neighbors are shown, being averaged with the neighbors lying in the lower

half. The size of the central particle is shown
by the smallest half circle. The sizes of the
other 4 concentric circles have been determined
to include roughly 6 neighboring particles each.
These circles have been further sectioned into
4 parts, as indicated by the lines, so as to have
a measure of direction relative to the velocity
vector of the central particle at zero time. The
arrows at the center of each of these sections
give the magnitude and direction of the average
velocity of the particles in the section at the
times indicated. The ordinary autocorrelation
function is represented by a vector located at
the central particle. The scale of velocity is
indicated by the solid arrow of the central particle which is about 0.02 of the initial velocity.
[ALDER and WAINWRIGHT, 1969].

Fig. 6. The velocity autocorrelation function, here denoted
by $\rho(s)$, versus s^{-1} at long times for hard disks at
$V/V_o = 2$, 3, and 5. The closed (N = 986) and open
(N = 504) triangles are molecular dynamics results
with a $(N-1)^{-1}$ correction added; the N = 504 points
at the largest values of s are perturbed by effects
of the finite period of the system. The solid lines
are drawn to fit the linear portion of the data at
long times. The circles and squares, and the dashed
curve drawn through them, are from the model numerical hydrodynamics calculations for two different
initial velocities of the fluid element, for $V/V_o=2$.
[ALDER and WAINWRIGHT, 1970].

Fig. 7. The velocity autocorrelation function, here denoted by $\rho(s)$, versus s for hard spheres at $V/V_o = 3$. The triangles are from a molecular dynamics calculation with $N = 500$ and with a $(N-1)^{-1}$ correction added. The circles are from the model numerical hydrodynamics calculations. The line is drawn with the slope 3/2. [ALDER and WAINWRIGHT, 1970].

Fig. 8. The normalized velocity autocorrelation function ρ_D versus s^{-1} (s = number of collisions per particle) for 1512 hard disks at $V/V_o = 2$, as calculated by ERPENBECK and WOOD [1972]. The upper straight line is ALDER and WAINWRIGHT's [1967] hydrodynamical relation $\rho_D = 0.205\ s^{-1}$ for an infinite system; the lower line is displaced downwards by their N^{-1} correction for $N = 1512$. The small vertical arrow indicates the calculated time for a sound wave to traverse a period of the finite system.

Fig. 9. The reduced wavelength and time-dependent self-diffusion coefficient D_k^* versus the time s in collisions per particle for $N = 1512$ hard disks at $V/V_o = 2$. The $k = 1$ and $k = 10$ points correspond to wavelengths of 223.4 and 22.34 mean free paths. [ERPENBECK and WOOD, 1972].

Fig. 10. The velocity autocorrelation function (here denoted by C) for the wind-tree model at $V/V_o = 2$ versus the time (t^*) in units of the Boltzmann mean-free time. NOV and OV denote the "nonoverlapping" and "overlapping" cases. The plotting symbols cover the range of several calculations with $N = 512$, 2048, and 8192. The line is the Boltzmann approximation $\exp(-t^*)$. [WOOD and LADO, 1971].

Fig. 11. The reduced mean square displacement function S vs t^* in the NOV and OV cases for $V/V_0 = 2$ and $N = 8192$. The estimated standard deviations are everywhere much smaller than the plotting symbols. [WOOD and LADO, 1971].

Fig. 12. The time-correlation functions $\rho_\eta^k - \rho_{\eta,E}^k$ and $\rho_\lambda^k - \rho_{\lambda,E}^k$ for hard disks at $V/V_0 = 2$ versus s^{-1}, where s is the time in collisions per particle and $\rho_{\eta,E}^k$ and $\rho_{\lambda,E}^k$ are the Enskog approximations. The lines are the theoretical predictions, using Enskog values of the transport coefficients. The molecular dynamics results are plotted as ×
(N = 504) and ⊙ (N = 1672) for ρ_λ^k and □ (N = 1672) for ρ_η^k. [WAINWRIGHT, ALDER, and GASS, 1971].

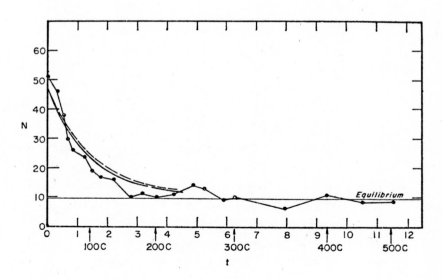

Fig. 1a

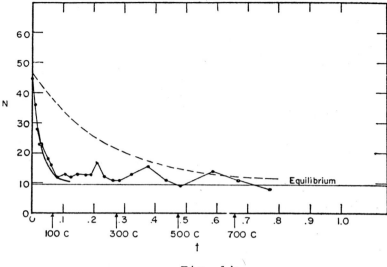

Fig. 1b

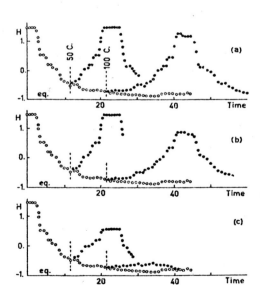

Fig. 2

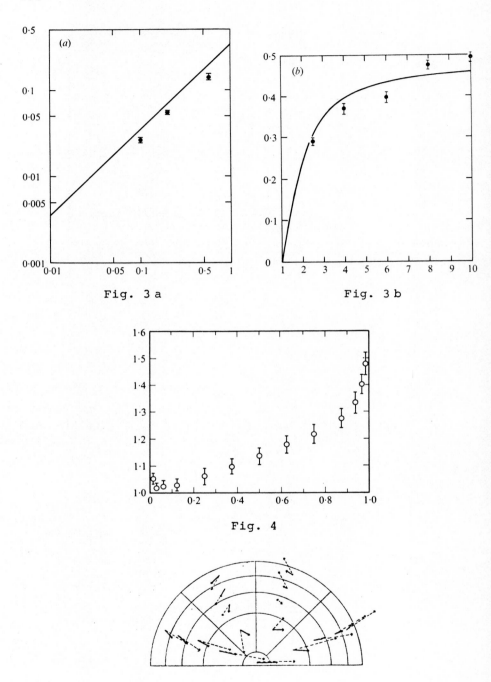

Fig. 3a

Fig. 3b

Fig. 4

Fig. 5

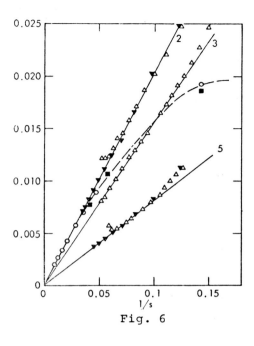

Fig. 6

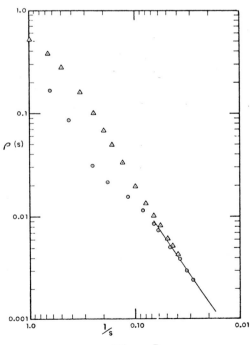

Fig. 7

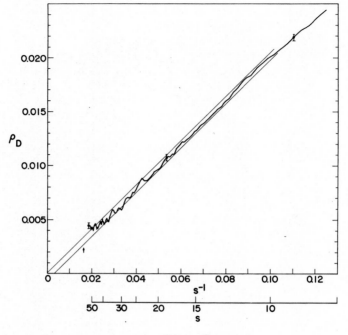

Fig. 8

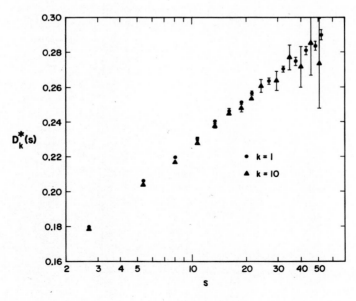

Fig. 9

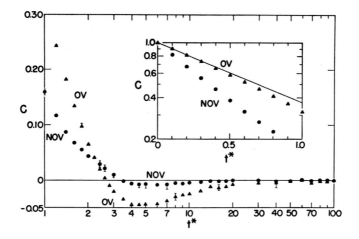

Fig. 10

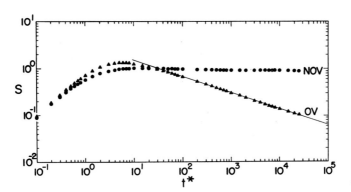

Fig. 11

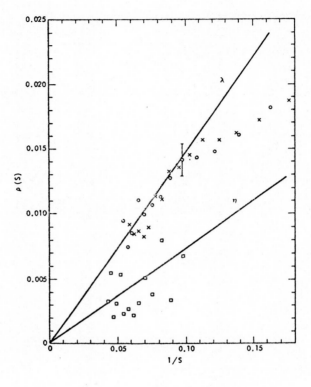

Fig. 12

A SURVEY OF NEUTRON TRANSPORT THEORY

IVAN KUŠČER

Department of Physics
University of Ljubljana, Yugoslavia

ABSTRACT

The basic ideas of neutron transport theory are compared with those of the kinetic theory of gases, and the essential differences are indicated. Typical idealized problems are briefly described, and the mathematical and computational methods reviewed.

1. Basic Premises

The hundredth anniversary of Boltzmann's discovery is a welcome occasion to review briefly the foundations of neutron transport theory and to present a selection of its principal methods and results. The interdisciplinary character of this conference suggests that the survey be addressed to workers in other fields, in particular in the kinetic theory of gases. Accordingly, the connections to, as well as the deviations from, gas theory will be indicated.

Inevitably the selection will be biassed, and I particularly apologize for not being able to offer more than a scant account of the great development in reactor calculations.

A historian might say that neutron transport theory should have had an easy life since many of the basic concepts and methods were well developed by Boltzmann and his followers long before the neutron was even discovered. The theory is in essence little else than a specialized branch of the kinetic theory of gases, so there is much common ground. Moreover, the mathematical development with respect to some idealized models has been anticipated in radiative transfer [1]. Yet in neutron transport there are also some highly specific and often complex aspects that have led to an independent development.

Though the first start was undertaken by Fermi and Wick already in 1936 [2,3], the field received wider attention only after the discovery of fission. It attracted men like Peierls, Placzek, Wigner, Bethe, Marshak and others. Much of the early work remained temporarily unpublished and altogether only about a dozen papers appeared before 1946 [2-10]. The lift of war-time secrecy triggered

a burst of activity that gained further momentum with the
development of reactor technology. A great deal of effort
had to be spent to meet the needs of reactor designers,
yet in parallel research of a more academic character
flourished also. Hardly anywhere else has the Boltzmann
equation found such extensive and diversified application,
and more than a hundred papers per year continue to appear,
according to recent bibliographies [11,12]. Neutron transport
theory is now well covered by a number of monographs
[13-26], and discussions of many details are collected in
proceedings of conferences [27-38].

A basic difference from the conventional theory of
gases is that we never consider a pure neutron gas but
neutrons moving in some kind of material, henceforth called
the medium. This may be a moderator, or the reactor shield,
or nuclear fuel, or a combination of different materials.
The state of the medium is that of thermodynamic equilibrium
(at least locally) at some temperature T. Even in a
high flux reactor the number of neutrons is 10^{11}-times
smaller than the number of atoms. Though the medium need
not be a gas we may for comparison think of a gas mixture,
with one of the components present only in small concentration.

Because of the high dilution, and since the cross
section for neutron-neutron collisions is of the nuclear
order of magnitude, these collisions may safely be neglected.
Hence for all practical purposes a Boltzmann equation
that is linear in the neutron distribution function may
be considered as exact. Nonlinearities only arise if thermal
and nuclear effects of the neutrons upon the properties
of the medium are taken into account, e.g. in reactor
kinetics [24,39].

For a macroscopically uniform and isotropic medium with time-independent properties, and if fission is absent, that equation is

$$[\frac{\partial}{\partial t} + \vec{v}\nabla + \nu(v)]f(\vec{r},\vec{v},t) = \int f(\vec{r},\vec{v}',t)K(\vec{v}'\to\vec{v})d^3v' + q(\vec{r},\vec{v},t), \quad (1)$$

where the notation is borrowed from gas theory. In terms of operators the equation may be shortened to

$$(\frac{\partial}{\partial t} + \hat{T})f = \hat{K}f + q \quad . \quad (1a)$$

The left hand side combines the substantial derivative and the extinction term, whereas the scattering and source terms appear on the right.

The Boltzmann equation is concerned only with the average number of neutrons per unit of phase space, as described by the single particle distribution function $f(\vec{r},\vec{v},t)$. As long as fission is absent, there is little interest in a study of random phenomena, such as fluctuations (neutron noise), which are trivially described by Poisson's distribution. In a reactor such problems are much less trivial, because more than one neutron may be born in a fission event, which is the source of interesting correlations. A new branch of neutron transport theory has evolved from such considerations [40-43] which, however, will not be included in the present survey.

The properties of the medium enter through the collision rate $\nu(v)$ and the scattering kernel $K(\vec{v}'\to\vec{v})$ (the differential scattering rate). While in gas theory sources and losses are seldomly considered, they naturally arise in neutron transport. In general, the collision rate contains an absorption term,

$$\nu(v) = \nu_s(v) + \nu_a(v) \quad,$$

$$\nu_s(v) = \int K(\vec{v} \to \vec{v}') d^3v' \quad.$$

The behavior of neutrons is simpler than that of gas molecules insofar as up to energies of 10^5 or 10^6 eV scattering by an isolated nucleus is isotropic in the center-of-mass system. However, the kinetic-theory way [44] of evaluating the scattering kernel from single-collisions, i.e., by transformation to the lab system and summation over a Maxwellian distribution for the nuclei, does not apply to slow neutrons in a condensed medium. In such case the kernel may be more complicated because of effects of chemical binding and of interference [20,23,45]. The kernel also may become mathematically more difficult by ceasing to represent a compact operator. Such is the case of a solid medium, where K contains a delta term describing elastic scattering.

A simple consequence of the thermal equilibrium of the medium and of time-reversal invariance is that all scattering kernels obey the reciprocity (or detailed balance) relation [46,47].

$$e^{-v'^2} K(\vec{v}' \to \vec{v}) = e^{-v^2} K(-\vec{v} \to -\vec{v}') \quad, \tag{2}$$

where we write v^2 for $mv^2/2kT$. Of course this relation is only useful for neutron energies in the thermal range. However, since fission energies are in the MeV range, we are also very much interested in the vast intermediate slowing-down interval. Most calculations for these energies completely neglect the thermal motion of the medium by taking T = 0, which in gas theory would be a very unusual assumption.

It should be mentioned that an equation like (1) also arises in the problem of a single gas if f is taken to denote a small departure from the equilibrium distribution. Contrary to the neutron problem, f may then have either sign, and the kernel K too is negative in some regions. The difference is best illustrated by quoting the kernel (K_H) for neutron transport in a monatomic hydrogen gas, and the one (K_s) for the hard-sphere single gas problem [25].

$$K_H(\vec{v}'\to\vec{v}) = \frac{1}{2\pi} \frac{\nu_s(0)}{|\vec{v}-\vec{v}'|} \exp[-(\frac{\vec{v}(\vec{v}-\vec{v}')}{|\vec{v}-\vec{v}'|})^2] \quad , \qquad (3)$$

$$K_s(\vec{v}'\to\vec{v}) = 2K_H(\vec{v}'\to\vec{v}) - \frac{1}{2\pi} \nu_s(0)|\vec{v}-\vec{v}'|e^{-v^2} \quad . \qquad (4)$$

Often the data for $\nu(v)$ and $K(\vec{v}'\to\vec{v})$ are taken from various simplified models. Neutron transport theory has been unusually prolific with inventions of this kind. They are said to represent approximations, but frequently their real origin is our longing for nice analytic solutions.

The monatomic gas, where the kernel is slightly more general than (3) to account for a mass ratio $A \neq 1$, has for some time also served as a model for non-gaseous media. An interesting special case is the heavy gas model (A >> 1) [20] which in kinetic theory is known as the Lorentzian gas mixture [48,49]. In this case it is of advantage to write $\vec{v} = v\vec{\Omega}$ and to introduce the angular flux $\phi(\vec{r},v,\vec{\Omega},t) = v^3 f(\vec{r},\vec{v},t)$. Let us also substitute $\nu(v) = v\Sigma(v)$ and $\nu_s(v) = v\Sigma_s(v)$. If the macroscopic scattering cross section $\Sigma_s(v)$ (the reciprocal mean free path for scattering) is a slowly varying function of v, an expansion of the kernel in reciprocal powers of A, truncated after A^{-1}, leads to the equation

$$[\frac{1}{v}\frac{\partial}{\partial t} + \vec{\Omega}\vec{\nabla} + \Sigma(v)]\Phi(\vec{r},v,\vec{\Omega},t) = \frac{1}{4\pi}\int d^2\Omega'[\Sigma_s(v)(1 + \frac{2}{A}\vec{\Omega}\vec{\Omega}')$$

$$+ \frac{1}{A}(1 - \vec{\Omega}\vec{\Omega}')\frac{\partial}{\partial v}\Sigma_s(v)(\frac{1}{2}\frac{\partial}{\partial v} - \frac{3}{2}v^{-1} + v)]\Phi(\vec{r},v,\vec{\Omega}',t) + v^2 q \ . \quad (5)$$

The differential operator on the right expresses the fact that for such a medium energy exchange in single collisions is infinitesimal.

In the slowing-down region (v >> 1 in our notation) the term with the second derivative and the $-\frac{3}{2}v^{-1}$ in parentheses may be neglected. After some further simplifications we obtain the age approximation [17,20,24].

For a very heavy medium one may neglect the terms of the order A^{-1} in the integrand altogether. This brings us to the extreme of such idealizations, the one-speed model with isotropic scattering [15,21]. Since speed reduces to a parameter we may write, after making variables dimensionless,

$$(\frac{\partial}{\partial t} + \vec{\Omega}\vec{\nabla} + 1)\Phi(\vec{r},\vec{\Omega},t) = \frac{c}{4\pi}\int \Phi(\vec{r},\vec{\Omega}',t)d^2\Omega' + q(\vec{r},\vec{\Omega},t) \ , \quad (6)$$

where c denotes the average number of secondary neutrons per collision. For time independent thermal problems such a model can also be justified with the assumption of constant cross sections [15], as is shown by integrating the Boltzmann equation over speed.

Somehow opposite to the heavy gas model is the separable kernel [20],

$$K(\vec{v}'\to\vec{v}) \propto e^{-v^2} \nu_s(v') \nu_s(v) \quad , \tag{7}$$

where energy exchange is maximal, so that full thermalization is achieved in a single collision. Degenerate kernels with several terms of the form (7) also are often considered. We recall that a special case of such a kernel is encountered in gas theory in the linearized version of the BGK model [25].

Various generalizations of the Boltzmann equation that are foreign to gas theory are of interest in neutron transport. Non-uniform media are easily included in the description by making ν and K dependent on $\vec{r}$. Of course such a modification greatly increases the difficulties in obtaining accurate solutions.

For a medium containing nuclear fuel one must take fission into account by adding to the scattering kernel another kernel K_f, usually assumed to be of the form $K_f(\vec{v}'\to\vec{v}) \propto \nu_f(v')\chi(v)$. Unlike the scattering kernel, K_f does not obey reciprocity, which introduces some difficulty in the mathematical treatment. In addition to Eq.(1) it often becomes necessary to consider the adjoint equation, where the variables in the kernel are interchanged, the signs of the derivatives reversed, and the boundary conditions accordingly modified [24,50]. Its solution is called the importance because it expresses the average contribution of a neutron born at $(\vec{r},\vec{v})$ to the total response of the detectors distributed according to a term q^+ in the equation.

More serious from a calculational point of view than the doubling of the effort implied in the introduction of the adjoint problem are the difficulties associated with the vast energy interval involved (up to $10^8 kT$), which by far exceeds anything encountered in gas problems. The problem of a thermal reactor looks more like a system of three problems: the diffusion of thermal neutrons, regeneration of fast neutrons by fission, and the slowing down from the fast to the thermal region.

In time dependent problems with fission, delayed neutrons play an essential role. The concentrations of the parent isotopes may then be included as additional unknowns, and we arrive at a system of equations [24,39].

The formulation of a neutron transport problem also includes the boundary conditions, which are inherently different from gas theory, because neutrons easily penetrate any material and cannot be held back by surfaces. In principle, therefore, every surrounding ought to be included in the calculation as part of the system. In practice, the condition is simplified by taking the medium of interest as surrounded by a perfect absorber or by empty space, and the neutron flux incident from the outside is given. This refers to "inner" problems, as they would be called in gas theory. "Outer" problems are rarely considered in neutron transport, though an important example is presented by the case of a totally absorbing object embedded in an extended medium.

It is sometimes an advantage to replace the integro-differential form (1) of the Boltzmann equation by the integral form in which the boundary condition is incorporated [24]. The time-independent version may be stated symbolically as

$$f = \hat{T}^{-1}[\hat{K}f + q] \quad . \tag{8}$$

If scattering is isotropic in the laboratory system, and in case of isotropic sources, we switch to the notation of Eq.(5) and integrate over the solid angle. The Peierls integral equation for the angle-integrated flux follows

$$\Phi(\vec{r},v) = \int[\int_0^\infty \Sigma_s(v'\to v)\Phi(\vec{r}',v')dv' + v^2 q(\vec{r}',v)]$$

$$\cdot \frac{\exp(-\Sigma(v)|\vec{r}-\vec{r}'|)}{|\vec{r}-\vec{r}'|^2} d^3r' \quad . \tag{9}$$

For any medium we may define both a time independent and a time dependent Green's function that describe point-source solutions. We recognize in the time-dependent solution $G(\vec{r}',\vec{v}'\to\vec{r},\vec{v};t)$ a transition probability density[*], or also a kernel expressing the time-evolution operator,

$$\hat{G}_t = \exp[-(\hat{T}-\hat{K})t] \quad . \tag{10}$$

If fission is absent, and if the medium has a uniform temperature (but not necessarily a uniform composition), the Green's functions obey a reciprocity relation similar to Eq.(2) [51,52,47], e.g. in the time-dependent case,

$$e^{-v'^2} G(\vec{r}',\vec{v}'\to\vec{r},\vec{v};t) = e^{-v^2} G(\vec{r},-\vec{v}\to\vec{r}',-\vec{v}';t) \quad . \tag{11}$$

---

[*] In general G is not normalized because the probability for the neutron being lost by absorption or leakage is not included.

More generally, the adjoint Green's function must also be considered, and we have the relation [24]

$$G(\vec{r}',\vec{v}'\to\vec{r},\vec{v};t) = G^{\dagger}(\vec{r},\vec{v}\gets\vec{r}',\vec{v}';t) \quad . \tag{12}$$

If delayed neutrons are absent, neutron diffusion in phase space (but not in coordinate space) is a Markoffian process. Hence the time-dependent Green's function of Eq.(1) obeys the Markoff (or Chapman-Smoluchowski) equation [53-55],

$$G(\vec{r}',\vec{v}'\to\vec{r},\vec{v};t) = \int\int G(\vec{r}',\vec{v}'\to\vec{r}'',\vec{v}'';\tau)G(\vec{r}'',\vec{v}''\to\vec{r},\vec{v},t-\tau)d^3r''d^3v'', \tag{13}$$

that expresses the semigroup property of the time-evolution operator,

$$\hat{G}_t = \hat{G}_\tau \hat{G}_{t-\tau} \quad . \tag{13a}$$

For an arbitrary solution of Eq.(1), if $q = 0$ for $t > 0$, we have

$$f(\vec{r},\vec{v},t) = \int\int f(\vec{r}',\vec{v}',t')G(\vec{r}',\vec{v}'\to\vec{r},\vec{v};t-t')d^3r'd^3v' \quad . \tag{14}$$

The Boltzmann equation (1) and its adjoint may be regarded as the corresponding forward and backward Kolmogorov-Feller equations, respectively [53].

2. Analytical Results

Before embarking upon solving the Boltzmann equation for specific data, it is desirable to make sure the existence and uniqueness of solutions. Though such questions enjoy little popularity, they are by no means trivial. Methods of functional analysis and, for time-dependent problems, of semigroup theory have to be engaged to derive strict answers. Investigations of Lehner and Wing [56], Vladimirov [57], Case and Zweifel [58], Marti [59], Hejtmanek [60], Vidav [61], and Mika et al. [62] may be mentioned in this connection. Similar considerations in linearized gas theory [63-66] turn out to be somewhat more involved, because of the different boundary conditions.

Another general task is the derivation of certain integral theorems. For instance integration of both sides of Eq.(1) over velocity space leads to the continuity equation. For a constant absorption rate it assumes the simple form

$$(\frac{\partial}{\partial t} + \nu_a) \, n(\vec{r},t) + \nabla \vec{j}(\vec{r},t) = q(\vec{r},t) \quad , \tag{15}$$

where n is the neutron density and $\vec{j}$ the net current.

Only a small selection of simple transport problems can be discussed here. Some of these problems appear quite unrealistic, yet they give valuable information about the general behavior of solutions of the linear Boltzmann equation. Moreover, they are the standard training ground for the development of methods of solution, and often they are used to test the accuracy of approximations.

Closest to gas theory are thermal problems, which

since more than a decade were thoroughly investigated by Nelkin, Corngold, Williams, and many others [20,23,28,29, 32,33,68]. Here fission is absent and the neutron distribution is confined to energies of the order kT. We first consider problems for a uniform sourceless infinite medium, which were mainly conceived to define and calculate the transport parameters and various eigenvalues that are characteristic for the medium.

In the same way as in gas theory we may calculate the diffusion coefficient D, and if the temperature is non-uniform also the thermal diffusion factor [69]. For a non-absorbing medium we postulate $f \propto e^{-v^2}[\text{const} + x - \Omega_x U(v)]$, in order to obtain an integral equation for $U(v)$, and then through Maxwellian averaging $D = \frac{1}{3}<vU(v)>$.

More generally, we may look for eigenmodes, i.e., for solutions of the form $f(\vec{r},\vec{v}) = \phi_L(\vec{v})e^{-x/L}$. An eigenvalue equation follows from which the eigenfunctions ϕ_L and the spectrum of relaxation lengths L are to be determined [70]. The spectrum represents the singularities of the reciprocal dispersion function $[\Lambda(z)]^{-1}$. There is a continuous spectrum consisting of the cut $[-1/\Sigma_{min}, 1/\Sigma_{min}]$. In addition, we may have one or several pairs of isolated real eigenvalues $\pm L_j$, the zeros of $\Lambda(z)$, outside the continuum. The largest one, L_o, is called the diffusion length. It corresponds to the only pair of non-negative solutions, and plays an important role in reactor calculations. For a small and constant absorption rate we find $L_o^2 = D/\nu_a$.

In an analogous manner we may ask for space-independent solutions that behave exponentially in time, $f = \psi_\lambda(\vec{v})e^{-\lambda t}$. The spectrum now covers the continuum $[\nu(0),\infty)$, and in all realistic cases there is at least one isolated eigenvalue $\lambda_o \in [0, \nu(0))$, corresponding to the unique non-negative eigenfunction ψ_{λ_o}. For typical media this problem has been studied extensively [20], and an application of the same methods to the analogous single-gas problem has led to surprisingly similar results [25].

By taking $f = \phi(\vec{v})\exp(-\lambda t + iBx)$ we formulate a combined problem that roughly simulates the behavior of neutrons in a bounded medium. We may study the eigenvalues λ as functions of the "buckling" B^2 or vice versa (Fig. 1). Calculations for particular models have been carried out in order to support the experimental investigations with neutron pulses. Analogous problems in gas theory appear less interesting.

Allowing for complex values of both λ and B, we are able to include various wave problems which have their counterpart in sound theory. For an idealized description of a neutron-wave experiment [25,31] we try $f = \phi(\vec{v}) \cdot \exp(-i\omega t + \kappa x)$, and look for complex eigenvalues $\kappa = ik - \beta$ as functions of the frequency ω.

In all of these problems it is possible to vary the discrete part of the spectrum by varying some parameter. Thus in the last case, with increasing ω the eigenvalues κ move toward the continuous part of the spectrum. After touching its boundary they apparently disappear. This phenomenon has caused considerable controversy, especially when unexpectedly clean exponential modes were observed where they should not exist. Several explanations were offered, for instance in terms of analytic continuation, with hidden eigenvalues appearing on another Riemann sheet for the dispersion function. Here an exchange of ideas between neutron transport and gas theory has greatly helped towards clarification [68,71].

Foremost among the solutions of the inhomogeneous equation (1) is the Green's function for an infinite medium. For a low-energy source this is relatively easy to obtain either by a Fourier transformation or by the equivalent method of singular eigenmode expansion. As expected, at large distances from the source the neutron density is proportional to $r^{-1} \exp(-r/L_o)$, and in the time-dependent problem for a nonabsorbing medium to $t^{-3/2} \exp(-r^2/4Dt)$.

Among the boundary-value problems, those for the uniform semi-infinite medium play a special role [15,21,25]. If a stationary source is in the far interior of the medium, we have the celebrated Milne problem, which was first considered in astrophysics and solved in more detail by Placzek et al. [72]. Only much later was it noticed that the slip problems of gas theory are closely analogous [25].

Except for a boundary layer, a few mean free paths thick, the neutron density in the Milne case may be expressed as $n \propto \sinh[(x+x_o)/L_o]$, where x_o is the extrapolation distance (e.g. $x_o = 0,710446$ m.f.p. for the one-speed model with isotropic scattering and no absorption). This is a useful parameter in the diffusion approximation, where it serves to modify the boundary condition in order to achieve an asymptotic best fit.

The reflection of neutrons by a semi-infinite medium is studied in the albedo problem. For the one-speed model and also for the separable kernel the reflected distribution may be factorized in terms of Chandrasekhar's H-function [1,67]. Similar results are found in problems with two adjacent semi-infinite media.

The solutions for the Milne and albedo problems

may be combined in the manner of asmptotic fitting to give a satisfactory approximation for the transmissivity of thick slabs, both for the one-speed and thermal problems.

A similar approximation applies to the homogeneous equation, encountered e.g. in the one-speed critical problem. Here the condition is sought for the existence of a stationary distribution in a multiplying slab. In the interior of a thick slab ($0 \leq x \leq h$) with weak multiplication ($c - 1 \ll 1$ in Eq.(6)) the neutron density is approximately described by $n \propto \sin[B(x+x_0)]$, where x_0 is taken from the Milne problem, and the "buckling" $B^2 = 3(c - 1)$ is such that $B(h + 2x_0) = \pi$.

In an elementary way many problems with spherical symmetry can be reduced to the same form as in plane geometry [15,21,24].

Formally similar to the critical problem is the time-eigenvalue problem for a bounded medium. However, the eigenvalue spectrum is more complicated now, since part of it covers a half-plane [56,61,73,74]. For comparison the problem of relaxation of a gas in a container may be mentioned [66].

The simplest slowing-down problems deal with time-independent, uniform, and isotropic neutron distributions in an infinite medium. We first consider the region well above kT and well below the source energy. If absorption is absent there, we have in velocity space a constant neutron flow, called the slowing-down density, directed towards the lower energy regions, where the neutrons are eventually absorbed. If scattering is elastic and isotropic in the center-of-mass system, the flux is given by

$$\phi(v) \propto 1/v \, \Sigma_s(v) \, . \tag{16}$$

The approach to thermal equilibrium as the ultimate stage of slowing-down has first been studied by Wigner and Wilkins [75,17]. If absorption is weak and confined to the thermal region, the result is an almost Maxwellian distribution with a slowing-down tail of the form (16).

At higher energies absorption is often confined to narrow resonances. The "slowing down density" then is no longer constant but makes a step down at each resonance. The flux takes a dip there, but recovers close to the undisturbed curve (16) farther below. For obvious reasons such calculations are essential in reactor design [19,20,76]. Even such a subtle effect as Doppler broadening of the resonances, caused by the thermal motion of the nuclei, can be of vital importance since through the temperature coefficient of reactivity it may affect the time behavior and thereby the safety of a reactor.

In the hypothetic case with constant cross sections, or with cross sections proportional to the same power of v, space dependent steady-state slowing down problems can be reduced to the form of a one-speed equation by applying a Mellin transformation [13,77]. The problem of a discrete source of fast neutrons in an infinite medium has been handled in this way.

At large distances from the source only a small flux of fast neutrons will remain, yet in designing reactor shields we must know its magnitude. This "deep penetration problem" has turned out to be quite intricate [78].

3. Methods

A specific success of neutron transport theory must be seen in the development of a multitude of ways of actually, not just formally, solving the linear Boltzmann equation or its approximations. Hence it would be unfair to conclude this survey without at least a brief, even if largely second-hand, account of this achievement.

Traditional classifications into exact and approximate methods, or alternatively into analytical and numerical ones, appear to miss the essential characteristics and will therefore be avoided. Consequently if, for the purpose of numerical work, some variable is made discrete, this will be regarded merely as a technical modification, rather than a new method.

Such a point of view may not be quite fair in cases where a relatively coarse discretization of the variables must be made, which calls for a judicious choice of weights and averaging. For instance, in the multigroup treatment of the speed (or energy) variable the group cross sections are determined as averages over the neutron spectrum, which requires partial previous knowledge of the solution or even iterative adaptation during the computation. There is undoubtedly more to such a technique than the mere application of standard recipes for numerical integration.

With neglect of this warning and with the inevitable risk of some overlap, most methods can be classified into the following types:
1) expansions and transformations;
2) direct or iterative solution of the differential version of the Boltzmann equation, or of its approximations;

3) direct or iterative solution of the integral version of the Boltzmann equation;
4) step-by-step integration of the differential form of the Boltzmann equation;
5) semigroup method
6) composition of response functions;
7) iterative solution of nonlinear integral equations for the response functions;
8) variational methods;
9) perturbation;
10) simulation of neutron life histories (Monte Carlo method).

Expansions and transformations are often two names for the same thing, and "moments methods" may be regarded as a third name, since moments are some expansion coefficients or linear combinations thereof. The principal step consists in the derivation of a transformed equation, or a system of equations, which is hopefully easier to solve than the original one. Examples are: eigenmode expansions, Laplace transformation with respect to the time variable, Fourier (or Laplace) transformation with respect to coordinates, spherical-harmonics expansion with respect to the angular variables, Mellin transformation with respect to speed, and expansions in polynomials of speed or energy.

In time independent problems with plane geometry (or in problems that can be reduced to such a form) the solution may be expanded in the eigenmodes $\phi_L(\vec{v}) \, e^{-x/L}$ mentioned before. A completeness property of the set of eigenfunctions $\phi_L(\vec{v})$ guarantees that this is always possible, provided that the singular eigenfunctions corresponding to values within the continuous spectrum of L are also included. Since the work of Case [79] this method has found wide application to a large number of idealized transport problems [21,67]. Cercignani [80] adapted the method to time dependent gas problems, where an area spectrum is encountered, and his generalization has in turn been taken over to neutron waves and to related problems [81,67].

For simple scattering kernels this method leads to closed-form solutions of problems both for an infinite and a semi-infinite medium. In the latter case we must rely upon a half-range completeness theorem. An essential step is the factorization by the Wiener-Hopf or Muskhelishvili procedures of the reciprocal dispersion function [1,67],

$$\Lambda^{-1}(z) = H(z) H(-z) \quad . \tag{17}$$

The factors $H(z)$ and $H(-z)$ are to inherit the singularities of Λ^{-1} in the left and right half-planes, respectively. The H-function also appears as a weight factor in half-range orthogonality relations that facilitate the calculation of the expansion coefficients.

While singular eigenfunctions hold little promise for general three dimensional problems, the method has greatly promoted our basic understanding of the linear Boltzmann equation. At least for a few simple cases the structure of the Boltzmann operator has been explicitly revealed.

In time-independent problems Fourier transformation with respect to coordinates leads to an integral equation. A method of solution starts with expanding the kernel according to a formula of Gegenbauer [82,83]. On the other hand, for infinite and semi-infinite media, and under restrictions as before, the transformed equation

permits closed-form analytical solutions [84]. This is the
approach that was developed by Wiener and Hopf [85] to
produce the first closed-form solution of a non-trivial
radiative-transfer problem - the Milne problem. Surprisingly, a closer inspection of this technique reveals not only
an overall equivalence with the method of eigenmode expansions but also an intimate relationship of the individual steps [67,84].

Expansions in spherical harmonics of the angular
variables are often used. Mostly such expansions are not
pushed to the exact limit but are truncated after a few
terms. This results in a system of integrodifferential
equations, representing a substitute for the Boltzmann
equation. The easiest and most widely used among such substitutes is the diffusion equation, derived from the P_1-approximation in combination with some relatively innocent
further simplifications [24]. A characteristic feature of
this equation is a speed-dependent diffusion coefficient.
In thermal problems, if absorption is weak, this coefficient may be substituted by a constant - the one mentioned
in Sec. 2. Only then does the diffusion equation coincide
with the one derived from Fick's law.

In reactor calculations the energy variable is
mostly made discrete. A combination with the previous
approximation leads to the multigroup diffusion (or P_1)
equation which is a system of second-order differential
equations. Presently this is still the most widely used
basis for computing the neutron distribution in a reactor
as a whole [86], while other methods are mostly used only
for subsidiary tasks or for special problems. One of the
preliminary steps is the homogenization of the reactor or

of its major parts, that is, a substitution of the reactor lattice by an equivalent homogeneous medium.

A typical reactor code [24] involves iteration at two levels. First an "inner" iteration within the highest energy group is performed, with some initial approximation substituted for the right hand side - the fission source. After proceeding in the same way successively through all the lower groups, the right hand side is improved through "outer" iteration, i.e., through repetition of the whole cycle. This corresponds to the stage of fission regeneration in the neutron life cycle.

Of course, in any such computation space is made discrete too, and derivatives replaced by difference ratios. To avoid overburdening the computer, in particular its fast memory, Kaper et al. [87] recommend a coarser space subdivision, in combination with more sophisticated difference formulas.

Once all variables are made discrete, both forms of the Boltzmann equation, as well as any of their approximations, reduce to systems of algebraic equations that in principle could be solved directly by matrix inversion. While with the mentioned improvement [87] and with further advance in computers this might become a real possibility, traditional procedures involve such a large number of mesh points that iteration is unavoidable, except for the simplest idealized problems.

For any subcritical system the integral version of the Boltzmann equation may be solved by iteration. An initial approximation, obtained by neglecting the unknown term in the brackets in Eqs.(8) or (9), is inserted into the integrand to yield the next approximation etc. The resulting Neumann series is also called the multiple collision expansion because the individual terms represent contributions from the uncollided neutrons and from those scattered once, twice etc. A number of the early iterative steps is saved by guessing a better initial approximation. The method can be extended to the critical problem and to other eigenvalue problems if the approximation is renormalized after each step.

In the collision probability method the Peierls integral equation (9) is iterated in such manner. A rather coarse discretization of space must be used, and averages taken over the cells.

In view of the computational burden involved in the evaluation of those averages it is sometimes of advantage to keep the angular variables and to iterate Eq.(8) instead of (9). The use of this equation becomes implicit if the computation starts with the differential form of the Boltzmann equation which is integrated along the neutron path, much in the same way as this is done with first-order differential equations.

For time-dependent problems Tavel et al. [88] carry out the integration steps in time according to the substantial derivative $(\partial/\partial t + \vec{v}\vec{\nabla})$ in Eq.(1), which is as if we were moving along with the neutrons. A motion picture of the time evolution of the neutron distribution can be produced in this way. The procedure is accelerated by first calculating the Green's function and then invoking the semigroup principle. That is, further computation proceeds according to Eq.(14) [60]. By letting the calculation run long enough, stationary solutions may be derived or eigenvalue problems solved.

The various S_N or discrete-ordinates procedures (where "ordinates" means angles) of Carlson and Lathrop [89] are more general insofar that they also apply to time-independent problems and that the integration path does not need to follow strictly the movement of the neutrons.

"Composition of response functions" shall be a label for the evaluation of the response of a large system to an injection of neutrons, if the response of its parts is known [90]. Though the idea has successfully been applied also to a two-dimensional problem [91], it is best known from problems in plane geometry where it is variously referred to as the layer-composition method or the interaction principle [92] or the transfer matrix method [93]. If only infinitesimal layers are added or removed, we obtain the principle of invariance [1] or of invariant imbedding [94]. Another special modification is the doubling method [95-97].

The response functions (reflectivity and transmissivity) of a plane layer satisfy non-linear integral equations that are derived either through invariance principles [1], or by other means [98]. Mostly these equations are solved by iteration.

Though applications of the Ritz-Galerkin principle to integral equations derived from the Boltzmann equation were known since the work of Chapman [48], neutron transport theory took the priority in developing variational methods for general space dependent problems [15,99,100]. Only later these methods were also applied in gas theory [101,102]. The integral as well as the differential form of the Boltzmann equation can be used as a starting point. Since usually many variables are involved, accurate approximations to the distribution function itself are difficult to reach, yet characteristic averages are obtained with satisfactory accuracy if they can be expressed with the stationary value of the variational functional. It is of advantage if the operators involved are self-adjoint, as is the case with the one-speed Peierls integral equation. If, moreover, the operator is positive definite, the exact value is an extremum of the functional. With suitable extensions the method applies equally well to one-speed problems in general, and also to thermal problems if the medium has a uniform temperature so that detailed balance can be invoked [103,104].

For more general operators, or if a greater arbitrariness of the functional is desired, the Roussopoulos method is applied where the original and adjoint equations are studied simultaneously, and trial functions for both solutions invented [105]. Here the results become somewhat less reliable, and one cannot hope for an extremum.

In an interesting manner variational principles have been applied by Kaplan [106] to synthesize solutions of three-dimensional problems from those in fewer dimensions. The solution, say, of the diffusion equation is approximated by a sum of products, $\phi(\vec{r}) = \sum H_j(x,y) \, Z_j(z)$, where solutions of corresponding two-dimensional problems are taken for the trial functions $H_j(x,y)$. By applying Galerkin's prescription we obtain a system of equations for the unknown functions Z_j involving one variable only.

The perturbation technique [17,107] is mainly used to calculate effects of small modifications in a reactor, e.g. of additions of fuel, or of movements of the regulating devices. The formalism is intimately related to that of the variational methods.

The Monte Carlo method [108-111] solves the Boltzmann equation without ever using it. It offers great flexibility, yet high accuracy is hard to reach, because statistical errors can only be cut down by simulating a large number of neutron life histories. One might argue that the method does more than we are asking for, since the complete record also contains information about fluctuations and correlations, and that this accounts for an excessive consumption of computer time. However, with all the multiple integrals involved, other methods do not fare that much better. Through clever modifications of the sampling procedures computing time in Monte Carlo can be cut down and the method becomes competitive, or even a necessity, at least in some cases, the deep penetration problem being a typical example. According to Askew [112], the future may be more in favor of Monte Carlo, in view of the rapid progress of computers and the relatively slow advance in numerical analysis.

4. Conclusions

Neutron transport theory can partly be regarded as a branch of theoretical physics, and partly as an engineering discipline. On the theoretical-physics side it is devoted to analytical investigations of the linear Boltzmann equation for conditions encountered in neutron diffusion, and to idealized problems that lead to clean mathematical results. The engineering branch consists of the art of inventing practical methods and computer programs for obtaining solutions in realistic problems of almost arbitrary complexity.

In spite of the difference in aims and spirit, both branches intimately depend upon each other. The idealized problems of the theoreticians, though far removed from direct engineering applications, often originate right there. (For instance, the thermal Milne problem stems from an idealized picture of the evaporation of neutrons from the surface of a thick flat shield.) On the other hand, the general theory is a prerequisite for the understanding and for the development of the computational methods.

The analytic achievements of neutron transport theory have led to a reasonably clear and consistent overall view. Moreover, in many cases, e.g. in problems of thermal neutron spectra, excellent agreement with experiments has been reached. In part this is due to the experimental advantage of neutrons, which in comparison to molecules are much more easily individually counted and their energy accurately measured.

All this satisfaction inevitably creates the feeling that this is becoming a mature subject where meaningful unsolved problems or possibilities for further general advance are increasingly harder to find.

On the engineering side the situation is entirely different. While the classical methods are well understood mathematically, the techniques incorporated in the computer codes for reactor calculations have in part been developed half-empirically. As a result of the fast development enforced by reactor technology, little time was left for details and for mathematical clean-up. In the opinion of Gelbard et al. [113] some of those techniques still "have no firm mathematical foundation, and the work of the applied mathematician has, in a sense, just begun". Much remains to be done about the comparison of the various methods and approximations, about their mathematical justification, optimization, and error estimates. Though to a surprising extent the engineering needs have been met by the relatively crude present methods, further advance of great value should be expected.

As was indicated in Sec. 2, many problems of neutron transport have their analogues in gas theory and in other fields of physics. In a number of cases the results and methods developed in one field were successfully transferred to another. However, communication was sometimes slow or only a one-way affair, as can be inferred from cases of delay or of duplication. Much of the computational art in neutron transport is virtually unknown to other users of the Boltzmann equation. Though the individual codes are far too specific to be translatable, it might be well worth to learn more about the built-in ideas and about the general approach.

The reciprocal relationship with mathematics must not be overlooked. Since a better mathematical foundation for some of the methods of neutron transport is surely needed, it is also true, as Greenspan et al. [114] are saying, that reactor physics is a prolific source of problems in numerical analysis. Indeed, the Monte Carlo technique owes much of its development to the application in neutron transport, and the same is true for the iterative methods dealing with large systems of algebraic linear equations where matrices of the order up to $10^5 \times 10^5$ have already been considered [115].

Acknowledgment

The author wishes to express his gratitude to Professors Milan Čopič, Norman J. McCormick and Paul F. Zweifel for their very substantial aid with detailed suggestions and criticism, to Professor Anton Moljk for a thorough examination of the manuscript and for valuable advice towards improving the presentation, and to Professor Noel Corngold for encouragement and helpful comments.

References

[1] S. Chandrasekhar, Radiative Transfer. Oxford Univ. Press, 1950.
[2] E. Fermi, Ricerca Sci. $\underline{7}$-II, 13 (1936).
[3] G.C. Wick, Rendiconti Accad. Lincei $\underline{23}$, 774 (1936); Zschr. Physik $\underline{121}$, 702 (1943).
[4] L.S. Ornstein and G.E. Uhlenbeck, Physica $\underline{4}$, 478 (1937).
[5] O. Halpern, R. Lueneburg and O. Clark, Phys. Rev. $\underline{53}$, 173 (1938).
[6] R.E. Peierls, Proc. Cambridge Phil. Soc. $\underline{35}$, 610 (1939).
[7] G. Placzek, Phys. Rev. $\underline{55}$, 1130 (1939); $\underline{60}$, 166 (1941).
[8] E.A. Schuchard and E.A. Uehling, Phys. Rev. $\underline{58}$, 611 (1940); E.A. Uehling, Phys. Rev. $\underline{59}$, 136 (1941).
[9] F. Adler, Phys. Rev. $\underline{60}$, 279 (1941).
[10] W. Bothe, Zschr. Physik $\underline{118}$, 401 (1941-42); $\underline{119}$, 493 (1942); $\underline{122}$, 648 (1944).
[11] T. Posescu, Roumanian Reports IFA-FR-48 and 51, (1966).
[12] W.L. Hendry, K.D. Lathrop, S. Vandervoort, and J. Wooten, U.S.AEC Report LA-4287-MS (1970).
[13] R.E. Marshak, Theory of the Slowing Down of Neutrons by Elastic Collisions with Atomic Nuclei. Rev. Mod. Phys. $\underline{19}$, 185 (1947).
[14] K.M. Case, F. de Hoffmann, and G. Placzek, Introduction to the Theory of Neutron Diffusion. Washington: U.S. Government Printing Office. 1953.
[15] B. Davison, Neutron Transport Theory. Oxford Univ. Press. 1957.
[16] A.D. Galanin, Teoriya yadernykh reaktorov na teplovykh neitronakh. Moscow: Atomizdat. 1957 - English translation, Thermal Reactor Theory, Oxford: Pergamon Press. 1960.

[17] A.M. Weinberg and E.P. Wigner, The Physical Theory of Neutron Chain Reactors. Univ. Chicago Press, 1958.
[18] G.I. Marchuk: Metody rascheta jadernykh reaktorov. Moscow: Gosatomizdat. 1961.
[19] J.H. Ferziger and P.F. Zweifel, The Theory of Neutron Slowing Down in Nuclear Reactors, Oxford: Pergamon Press. 1966.
[20] M.M.R. Williams, The Slowing Down and Thermalization of Neutrons. Amsterdam: North Holland. 1966.
[21] K.M. Case and P.F. Zweifel: Linear Transport Theory. Reading, Mass.: Addison-Wesley. 1967.
[22] Computing Methods in Reactor Physics (H. Greenspan, C.N. Kelber, and D. Okrent, eds.). New York: Gordon and Breach. 1968.
[23] D.E. Parks, M.S. Nelkin, J.R. Beyster, and N.F. Wikner, Slow Neutron Scattering and Thermalization. New York: Benjamin. 1970.
[24] G.I. Bell and S. Glasstone, Nuclear Reactor Theory. New York: Van Nostrand Reinhold. 1970.
[25] M.M.R. Williams, Mathematical Methods in Particle Transport Theory. London: Butterworths. 1971.
[26] G.I. Marchuk, V.I. Lebedev, Chislennye metody v teorii perenosa neitronov. Moscow: Atomizdat. 1971.
[27] Proc. Symposia in Appl. Math. Vol.11: Nuclear Reactor Theory. Providence: Amer. Math. Soc. 1961.
[28] Proc. Brookhaven Conf. on Neutron Thermalization. U.S. AEC Report BNL 719 (C-32), 1962.
[29] Pulsed Neutron Research (Proc. Symposium, Karlsruhe 1965). Vienna: IAEA. 1965.
[30] Developments in Transport Theory (E. Inönü and P.F. Zweifel, eds.). New York: Academic Press. 1967.

[31] Neutron Noise, Waves, and Pulse Propagation (Proc. Symposium, Gainesville, 1966). U.S. AEC Symposium Series, No.9 (1967).

[32] Reactor Physics in the Resonance and Thermal Regions (ANS Meeting, San Diego 1966; A.J. Goodjohn and G.C. Pomraning, eds.). Cambridge, Mass.: MIT Press. 1966.

[33] Neutron Thermalization and Reactor Spectra (Proc. Symposium, Ann Arbor 1967). Vienna: IAEA. 1968.

[34] SIAM-AMS Proc., Vol. 1: Transport Theory (Symposium, New York 1967; R. Bellman, G. Birkhoff and I. Abu-Shumays, eds.). Providence: Amer. Math. Soc. 1969.

[35] Vychislitel'nye metody v teorii perenosa (G.I. Marchuk, ed.). Moscow: Atomizdat. 1969.

[36] Neutron Transport Theory Conf. (Blacksburg 1969). U.S. AEC Report ORO-3858-1 = CONF-690108 (1969).

[37] Transport Theory (Atlas Symposium No. 3., Oxford 1970; G.E. Hunt, ed.). J. Quant. Spectroscopy Radiative Transfer $\underline{11}$, 511-1033 (1971).

[38] Second Conference on Transport Theory (Los Alamos 1971). U.S. AEC Report CONF 710107 (1971).

[39] G.R. Keepin, Physics of Nuclear Kinetics. Reading, Mass.: Addison-Wesley. 1965.

[40] L. Pal, Nuovo Cimento Suppl. $\underline{7}$, 25 (1958); Acta Phys. Hung. $\underline{14}$, 345, 357, 369 (1962).

[41] G.I. Bell, Nucl. Sci. Eng. $\underline{21}$, 390 (1965); and in Ref. 34, p. 181.

[42] W. Matthes, Nukleonik $\underline{8}$, 87 (1966).

[43] M.M.R. Williams, J. Nucl. Energy $\underline{22}$, 153 (1968).

[44] L. Waldmann, in "Encyclopedia of Physics", Vol. 12, p. 295. Berlin: Springer. 1958.

[45] V.F. Turchin, Medlennye neitrony. Moscow. 1963. - Engl. Translation, Slow Neutrons. New York: Davey & Co. 1965.

[46] H. Hurwitz, Jr., M.S. Nelkin, and G.J. Habetler, Nucl. Sci. Eng. $\underline{1}$, 280 (1956).
[47] I. Kuščer and G.C. Summerfield, Phys. Rev. $\underline{188}$, 1445 (1969).
[48] S. Chapman and T.G. Cowling, The Mathematical Theory of Non-Uniform Gases. Cambridge Univ. Press. 1939.
[49] K. Andersen and K.E. Shuler, J. Chem. Phys. $\underline{40}$, 633 (1964).
[50] J. Lewins, Importance, The Adjoint Function. Oxford: Pergamon Press. 1965.
[51] K.M. Case, Rev. Mod. Phys. $\underline{29}$, 651 (1957).
[52] I. Kuščer and N.J. McCormick, Nucl. Sci. Eng. $\underline{26}$, 522 (1966).
[53] B.V. Gnedenko, Kurs teorii veroyatnostei. Moscow: Gostehizdat. 1954.
[54] W. Feller, An Introduction to Probability Theory and its Application, Vol. 2. New York: Wiley & Sons, 1966.
[55] E.V. Tolubinskii, Teoriya processov perenosa. Kiev: Naukova dumka. 1969.
[56] G.M. Wing, An Introduction to Transport Theory. New York: Wiley & Sons. 1962.
[57] V.S. Vladimirov, Matematicheskye zadachi odnoskorostnoi teorii perenosa chastits. Trudy matem. Inst. im. Steklova $\underline{61}$, (1961).
[58] K.M. Case and P.F. Zweifel, J. Math. Phys. $\underline{4}$, 1376 (1963).
[59] J.T. Marti, Nukleonik $\underline{8}$, 159 (1966); Zschr. angew. Math. Phys. $\underline{18}$, 247 (1967).
[60] H. Hejtmanek, Nukleonik $\underline{12}$, 145 (1969); J. Math. Phys. $\underline{11}$, 995 (1970); Transport Calculations by the Semigroup Method (unpublished).

[61] I. Vidav, J. Math, Anal. Applic. **22**, 144 (1968).

[62] J. Mika, Studia Mathematica (Warsaw), **37**, 213 (1970); J. Mika and R. Stankiewicz, Transport Theory Statist. Phys. **2**, 55 (1972).

[63] G. Scharf, Helv. Phys. Acta **40**, 929 (1967); **42**, 5 (1969).

[64] C. Cercignani, J. Math. Phys. **9**, 633 (1968).

[65] J.-P. Guiraud, Comptes rendus **266A**, 671 (1968); **268A**, 1300 (1969).

[66] I. Kuščer, in Ref. 36, p. 380.

[67] N.J. McCormick and I. Kuscer, Adv. Nucl. Sci. Technol. **7** (to be published).

[68] N. Corngold, in Ref. 34, p. 79, and in Ref. 37, p.851.

[69] I. Kuščer, in Ref. 30, p. 243.

[70] I. Kuščer and I. Vidav, J. Math. Anal. Applic. **25**, 80 (1969).

[71] J. Dorning, B. Nicolaenko and J.K. Thurber, in Ref. 36, pp. 1, 36, 76, and in Ref. 37, p. 1007; B. Nicolaenko and J.K. Thurber, in Ref. 38, pp. 96, 115, and in "The Boltzmann Equation" (Lecture Notes of Seminar, 1970-71, F.A. Grünbaum, ed.), pp. 125, 173, 211. New York: Courant Institute, NYU. 1972.

[72] G. Placzek and W. Seidel, Phys. Rev. **72**, 550 (1947); G. Placzek, Phys. Rev. **72**, 556 (1947); C. Mark, Phys. Rev. **72**, 558 (1947); J. LeCaine, Phys. Rev. **72**, 564 (1947).

[73] S. Albertoni and B. Montagnini, in Ref. 29, p. 239.

[74] M. Borysiewicz and J. Mika, J. Math. Anal. Applic. **26**, 461 (1970).

[75] E.P. Wigner and J.E. Wilkins, U.S. Report AECD 2275 (1944); J.E. Wilkins, U.S. Report CP-2481 (1944).

[76] L. Dresner, Resonance Absorption in Nuclear Reactors. New York: Pergamon Press. 1960.

[77] J.J. McInnerney, Nucl. Sci. Eng. 22, 215 (1969).

[78] F.H. Clark, Adv. Nucl. Sci. Technol. 5, 95 (1969).

[79] K.M. Case, Ann. Phys. (N.Y.) 9, 1 (1960).

[80] C. Cercignani, Ann. Phys. 40, 454 (1966).

[81] H.G. Kaper, J.H. Ferziger and S.K. Loyalka, in Ref. 33, p. 95.

[82] T. Asaoka, J. Nucl. Energy 22, 99 (1968); Nucl. Sci. Eng. 34, 122 (1968).

[83] M. Čopič, Nucl. Sci. Eng. (to be published).

[84] M.M.R. Williams, Adv. Nucl. Sci. Technol. 7 (to be published).

[85] N. Wiener and E. Hopf, Ber. Akad. Berlin, Math. Phys. Kl. (1931), p. 696.

[86] E.L. Wachpress, Iterative Solution of Elliptic Systems, and Applications to the Neutron Diffusion Equations in Reactor Physics. Englewood Cliffs, N.J.: Prentice-Hall. 1966.

[87] H.G. Kaper, G.K. Leaf and A.J. Lindeman, U.S. Report ANL-7925 (1972).

[88] M.A. Tavel and M.S. Zucker, Transport Theory Statist. Phys. 1, 115 (1971).

[89] B.G. Carlson and K.D. Lathrop, in Ref. 22, p. 171.

[90] M. Ribarič, Arch. Rational Mech. Anal. 8, 381 (1961); 15, 54 (1964); 16, 196 (1964).

[91] K. Aoki and A. Shimizu, J. Nucl. Sci. Technol. 2, 149 (1965).

[92] G.E. Hunt, in Ref. 37, p. 655.

[93] R. Aronson, Nucl. Sci. Eng. 27, 271 (1967); Transport Theory Statist. Phys. 1, 209 (1971).

[94] R. Bellman, R. Kalaba, and G.M. Wing, J. Math. Phys. **1**, 280 (1960).

[95] H.C. van de Hulst and K. Grossman, in "The Atmospheres of Venus and Mars" (J.C. Brandt and M.B. McElroy, eds.) p. 35. New York: Gordon & Breach, 1968.

[96] H.C. van de Hulst, J. Computational Phys. **3**, 291 (1968).

[97] W. Pfeiffer and J.L. Shapiro, Nucl. Sci. Eng. **34**, 336 (1968).

[98] S. Pahor and P.F. Zweifel, J. Math. Phys. **10**, 581 (1969).

[99] G.C. Pomraning and M. Clark, Nucl. Sci. Eng. **16**, 147 (1963); G.C. Pomraning, ibid. **29**, 220 (1967).

[100] S. Kaplan, Adv. Nucl. Sci. Technol. **5**, 185 (1969).

[101] S.K. Loyalka and H. Lang, Rarefied Gas Dynamics, 7th Symposium (in press).

[102] C. Cercignani, in Ref. 37, p. 973.

[103] M.S. Nelkin, Nucl. Sci. Eng. **7**, 552 (1960).

[104] R. Kladnik and I. Kuščer, Nucl. Sci. Eng. **11**, 116 (1961); **13**, 149 (1962).

[105] T. Kahan, G. Rideau et P. Roussopoulos, Les méthodes d'approximation variationnelles dans la théorie des collisions atomiques et dans la physique des piles nucléaires. Mém. Sci. Math. **134**. Paris: Gauthier-Villars. 1956.

[106] S. Kaplan, Adv. Nucl. Sci. Technol. **3**, 233 (1966).

[107] J. Lewins, Adv. Nucl. Sci. Technol. **4**, 309 (1968).

[108] G. Goertzel and M.H. Kalos, Monte Carlo in Transport Problems, in Progress in Nuclear Energy, Ser. I: Physics and Mathematics. London: Pergamon Press. 1958.

[109] E.D. Cashwell and C.J. Everett, The Monte Carlo Method for Random Walk Problems. New York: Pergamon Press. 1959.

[110] Metod Monte Karlo v probleme perenosa izluchenii, (G.I. Marchuk, ed.). Moscow: Atomzidat. 1967.
[111] J. Spanier and E.M. Gelbard, Monte Carlo Principles and Neutron Transport Problems. Reading, Mass.: Addison-Wesley, 1969.
[112] J.R. Askew, in Ref. 37, p. 905.
[113] E.M. Gelbard, J.A. Davis, and L.A. Hageman, in Ref. 34, p. 157.
[114] H. Greenspan, C.N. Kelber and D. Okrent, foreword to Ref. 22.
[115] R.S. Varga, Matrix Iterative Analysis, p. 1. Englewood Cliffs, N.J.: Prentice-Hall. 1962.

Figure Caption

Fig. 1. A threedimensional picture of typical λ vs. B^2 eigenvalue spectra for real $B^2 = -1/L^2$. Thick lines ... loci of eigenvalues $\lambda_i(B^2)$. Shaded area ... locus of line spectra. Shaded solid ... locus of area spectra.

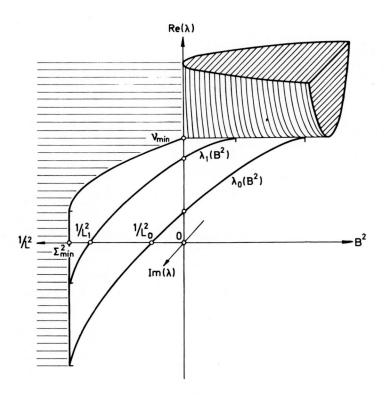

Fig. 1

RELATIVISTIC BOLTZMANN THEORY

S.R. DE GROOT

Institute of Theoretical Physics
University of Amsterdam, Amsterdam, The Netherlands

SUMMARY

In the last thirty years a relativistic kinetic theory of transport processes has been developed, which is based on an appropriate generalization of Boltzmann's celebrated equation. The relevant macroscopic laws could be derived in the framework of this theory. Moreover transport coefficients could be calculated, according to various procedures:
  (a) the eigenvalue method,
  (b) the simplified collision description,
  (c) the moment method and
  (d) the orthogonal functions method.

The last - new - method is free from the limitations, inherent in the first three. In fact it allows to find values for the relativistic transport coefficients of systems of particles with <u>arbitrary cross-sections</u> in <u>successive approximations</u>. Results are presented for the heat conductivity, the viscosities and the coefficient of sound absorption of simple relativistic gases with different cross-sections, in various approximations.

1. Introduction

The first paper on relativity, entitled "Zur Elektrodynamik bewegter Körper", was published one year before the death of Ludwig Boltzmann. The generalization of Boltzmann's transport equation, in such a way that it satisfied the requirements of relativity theory, was given more than thirty years later. Yet the title of the subject, on which I have been asked to give an outline here, namely "Relativistic Boltzmann Theory", is no misnomer, since the relativistic version of the kinetic theory of transport processes leans on Boltzmann's fundamental ideas.

The purpose of this review is threefold. First (in section 2) it is shown how one may formulate a relativistic transport equation. Then (in section 3) the macroscopic laws (conservation laws, entropy law, phenomenological laws) are derived from the kinetic basis, together with statistical expressions for all relevant macroscopic quantities. Finally (in section 4) it is indicated how one may obtain the properties of the transport coefficients, in particular their symmetries and magnitudes. The latter have been calculated by various methods: (a) the eigenvalue method, (b) the simplified collision description, (c) the moment method, and quite recently (d) the orthogonal functions method. The __orthogonal functions method__ is free from the limitations to which the other procedures are subject. In fact this method permits to find transport coefficients for systems of particles with __arbitrary cross-section__ in any __desired approximation__ (just as does its non-relativistic counterpart, due to Chapman and Enskog).

2. Relativistic Transport Equation

The transport equation for a relativistic gas is a direct generalization of Boltzmann's integro-differential equation of 1872. It was written down originally by A. Lichnerowicz and R. Marrot in 1940 [1-5]. The equation describes the behaviour of a relativistic distribution function $f_i(x, p_i)$, where $i = 1, 2, \ldots, n$ indicates the component of mixture of n species, x the vector x^α ($\alpha = 0, 1, 2, 3$) with time and space components $(ct, \vec{r})$ and p_i the energy-momentum vector p_i^α with components $(E_i/c, \vec{p}_i)$. This function is defined in such a way that

$$p_i^\alpha \, d\sigma_\alpha \, f_i(x, p_i) d^3p_i / p_i^0 \qquad (1)$$

gives the number of particle world lines with the energy-momentum four-vector in the range $(p_i, p_i + d^4p_i)$ which cross a spacelike three-surface $d\sigma_\alpha$ ($d\sigma_0 > 0$) situated at x. In view of the relation $p_{i\alpha} p_i^\alpha = -(p_{i0})^2 + (\vec{p}_i)^2 = -(m_i c)^2$ (with m_i the mass of a particle of species i), the function f_i depends on x and $\vec{p}_i$ only.
(Metric: $g^{00} = -1$, $g^{kk} = 1$ for $k = 1, 2, 3$, other $g^{\alpha\beta}$ zero.)
It has been proved [6-9] that $f(x, p_i)$ is a scalar independent of the orientation of $d\sigma_\alpha$. This implies that $f(x, p_i) d^3p_i$ is the number density at $\underline{x}$ of particles of species i with momenta in the range $(p_i, p_i + dp_i)$.

The distribution function satisfies the relativistic transport equation, referred to in the beginning of this section, and which reads for the species i:

$$p_i^\alpha \partial_\alpha f_i(x, p_i) = \sum_{j=1}^{n} \iiint (f_i' f_j' W_{ij}' - f_i f_j W_{ij})$$

$$\frac{d^3p_j}{p_j^0} \frac{d^3p_i'}{p_i'^0} \frac{d^3p_j'}{p_j'^0} \equiv \sum_{j=1}^{n} C(f_i, f_j) \quad , \tag{2}$$

where the abbreviations $f(x,p_i) \equiv f_i$ and $f(x,p_i') \equiv f_i'$ have been used and where the collision rate W_{ij} is related to the differential cross-section σ_{ij} for the collision $(p_i, p_j \to p_i', p_j')$ as [10]

$$W_{ij} = \sigma_{ij} s_{ij} \delta^{(4)}(p_i^\alpha + p_j^\alpha - p_i'^\alpha - p_j'^\alpha) \quad . \tag{3}$$

The cross-section σ_{ij}, which is defined in the centre-of-mass frame, depends only upon the relative velocity and the scattering angles. The parameter s_{ij} is the "Mandelstam variable"

$$s_{ij} \equiv P_{ij}^2 = -P_{ij}^\alpha P_{ij\alpha} \quad , \tag{4}$$

which represents the square of the magnitude P_{ij} of the total four-momentum

$$P_{ij}^\alpha \equiv p_i^\alpha + p_j^\alpha = p_i'^\alpha + p_j'^\alpha \equiv P_{ij}'^\alpha \tag{5}$$

of two colliding particles. This latter energy-momentum conservation law is responsible for the appearance of the four-dimensional delta function in the expression (3). The collision rate W_{ij}' is related in the same way to the cross-section σ_{ij}' of the inverse collision $(p_i', p_j' \to p_i, p_j)$:

$$W_{ij}' = \sigma_{ij}' s_{ij} \delta^{(4)}(p_i^\alpha + p_j^\alpha - p_i'^\alpha - p_j'^\alpha) \quad . \tag{6}$$

It may be noted here that the discussion of quantum effects and the influence of phenomena of general relativity have been left out in the present discussion.

3. Macroscopic Covariant Laws

The systems studied here are characterized by a number of macroscopic laws: (i) the conservation laws of mass and of energy-momentum, (ii) the entropy law and (iii) the phenomenological laws. An important purpose of relativistic kinetic theory is the (covariant) derivation of these macroscopic laws, together with expressions for the macroscopic quantities which appear in them as integrals involving the distribution functions and relevant microscopic quantities. This programme has been carried out both for a one-component system [5] and for a mixture [11]. In this review the laws will be stated for a mixture of non-reacting species $i = 1,2,\ldots,n$.

(i) *The conservation laws*. The mass four-flow and the energy-momentum tensor may be defined as the following averages:

$$M_i^\alpha \equiv m_i c \int p_i^\alpha f_i \frac{d^3 p_i}{p_i^0}, \quad (\alpha = 0,1,2,3), \quad (i = 1,2,\ldots,n) \tag{7}$$

$$T^{\alpha\beta} \equiv c \sum_{i=1}^{n} \int p_i^\alpha p_i^\beta f_i \frac{d^3 p_i}{p_i^0}. \quad (\alpha,\beta = 0,1,2,3). \tag{8}$$

With the help of the transport equation (3) one derives the conservation laws of mass and of energy-momentum:

$$\partial_\alpha M_i^\alpha = 0, \quad (i = 1,2,\ldots,n) \tag{9}$$

$$\partial_\beta T^{\alpha\beta} = 0, \quad (\alpha = 0,1,2,3) \tag{10}$$

where $\partial_\alpha \equiv (\partial_0, \nabla)$ with $\partial_0 \equiv \partial/\partial(ct)$. One may notice that the time component $\alpha = 0$ of (7) represents a mass density of component i, while the space components $\alpha = 1,2,3$ represent a three-flow of component i; thus (9) is simply the covariant generalization of the familiar continuity equation of mass. Similarly from (8) it is apparent that T^{oo} is an energy density, cT^{ok} ($k = 1,2,3$) an energy flow, $c^{-1}T^{ko}$ ($k = 1,2,3$) a momentum density and $T^{k\ell}$ ($k,\ell = 1,2,3$) a momentum flow (pressure), so that the cases $\alpha = 0$ and $\alpha = 1,2,3$ of (10) represent indeed the covariant laws of energy and momentum conservation respectively. The total mass flow

$$M^\alpha \equiv \sum_{i=1}^{n} M_i^\alpha$$

is used to define the covariant mass density ρ as $c^{-1}(-M_\alpha M^\alpha)^{1/2}$ and the hydrodynamic four-velocity U^α as $\rho^{-1}M^\alpha$, so that $U_\alpha U^\alpha = -c^2$. The density ρ_i of the species i is subsequently defined as $-M_i^\alpha U_\alpha/c^2$. (In this way in the "local rest frame", i.e. the frame in which U^α is purely timelike, the density ρ_i is equal to the zero component of M_i^α divided by c.) The specific energy e, the energy flow I_q^α, the diffusion flow I_i^α (of species i) and the pressure tensor $P^{\alpha\beta}$ are given by the covariant expressions:

$$e \equiv U^\alpha T_{\alpha\beta} U^\beta/\rho c^2 , \tag{11}$$

$$I_q^\alpha \equiv -U^\beta T_{\beta\gamma} \Delta^{\gamma\alpha} , \tag{12}$$

$$I_i^\alpha \equiv M_i^\alpha - \rho_i U^\alpha , \tag{13}$$

$$P^{\alpha\beta} \equiv \Delta^{\alpha\gamma} T_{\gamma\varepsilon} \Delta^{\varepsilon\beta} , \tag{14}$$

with the "projector"

$$\Delta^{\alpha\beta} \equiv g^{\alpha\beta} + U^{\alpha}U^{\beta}/c^2 \ . \tag{15}$$

(The quantities are all seen to be related in a simple way to the components of $T^{\alpha\beta}$ in the local rest frame, where $U^{\alpha} = (c,0,0,0)$.) All physical quantities introduced depend on the first and second "moments" of the distribution functions, written down in (7) and (8).

(ii) *The entropy law*. In contrast to the mass four-flow and the energy-momentum tensor, the entropy is not an average of a microscopic quantity, but an expression involving an integral over the logarithm of f_i multiplied by f_i itself. Generalizing Boltzmann's definition one writes down an entropy four-vector given as

$$S^{\alpha} = -kc \sum_i \int p_i^{\alpha} (\ln h^3 f_i - 1) f_i \frac{d^3 p_i}{p_i^0} \tag{16}$$

(with h a constant such that $h^3 f_i$ is dimensionless), where again the zero component is a density, of the form of Boltzmann's original entropy expression, and the space components represent a flow, in this case the entropy flow. The covariant specific entropy s is now defined as the quantity $-S^{\alpha}U_{\alpha}/\rho c^2$. By differentiation with respect to the space-time coordinates one obtains, with the help of the mass law, a so-called balance equation which expresses the fact that the entropy density not only changes through entropy flow but also due to a local entropy production given by

$$\sigma \equiv \partial_{\alpha} S^{\alpha} \ . \tag{17}$$

One may characterize the equilibrium situation by $\sigma = 0$. From the transport equation, together with the condition of "bilateral normalization" [12-13] of the collision rates:

$$\iint W_{ij} \frac{d^3 p'_i}{p'^0_i} \frac{d^3 p'_j}{p'^0_j} = \iint W'_{ij} \frac{d^3 p'_i}{p'^0_i} \frac{d^3 p'_j}{p'^0_j} \tag{18}$$

(in quantum language: unitarity of the scattering matrix), one may derive [11] the relativistic H-theorem, which expresses the fact that $\sigma \geq 0$. The state of non-equilibrium is thus characterized by a positive entropy production σ.

The equilibrium distribution function f_i^{eq} follows from the condition of vanishing σ (17) together with (18). It turns out to be the Jüttner function [14], derived already in 1911, from relativistic equilibrium statistics:

$$f_i^{eq} = h^{-3} \exp(a_i m_i + b_\alpha p_i^\alpha) \tag{19}$$

with constants a_i and b_α. The length b of the last four-vector may be used to define the temperature T:

$$kT = c/b \quad , \quad (b^2 = -b_\alpha b^\alpha) \quad . \tag{20}$$

The constants mentioned then follow with the help of the physical quantities defined in the foregoing in such a way that (19) gets the form

$$f_i^{eq} = h^{-3} \exp(\frac{m_i \mu_i + p_i^\alpha U_\alpha}{kT}) \quad , \tag{21}$$

where μ_i is the chemical potential (partial specific Gibbs function) of species i, which is a function of the temperature and the density ρ_i of species i

$$\mu_i = \frac{kT}{m_i} \ln \frac{h^3 \rho_i}{4\pi m_i^3 ckT K_2(m_i c^2/kT)} \quad , \tag{22}$$

where K_n is the modified Bessel function of the second kind of order n (also referred to as Kelvin function). With the help of the equilibrium distribution function (21-22) the macroscopic quantities mentioned may be calculated. One finds thus for the energy density ρe (11) and the pressure tensor (14-15), if use is made of (8) and (21-22):

$$\rho e^{eq} = \sum_{i=1}^{n} \rho_i \{ c^2 \frac{K_3(m_i c^2/kT)}{K_2(m_i c^2/kT)} - \frac{kT}{m_i} \} , \qquad (23)$$

$$P_{\alpha\beta}^{eq} = p \, \Delta_{\alpha\beta} , \qquad p \equiv n k T \qquad (24)$$

with $n = \sum_{i=1}^{n} \rho_i/m_i$ the total particle number density.

The energy density (23) is thus a peculiar function of the temperature: for moderately relativistic systems it is equal to ρc^2, the rest mass energy, plus the non-relativistic term $\frac{3}{2}nkT$, and corrections of order c^{-2}; for the ultra-relativistic case of $kT \gg m_i c^2$ the result is $3nkT$. The pressure on the other hand is still formally the same as given by Boyle-Gay Lussac's law, as (24) shows. The heat flow I_q^α (12), the diffusion flows I_i^α (i = 1,2,...,n) (13), and the viscous pressure tensor, defined as

$$\Pi_{\alpha\beta} = P_{\alpha\beta} - p \, \Delta_{\alpha\beta} \qquad (25)$$

all vanish at equilibrium, as follows if (21-22) is inserted into (11-15) with (7-8).

Outside equilibrium, however, these flows are different from zero. The situation may then be described by means of distribution functions of the form

$$f_i(x,p_i) = f_i^{(o)}(x,p_i)\{1+\phi_i(x,p_i)\} \quad , \quad (i = 1,2,\ldots,n) \quad . \tag{26}$$

Here the so-called "zero order distribution function" $f_i^o(x,p_i)$ has the same form as the equilibrium function f_i^{eq} (21-22), but with "local" quantities $\rho_i(x)$, $T(x)$ and $U^\alpha(x)$, i.e. functions of the space-time coordinates x rather than uniform constants, as they are at equilibrium. The temperature is defined by the requirement that the connexion with the local specific energy and the local density is the same as in equilibrium, cf. expression (23). It may be remarked here that this definition of temperature in a local frame suffices for the present theory. One finds all equilibrium as well as non-equilibrium properties in terms of this temperature. (A discussion on the problem of how the thermodynamic quantities transform from one frame to the other is therefore superfluous in the framework of the present theory.) The expression (26) contains the function $\phi_i(x,p_i)$, which is a measure for the deviation from "local equilibrium", described by the zero order distribution function $f_i^{(o)}(x,p_i)$. The function $\phi_i(x,p_i)$ will be considered in "first Enskog order" $\phi_i^{(1)}(x,p_i)$ only. This means that it will be solved from an equation which is obtained from the transport equation (2) by the following procedure. One substitutes (26), neglects the influence of $\phi_i(x,p_i)$ itself in the left-hand side and of second-order terms in the right-hand side. In this way, using moreover (5), (18) and (21-22) for $f_i^{(o)}(x,p_i)$ one obtains for $\phi_i^{(1)}(x,p_i)$ the "linearized transport equation"

$$p_i^\alpha \partial_\alpha f_i^{(o)} = - f_i^{(o)} \sum_{j=1}^{n} \iiint f_j^{(o)} (\phi_i^{(1)} + \phi_j^{(1)} - \phi_i'^{(1)} - \phi_j'^{(1)})$$

$$W'_{ij} \frac{d^3p_j}{p_j^{\circ}} \frac{d^3p'_i}{p_i'^{\circ}} \frac{d^3p'_j}{p_j'^{\circ}} \equiv -f_i^{(o)} \sum_{j=1}^{n} L[\phi_{ij}^{(1)}]. \quad (27)$$

The results obtained with the help of this equation are said to be of the "first Enskog order", which will be used throughout the remainder of this review.

With the help of these linearized transport equations and the definitions of the macroscopic quantities given above it may be shown [11] that a relativistic Gibbs relation of the form:

$$T \partial_\alpha s = \partial_\alpha e + p \partial_\alpha v - \sum_{i=1}^{n} \mu_i \partial_\alpha c_i \quad (28)$$

(with $v = \rho^{-1}$ and $c_i = \rho_i/\rho$) remains valid outside equilibrium in the first Enskog order. The entropy production (17) is then found [11] to have the source strength:

$$\sigma = -\Pi \frac{\partial_\alpha U^\alpha}{T} - I_q^\alpha (\frac{\partial_\alpha T}{T^2} + \frac{DU_\alpha}{c^2 T}) - \sum_{i=1}^{n} I_i^\alpha \partial_\alpha (\frac{\mu_i}{T}) - \overset{\circ}{\Pi}{}^{\alpha\beta} \frac{<\partial_\alpha U_\beta>}{T}. \quad (29)$$

It contains the "flows" (heat flow I_q^α, diffusion flows I_i^α, and a trace Π and a traceless part $\overset{\circ}{\Pi}$ of the viscous pressure tensor $\Pi^{\alpha\beta}$), different from zero outside equilibrium, and as factors corresponding "thermodynamic forces". The latter contain gradients of the temperature T, the chemical potentials μ_i and the hydrodynamic velocity U^α (the brackets in the last term indicate a space-space like, symmetric and traceless part of the tensor enclosed); moreover in the heat term a relativistic thermodynamic force arises which contains accelerations (D stands for $U^\alpha \partial_\alpha$) and which was predicted on thermodynamic grounds [15], just as the complete expression [16]. If reacting mixtures are considered [17-18], one obtains in the entropy source strength an additional term which is the product of the chemical reaction rate and the chemical affinity.

(iii) The phenomenological laws. Again within the framework of the linearized kinetic theory relationships have been derived for single systems [5] and for mixtures [11] which connect the fluxes to the thermodynamic forces and which are known as the "phenomenological laws". They may be found in the following way. If in the linearized transport equations (27) the conservation laws are employed to eliminate the time derivatives from the left hand sides, one notices that general solutions exist, which are linear combinations of the thermodynamic forces of the form

$$\phi_i^{(1)} = A_i \partial_\alpha U^\alpha - B_i^\alpha \Delta_{\alpha\beta} (\frac{\partial^\beta T}{T} + \frac{DU^\beta}{c^2})$$

$$- \sum_{j=1}^{n} B_{ij}^\alpha \Delta_{\alpha\beta} T \partial^\beta (\frac{\mu_j}{T}) + C_i^{\alpha\beta} <\partial_\alpha U_\beta> \qquad (30)$$

with coefficients A_i, B_i^α, B_{ij}^α and $C_i^{\alpha\beta}$ still to be determined from the equations. If these expressions are used in (26) for ϕ_i and then the resulting distribution function introduced into the definitions of the "flows", given above, one finds "phenomenological laws", which relate these flows to the thermodynamic forces:

$$\Pi = - n_v \partial_\alpha U^\alpha , \qquad (31)$$

$$I_q^\alpha = - L_{qq} \Delta^{\alpha\beta} (\frac{\partial_\beta T}{T} + \frac{DU_\beta}{c^2}) - \sum_{j=1}^{n} L_{qj} \Delta^{\alpha\beta} T \partial_\beta (\frac{\mu_j}{T}) , \qquad (32)$$

$$I_i^\alpha = - L_{iq} \Delta^{\alpha\beta} (\frac{\partial_\beta T}{T} + \frac{DU_\beta}{c^2}) - \sum_{j=1}^{n} L_{ij} \Delta^{\alpha\beta} T \partial_\beta (\frac{\mu_j}{T}) , \qquad (33)$$

$$\overset{o}{\Pi}{}^{\alpha\beta} = - 2\eta <\partial^\alpha U^\beta> . \qquad (34)$$

The quantities n_v, L_{qq}, L_{qi}, L_{iq}, L_{ij} and η which appear here are the "transport coefficients". The first is the volume viscosity, the second is related to the heat conductivity λ, the third and fourth describe cross-effects between heat conduction and diffusion, the fifth represents the diffusion phenomena, and η is the shear viscosity. They depend upon the coefficients A_i, B_i^α, B_{ij}^α and $C_i^{\alpha\beta}$ which appeared in (30). Their properties will be studied in the next section.

It may be noticed that (31) describes a scalar phenomenon, since the flux and the thermodynamic force are scalars, while (32) and (33) contain vectors and (34) has tensor character.

If a chemical reaction (or a relaxation phenomenon) can take place inside the system, one finds an additional scalar flow, namely the chemical reaction rate I_r, and a corresponding thermodynamic force, the chemical affinity A. The phenomenological law (31) is now replaced by two coupled equations [17-18] containing scalar flows and thermodynamic forces:

$$\Pi = - n_v \, \partial_\alpha U^\alpha - L_{vr} A \qquad (35)$$

$$I_r = - L_{rv} \, \partial_\alpha U^\alpha - L_r A \quad . \qquad (36)$$

Besides the volume viscosity n_v, it contains two transport coefficients L_{vr} and L_{rv}, which represent cross-effects between the volume viscous flow and the chemical reaction. Finally the transport coefficient L_r characterizes the chemical reaction.

4. The Transport Coefficients

The transport coefficients, which appear in the phenomenological laws of the preceding sub-section, have still to be determined. Various methods, by which their character was studied, will be outlined here. In particular their general symmetry properties and their magnitudes will be determined.

Their general expressions follow from the preceding formalism, based on the linearized transport equation. To show this, let us consider the heat conductivity L_{qq} as a characteristic example. In the phenomenological law (32) this quantity stands for the following integral:

$$L_{qq} = -\frac{c}{3} \Delta_{\alpha\beta} \sum_{i=1}^{n} \int B_i^\alpha \, p_i^\beta (p_i^\gamma U_\gamma + mh) f_i^{(o)} \frac{d^3 p_i}{p_i^o} \qquad (37)$$

with $h = e + pv$ the specific enthalpy. The coefficient B_i^α appeared in the general solution (30) of the linearized transport equation (27). Since in (30) the thermodynamic force pertaining to the heat conduction is independent from the other thermodynamic forces the coefficient B_i^α must itself satisfy the transport equation (30), <u>i.e.</u>, we have

$$-(kT)^{-1}(p_i^\gamma U_\gamma + mh)\Delta^{\alpha\beta} p_{i\beta} = \sum_{j=1}^{n} L[B^\alpha_{(ij)}] \quad , \qquad (38)$$

where L indicates the collision operator defined in the last equality of (27). It follows from the linearity and the Lorentz covariance of this operator that the solution of the last equation must have the general form

$$B_i^\alpha = B_i \Delta^{\alpha\beta} P_{i\beta} , \qquad (39)$$

so that a scalar function B_i remains to be determined, if one wants to find the heat conductivity (37).

Similar expressions may be derived for the other transport coefficients. The complete set of transport coefficients contains the coefficients A_i, B_i^α, B_{ij}^α and $C_i^{\alpha\beta}$, which should be solved from equations like (38). They will depend on the collision rates W_{ij} and W'_{ij} and thus on the cross-sections σ_{ij} and σ'_{ij} of the particle collisions, since these quantities are contained in the collision operator, as was shown in section 1. For the study of the symmetry properties (subsection i) of the transport coefficients, which follow from symmetries of the collision rates or cross-sections, the formalism developed so far suffices. More detailed calculations will be needed, if we want to calculate (subsection ii) actual magnitudes of the transport coefficients.

(i) *Symmetry properties*. From space symmetries (Lorentz invariance), that is, in the case studied here, from the isotropy of the fluid system, it followed that the transport coefficients were all scalars and that phenomena of different tensorial order did not couple, as the laws (31-36) show. This statement constitutes the relativistic generalization [11] of the so-called Curie principle, originally formulated by Pierre Curie [19] and later proved with the help of the theory of invariants [20] in non-relativistic theory of continuous media.

An important general property, which electromagnetic interactions are supposed to obey, is the time (better: motion) reversal invariance of the microscopic equations of motion. This implies the following property to be valid for the collision rates:

$$W_{ij}(\vec{p}_i,\vec{p}_j|\vec{p}_i',\vec{p}_j') = W_{ij}(-\vec{p}_i',-\vec{p}_j'|-\vec{p}_i,-\vec{p}_j) \quad , \tag{40}$$

where only the space components of the particle momenta are indicated. (The time components do not change upon reversal of motion: they are the energies p_i^0 etc..) This invariance entails for the transport coefficients of (32), (33), (35) and (36) reciprocal relations between them of the form

$$L_{iq} = L_{qi} \quad , \quad L_{ij} = L_{ji} \quad (i,j = 1,2,\ldots,n) \tag{41}$$

$$L_{vr} = - L_{rv}$$

as was proved [10,17-18] from the scheme outlined above. These reciprocities constitute the relativistic generalization of the Onsager relations [21]. The minus sign in (42) arises because the two thermodynamic forces involved have different motion parity: in fact the affinity A is even and the velocity derivative $\partial_\alpha U^\alpha$ is odd in the particle velocities. The possibility of the occurence of such a minus sign was first pointed out by Casimir [22].

If macroscopic electromagnetic fields are present in the systems, many more transport coefficients are needed for the description (Hall effects etc.) of the irreversible processes. The quantities involved possess different character under time and space reflexions. This gives rise to a whole array of new reciprocal relations [8] provided with plus or minus signs depending on the various parities.

If motion reversal invariance (40) would not be strictly satisfied for the interactions present in the systems studied, as has been suggested [23], deviations from the reciprocal relations should be expected. However, in practice such deviations can hardly be detected in ordinary fluid systems, where pure electromagnetic interactions are preponderant.

(ii) **Magnitudes**. The final problem to be solved is how to determine numerical values for the transport coefficients if the collision cross-sections, which characterize the system, are given. In this subsection the various methods, proposed to deal with this task in relativistic kinetic theory, are reviewed, roughly in historical order.

a. The eigenvalue method. This method is a generalization of a non-relativistic theory [24-25] in which one seeks eigenfunctions of the collision operator of the linearized transport equation. Originally it was applied for a single gas [5]. Its heat conductivity λ and volume viscosity η_v were calculated. It turned out that the latter did not vanish, as is the case in non-relativistic theory, but has a magnitude of the order of c^{-4}. The method was later extended [26,10] to include the shear viscosity η and the transport properties of mixtures (so that values for the diffusion and thermal diffusion coefficients were also obtained). The method possesses the same limitation as the first non-relativistic treatments: it is confined to the study of one single model with a form of the particle collision cross-section chosen in such a way that the eigenvalue problem is solvable. In the non-relativistic theory this means that $\sigma(g,\Theta)$ (with g the momentum transfer and Θ the scattering angle, both defined in the centre-of-mass system of the colliding particles) has the form of a product of g^{-1} and a function $\Gamma(\theta)$. This situation corresponds to a fifth power repulsion law. Israel [5] was the first to show that the relativistic eigenvalue problem (for η_V and λ) is solvable if an extra factor $\{1 + (g/mc)^2\}^{-1}$ is added to the collision cross-section. We shall refer to particles with such a cross-section as "Israel particles".

The answers obtained in this case by calculating the volume viscosity [5], the heat conductivity [5] and the shear viscosity [26,10] correspond to the values for $[\eta_V]_1$, $[\lambda]_1$ and $[\eta]_1$ listed at the end of this subsection under Israel particles.

b. *Method with simplified collision term*. In contrast to the preceding and the two following methods, in this procedure the collision integral is simplified. In fact it is replaced by a simple difference, involving a relaxation time, which is left unspecified. The method generalizes a non-relativistic one [27] and has been employed [28] in combination with the third method. Owing to the appearance of an undetermined parameter its results cannot be compared with the expressions, obtained by means of the other formalisms.

c. *The moment method*. A method, in principle free from the limitation of being applicable only to the case of Israel particles, may be developed as a relativistic version of the moment method introduced in non-relativistic theory [29]. The general formalism was set up and it turned out that the same results as obtained in method *a* could be found for Israel molecules [30], after initial considerations of a similar nature [3,28,31-33] had been given.

An assessment of the range of validity of this method is possible within the framework of the formalism to be developed in the next point. It will then turn out that only for Israel particles the moment method will yield the first relativistic corrections completely.

    d. The orthogonal functions method. All three methods outlined above contain limitations. Moreover their range of validity as procedures to solve the linearized transport equation is hard to assess. In a paper [34] on an astrophysical subject Charles W. Misner wrote in 1967 on the desirability of obtaining numerical values for the relativistic transport coefficients: "... which would in any case require a Chapman-Enskog solution of the relativistic Boltzmann equation in order to be reliably computed."

    Such a relativistic theory, which generalizes the classical Chapman-Enskog [35-36] method, has been worked out recently [37]. It allows one to calculate successive approximations for the values of the transport coefficients as they occur in the linearized theory in a manner analogous to the non-relativistic method. Its principle will be explained in the following treatment of a characteristic case.

    Let us consider a one-component system of particles with a collision cross-section σ. (For particles without internal degrees of freedom, which case is considered here, the cross-sections σ and σ' of section 1 for a collision and its inverse are equal and thus also the collision rates W and W'.) Then the general expression (37) for the heat conductivity λ ($\equiv L_{qq}/T$) reads:

$$\lambda = - \frac{c}{3T} \Delta_{\alpha\beta} \int B \, p^\alpha \, p^\beta \, (p^\gamma U_\gamma + mh) f^{(o)} \frac{d^3p}{p^o} \qquad (43)$$

with a coefficient B which follows from the equation (38-39):

$$L[B \Delta^{\alpha\beta} P_\beta] = - (kT)^{-1} (p^\gamma U_\gamma + mh) \Delta^{\alpha\beta} P_\beta . \qquad (44)$$

The collision operator L occurs in the linearized transport equation (27), which in the present case may be written as:

$$p^\alpha \partial_\alpha f^{(o)} = -f^{(o)} \int f_1^{(o)} (\phi^{(1)} + \phi_1^{(1)} - \phi'^{(1)} - \phi_1'^{(1)}) W \frac{d^3p_1}{p_1^o} \frac{d^3p'}{p'^o} \frac{d^3p_1'}{p_1'^o} \equiv - f^{(o)} L[\phi^{(1)}] . \qquad (45)$$

(As the notation shows, collisions are now considered between a particle "sin nombre" and a particle with properties characterized by a lower index 1. Their properties after collision are indicated with the help of extra, upper, dashes.)

In the procedure of the "orthogonal functions method" the heat conductivity λ (43) is found from (44) with (45) in the following way. First the coefficient B is developed into a series of orthogonal functions. In the present case of the "vectorial" phenomenon of heat conduction the appropriate functions [38] are associated Laguerre polynomials of order 3/2. Thus, we write [37]

$$B(\tau) = \sum_{m=0}^{\infty} b_m L_m^{3/2}(\tau) , \qquad (46)$$

where the argument

$$\tau = - \frac{p_\alpha U^\alpha + mc^2}{kT} \qquad (47)$$

is the kinetic energy of a particle with four-momentum p^α in the proper frame of U^α, divided by kT. We must now find the coefficients b_m from the equation (44) with (46) inserted. To this end the equation is multiplied by $L_n^{3/2}(\tau)p^\alpha f^{(o)}(p^o)^{-1}$ and then integrated over the momentum $\vec{p}$. In this way we obtain - instead of the integral equation for B - an infinite set of algebraic equations for the coefficients b_m:

$$\sum_{m=0}^{\infty} b_m b_{mn} = \rho^{-1} \beta_n , \quad (n = 0,1,\ldots) , \qquad (48)$$

where the following abbreviations have been introduced:

$$\beta_n \equiv -\frac{mc}{\rho k^2 T^2} \Delta_{\alpha\beta} \int L_n^{3/2}(\tau)(p_\gamma U^\gamma + mh) p^\alpha p^\beta f^{(o)} \frac{d^3p}{p^o} , \qquad (49)$$

$$(n = 0,1,\ldots) ,$$

$$b_{mn} \equiv \frac{1}{mkT} \Delta_{\alpha\beta}[L_m^{3/2}(\tau)p^\alpha, L_n^{3/2}(\tau)p^\beta] , \quad (m,n = 0,1,\ldots) . \qquad (50)$$

In the last expression a "bracket symbol", containing the collision rate W, has been employed; it is defined as:

$$[F,G] \equiv \frac{1}{4}\frac{m^2c}{\rho^2} \int\int\int f^{(o)} f_1^{(o)} \, W \, \delta(F)\delta(G) \, \frac{d^3p}{p^o}\frac{d^3p}{p_1^o}\frac{d^3p'}{p'^o}\frac{d^3p_1'}{p_1'^o} , \qquad (51)$$

where delta symbols of the form

$$\delta\{F(p)\} \equiv F(p) + F(p_1) - F(p') - F(p_1') \qquad (52)$$

are used. With the last two expressions it follows that b_{mn} (50) is symmetrical in m and n; the components $b_{no} = b_{on}$ vanish, because the Laguerre polynomial of degree zero ($n = 0$) is equal to unity, and then in (52) the function $F(p)$ is p itself, in which case the right-hand side of (50) vanishes as a result of the conservation of four-momentum in a collision. The coefficient β_o (49) vanishes also because the Laguerre polynomial of degree zero is equal to unity. Therefore the set (48) may be written as

$$\sum_{m=1}^{\infty} b_m b_{mn} = \rho^{-1} \beta_n , \quad (n = 1, 2, \ldots) . \tag{53}$$

An approximate solution $b_m^{(r)}$ for the coefficients b_m (with $m = 1, 2, \ldots r$) is obtained by limiting the infinite number of equations to a finite set of r equations:

$$\sum_{m=1}^{r} b_m^{(r)} b_{mn} = \rho^{-1} \beta_n . \tag{54}$$

The heat conductivity (43) may be written in terms of the quantities β_m and b_m if use is made of (46) and (49):

$$\lambda = \frac{\rho k^2 T}{3m} \sum_{m=1}^{\infty} b_m \beta_m . \tag{55}$$

We now define as the <u>r-th approximation</u> $[\lambda]_r$ to the heat conductivity (55) the result which is obtained by limiting this sum to the first r terms and using the solutions $b_m^{(r)}$ ($m = 1, 2, \ldots, r$) of (54). In this way, for values of $r = 1, 2$, etcetera, successive approximations are obtained. The first and second approximations to the heat conductivity become for instance:

$$[\lambda]_1 = \frac{1}{3} \frac{k^2 T}{m} \frac{\beta_1^2}{b_{11}} , \tag{56}$$

$$[\lambda]_2 = \frac{1}{3} \frac{k^2 T}{m} \frac{\beta_1 b_{22} - 2\beta_1 \beta_2 b_{12} + \beta_2^2 b_{11}}{b_{11} b_{22} - b_{12}^2} . \tag{57}$$

The coefficients β_n and b_{mn} need thus to be calculated up to certain values of m and n to find the heat conductivity up to a certain desired approximation.

From the properties of the Laguerre polynomials and the values of the moments of the distribution function $f^{(o)}$ one may find expressions for β_n, defined by (49). The result for the case n = 1 is then

$$\beta_1 = -3\frac{m}{k}\frac{dh}{dT} \qquad (58)$$

with h the specific enthalpy. Series expansion in powers of kT/mc^2 of this expression yields:

$$\beta_1 = -\frac{15}{2}(1 + \frac{3}{2}\frac{kT}{mc^2}) + O(c^{-4}) \quad . \qquad (59)$$

The coefficients β_2 and β_3 have also been calculated. Their leading terms are of order c^{-2} and c^{-4} respectively. (In fact in the non-relativistic limit the coefficients β_n for $n \geq 2$ all vanish. This reduces the set of equations (53) to a simpler form, known from non-relativistic theory.)

The coefficients b_{mn} (50) are twelve-fold integrals, as (51) shows. Ten of these integrations may be performed. One then obtains linear combinations of quantities, which are two-fold integrals extended over the momentum transfer g and the scattering angle θ. These quantities, which we dubbed "relativistic omega integrals", are defined as

$$\omega_i^{(s)}(T) \equiv \frac{2\pi}{m^5 c(kT)^3 K_2^2(mc^2/kT)} \int_0^\infty dg\, g^7 x^{i-1} K_j(\frac{2xmc^2}{kT}) \cdot$$

$$\cdot \int_0^\pi d\theta\, \sin\theta(1-\cos^s\theta)\sigma(g,\theta); \qquad x \equiv \sqrt{1+(\frac{g}{mc})^2} \qquad (60)$$

with $s = 2,4,\ldots$, $i = 0, \pm 1,\ldots$, while $j = 2$ if i is odd and $j = 3$ if i is even. Besides the scattering cross-section $\sigma(g,\theta)$, they contain modified Bessel functions K_n of the second kind. As an example we give b_{mn} (50) for the case $m = n = 1$:

$$b_{11} = 8(\omega_1^{(2)} + \frac{kT}{mc^2}\omega_0^{(2)}) . \qquad (61)$$

If the relativistic omega integrals $\omega_i^{(s)}$ are developed as series of powers of kT/mc^2, one finds as coefficients the omega integrals:

$$\Omega_i^{(s)}(T) \equiv \frac{2\sqrt{\pi}}{m}\int_0^\infty dg\, e^{-g^2/mkT}(\frac{g}{\sqrt{mkT}})^{2i+3}$$

$$\cdot \int_0^\pi d\theta\, \sin\theta(1 - \cos^s\theta)\sigma(g,\theta) \qquad (62)$$

well-known in non-relativistic theory [39]. In this way (61) becomes

$$b_{11} = 8\Omega_2^{(2)} - \frac{kT}{mc^2}(\frac{29}{2}\Omega_2^{(2)}+2\Omega_3^{(2)}-2\Omega_4^{(2)}) + O(c^{-4}) , \qquad (63)$$

where the first two terms of the series expansion are given explicitly. (In the non-relativistic approximation only the first term survives.) With the help of the knowledge of β_1 and b_{11} one may find the first approximation $[\lambda]_1$ (56) to the heat conductivity. The coefficients b_{12}, b_{13}, b_{22}, b_{23} and b_{33} have also been calculated, so that, together with the values of β_2 and β_3, the heat conductivity may be obtained in second and third approximation [37].

We are now in a position to indicate which is the limitation of the relativistic moment method. In fact it turns out that the results obtained with the help of that method are identical with the first approximation of the present orthogonal functions method. (This is thus a situation similar to the connexion of the non-relativistic version of both methods.)

A number of results obtained with the help of the orthogonal functions method, as described above, will now be listed. They all refer to the case of a simple gas. Not only the heat conductivity λ, but also the shear viscosity η and the volume viscosity η_v will be presented in different approximations. Three different models, i.e., three different choices of the collision cross-section $\sigma(g,\theta)$, will be used in this illustration. The results have been calculated by introducing these particular choices into the general expressions valid for arbitrary collision cross-sections.

Israel particles: collision cross-section:

$$\sigma(g,\theta) = \frac{m}{2g\{1 + (g/mc)^2\}} \Gamma(\theta) \qquad (64)$$

with g the momentum transfer and $\Gamma(\theta)$ a function of the scattering angle θ. The r-th approximations $[\eta_v]_r$ to the volume viscosity, $[\lambda]_r$ to the heat conductivity and $[\eta]_r$ to the shear viscosity were calculated [37]. In all cases the two leading terms of a c^{-2}-expansion are given explicitly:

$$[\eta_v]_1 = \frac{5}{12}(\frac{kT}{mc^2})^2 \frac{2kT}{\pi a}(1 - \frac{29}{4}\frac{kT}{mc^2}) + O(c^{-8}) , \qquad (65)$$

$$[\eta_v]_r = [\eta_v]_1 + O(c^{-8}) , \qquad (r \geq 2) \qquad (66)$$

$$[\lambda]_1 = \frac{5}{4}\frac{k^2T}{\pi ma}(1 + \frac{21}{4}\frac{kT}{mc^2}) + O(c^{-4}) , \qquad (67)$$

$$[\lambda]_r = [\lambda]_1 + O(c^{-4}) \,, \quad (r \geq 2) \tag{68}$$

$$[\eta]_1 = \tfrac{1}{3}\tfrac{kT}{\pi a}(1 + \tfrac{25}{4}\tfrac{kT}{mc^2}) + O(c^{-4}) \,, \tag{69}$$

$$[\eta]_r = [\eta]_1 + O(c^{-4}) \,. \quad (r \geq 2) \tag{70}$$

Here a stands for $\tfrac{1}{2}\int_0^\pi d\theta \sin^3\theta \, \Gamma(\theta)$.

The volume viscosity is a relativistic effect, starting with a term of order c^{-4}; the heat conductivity and the shear viscosity show the well-known non-relativistic terms (of order c^0) [38].

The results (65), leading term [5], (67) [5,10,27] and (69) [10,26,40] have been obtained also with the eigenvalue method, which allows to treat Israel particles only.

<u>Fifth power repulsion</u>. The collision cross-section of relativistic particles with a fifth power repulsion law has the form

$$\sigma = \tfrac{m}{g}\,\Gamma(\theta) \,. \tag{71}$$

The results [37] of the calculations for the transport coefficients are:

$$[\eta_v]_1 = \tfrac{5}{12}(\tfrac{kT}{mc^2})^2\tfrac{kT}{\pi a}(1 - \tfrac{43}{4}\tfrac{kT}{mc^2}) + O(c^{-8}) \,, \tag{72}$$

$$[\eta_v]_r = [\eta_v]_1 + O(c^{-8}) \,, \quad (r \geq 2) \tag{73}$$

$$[\lambda]_1 = \tfrac{5}{4}\tfrac{k^2 T}{\pi ma}(1 + \tfrac{7}{4}\tfrac{kT}{mc^2}) + O(c^{-4}) \,, \tag{74}$$

$$[\lambda]_r = [\lambda]_1 + O(c^{-4}) \,, \quad (r \geq 2) \tag{75}$$

$$[\eta]_1 = \tfrac{1}{3}\tfrac{kT}{\pi a}(1 + \tfrac{11}{4}\tfrac{kT}{mc^2}) \,, \tag{76}$$

$$[\eta]_r = [\eta]_1 + O(c^{-4}) \,. \quad (r \geq 2) \tag{77}$$

Hard spheres. Their collision cross-section is:

$$\sigma = \frac{1}{4} \pi d^2 \qquad (78)$$

where d is the diameter of the particles. The relativistic transport coefficients become in first and second approximation [37]:

$$[n_v]_1 = \frac{25}{64}\left(\frac{kT}{mc^2}\right)^2 \sqrt{\frac{mkT}{\pi}} \frac{1}{d^2}\left(1 - \frac{183}{16}\frac{kT}{mc^2}\right) + O(c^{-8}) , \qquad (79)$$

$$[n_v]_2 = [n_v]_1 + \left(\frac{1}{30} + \frac{259}{900}\frac{kT}{mc^2}\right)[n_v]_1 + O(c^{-8}) , \qquad (80)$$

$$[\lambda]_1 = \frac{75}{64}\sqrt{\frac{k^3T}{\pi m}}\frac{1}{d^2}\left(1 + \frac{13}{16}\frac{kT}{mc^2}\right) + O(c^{-4}) , \qquad (81)$$

$$[\lambda]_2 = [\lambda]_1 + \left(\frac{1}{44} + \frac{763}{3872}\frac{kT}{mc^2}\right)[\lambda]_1 + O(c^{-4}) , \qquad (82)$$

$$[\eta]_1 = \frac{5}{16}\sqrt{\frac{mkT}{\pi}}\frac{1}{d^2}\left(1 + \frac{25}{16}\frac{kT}{mc^2}\right) + O(c^{-4}) , \qquad (83)$$

$$[\eta]_2 = [\eta]_1 + \left(\frac{3}{202} + \frac{2695}{4444}\frac{kT}{mc^2}\right)[\eta]_1 + O(c^{-4}) . \qquad (84)$$

From these expressions it is seen that, just as the non-relativistic values change by small amounts only if one passes from the first to the second approximation, the relativistic values do the same in the weakly relativistic case studied. While the first point is well-known from non-relativistic theory [39] the second point means that the behaviour of the relativistic terms is satisfactory in a similar fashion.

Sound absorption. The relativistic theory of sound propagation in a simple (dissipative) gas has been worked out [40-41]. One then obtains an expression for the absorption coefficient α, which shows the influence of the heat conductivity λ and the two viscosities η and η_v on the attenuation of sound:

$$\alpha = \frac{\omega^2}{2\rho c_s^3}\{\frac{\lambda}{c_v}(\frac{c_p-c_v}{c_p} - \frac{2}{h}\frac{T\gamma}{\rho\beta_T} + \frac{1}{h^2}\frac{Tc_p}{\rho\beta_T}) + \frac{c^2}{h}(\frac{4}{3}\eta + \eta_v)\}. \quad (85)$$

In this formula ω is the angular frequency of the sound wave, c_s is its speed of propagation $(\rho\beta_s h/c^2)^{-1/2}$, while ρ is the density, h the specific enthalpy (including the specific rest-mass density c^2), T the temperature, β_s and β_T the isentropic and isothermal compressibilities and γ the expansion coefficient. This expression contains two purely relativistic terms (with h^{-1} and h^{-2}) and furthermore the relativistic transport coefficients. If the first approximations found above are introduced for the latter coefficients, and if also the other quantities are calculated for the relativistic gas, one obtains simple formulae for the three models studied in the preceding. The results are [40]:

$$\alpha = \frac{7}{30\pi}(\frac{3}{5})^{1/2}\frac{\omega^2 m^{3/2}}{(kT)^{1/2}\rho a}(1 + \frac{81}{41}\frac{kT}{mc^2}) , \quad (86)$$

$$\alpha = \frac{7}{30\pi}(\frac{3}{5})^{1/2}\frac{\omega^2 m^{3/2}}{(kT)^{1/2}\rho a}(1 + \frac{16}{7}\frac{kT}{mc^2}) , \quad (87)$$

$$\alpha = \frac{7}{32}(\frac{3}{5\pi})^{1/2}\frac{\omega^2 m^2}{kT\rho d^2}(1 + \frac{135}{112}\frac{kT}{mc^2}) , \quad (88)$$

for Israel particles, fifth power law particles and hard spheres respectively.

The formula (85) was derived in view of its astrophysical importance [40-41]. Several other authors [32,34, 42-47] have expressed their interest in relativistic transport coefficients in view of possible applications in astrophysical problems.

5. Conclusion

We have thus seen that it is possible to build up a Boltzmann-type kinetic theory of transport processes in the classical framework of special relativity. The theory pertains to the Navier-Stokes regime and to elastic collisions of the particles. In particular the macroscopic laws could be derived and general properties of the transport coefficients, such as reciprocal relations, could be established. Moreover, values for the transport coefficients have been calculated according to different procedures: the "eigenvalue method", suited for Israel particles only, the "moment method", which enables one to obtain readily first approximations, and finally the "<u>orthogonal functions method</u>", constituting the relativistic generalization of the Chapman-Enskog procedure. The latter method yields the magnitudes of the transport coefficients for relativistic systems of particles with <u>arbitrary</u> <u>collision cross-sections</u> in <u>successive</u> <u>approximations</u>.

Finally it should be remarked that at high temperatures not only elastic collisions occur, but also inelastic processes, such as ionization, radiation and particle production, may take place. In the present stage of relativistic kinetic theory of transport processes their influence has not yet been accounted for. This remains a task to be carried out.

The author is greatly indebted to Mr. J. Guichelaar, Miss A. Kitselar, Mr. A.J. Kox, Dr. W.A. van Leeuwen, Mr. P.H. Polak and Dr. Ch.G. van Weert for numerous useful remarks on the contents and presentation of results in this review.

This work is based in part on investigations carried out by the "Stichting voor fundamenteel onderzoek der materie (F.O.M.)", which is financially supported by the "Organisatie voor zuiver wetenschappelijk onderzoek (Z.W.O.)".

References

[1] A. Lichnerowicz and R. Marrot, Compt. Rend. Acad. Sc., Paris 210 (1940) 759; 211 (1940) 177; 219 (1944) 270.
[2] R. Marrot, Journ. Math. pures et appliquées 25 (1946) 12.
[3] N.A. Chernikov, Sov. Phys. Doklady 2 (1957) 103, 248; 5 (1960) 764; Naukn. Dokl. Vyssh. Shkoly, Ser. Fiz.-Mat. 1 (1959) 168.
[4] G.E. Tauber and J.W. Weinberg, Phys. Rev. 122 (1961) 1342.
[5] W. Israel, J. Math. Phys. 4 (1963) 1163.
[6] R. Hakim, J. Math. Phys. 8 (1967) 1315, Phys. Rev. 162 (1967) 128.
[7] N.G. van Kampen, Physica 43 (1969) 244.
[8] Ch.G. van Weert, Proc. Kon. Ned. Akad. Wet., Amsterdam B 73 (1970) 381, 397, 500, 517 (thesis Amsterdam).
[9] S.R. de Groot and L.G. Suttorp, Foundations of electrodynamics (North-Holland) Amsterdam, 1972 (chapter V, section 2a).
[10] W.A. van Leeuwen, Proc. Kon. Ned. Akad. Wet., Amsterdam B 74 (1971) 122, 134, 150, 269, 276 (thesis Amsterdam).
[11] S.R. de Groot, Ch.G. van Weert, W.Th. Hermens and W.A. van Leeuwen, Physica 40 (1968-1969) 257, 581; 42 (1969) 309; Phys. Lett. 26A (1968) 345, 439.
[12] E.C.G. Stueckelberg, Helv. Phys. Acta 25 (1952) 577.
[13] W. Heitler, Ann. Inst. Henri Poincaré 15 (1967) 67.
[14] F. Jüttner, Ann. Physik 34 (1911) 856.
[15] C. Eckart, Phys. Rev. 58 (1940) 919.
[16] G.A. Kluitenberg, S.R. de Groot and P. Mazur, Physica 19 (1953) 689, 1079; 20 (1954) 199; 21 (1955) 148.
[17] W.Th. Hermens, Proc. Kon. Ned. Akad. Wet., Amsterdam B 74 (1971) 376, 387, 461, 478 (thesis Amsterdam).

[18] W.Th. Hermens, W.A. van Leeuwen, Ch.G. van Weert and S.R. de Groot, Physica 60 (1972) 472
[19] P. Curie, Oeuvres, p. 118, Paris, 1908.
[20] S.R. de Groot and P. Mazur, Non-equilibrium thermodynamics (North-Holland) Amsterdam, 1962.
[21] L. Onsager, Phys. Rev. 37 (1931) 405; 38 (1931) 2265.
[22] H.B.G. Casimir, Rev. Mod. Phys. 17 (1945) 343.
[23] J. Bernstein, G. Feinberg and T.D. Lee, Phys. Rev. 139B (1965) 1650.
[24] C.S. Wang (Chang) and G.E. Uhlenbeck, Rep. M 999, Eng. Res. Inst., Univ. of Michigan, 1952.
[25] L. Waldmann, Z. Naturf. 11A (1956) 523; Encycl. Phys. 12 (1958) 370.
[26] W.A. van Leeuwen and S.R. de Groot, Physica 51 (1971) 1, 16, 32.
[27] E. Gross, D. Bhatnager and M. Krook, Phys. Rev. 94, (1954) 511.
[28] C. Marle, Ann. Inst. Henri Poincaré 10 (1969) 67, 127.
[29] H. Grad, Comm. Pure Appl. Math. 2 (1949) 331; Phys. Fluids 6 (1963) 147.
[30] W. Israel and J.N. Vardalas, Lett. N. Cim. 4 (1970) 887.
[31] B. Vignon, Ann. Inst. Henri Poincaré 10 (1969) 31.
[32] J.L. Anderson, Proc. 1969 Relativity conference in the Mid-West (Plenum) New York, 1970.
[33] J.M. Stewart, Non-equilibrium relativistic kinetic theory (Springer), Berlin-Heidelberg-New York, 1971.
[34] C.W. Misner, Astrophys. J. 151 (1968) 431.
[35] S. Chapman, Phil. Trans. Roy. Soc. A 216 (1916) 279; A 217 (1917) 115.
[36] D. Enskog, Dissertation, Uppsala, 1917.
[37] W.A. van Leeuwen, P.H. Polak and S.R. de Groot, Phys. Lett. 37A (1971) 323; C.R. Acad. Sc., Paris 274 (1972) 431; Physica, to be published.

[38] D.C. Kelly, Unpublished report, Miami University, Oxford, Ohio, 1963. A different set of orthogonal functions was proposed here, but since it is incomplete, no method could be developed from it.

[39] S. Chapman and T.G. Cowling, The mathematical theory of non-uniform gases, Cambridge University Press, 1939 (third edition, 1970).

[40] J. Guichelaar, W.A. van Leeuwen and S.R. de Groot, Physica 59 (1972) 97.

[41] S. Weinberg, Astrophys. J. 168 (1971) 175.

[42] T.D. Lee, Astrophys. J. 111 (1950) 625.

[43] M. Ruderman, Rep. Progr. Phys. 28 (1965) 411.

[44] M. Ruderman, Acc. Naz. Lincei 368 (1971) 1.

[45] T. de Graaf, Acc. Naz. Lincei 368 (1971) 81.

[46] J. Ehlers, Rendiconti Scuola Int. di Fisica "Enrico Fermi" 47 (1971) 1.

[47] W. Israel, in General Relativity, Papers in honour of J.L. Synge (Clarendon Press) Oxford, 1972.

KINETIC EQUATION FOR ELEMENTARY EXCITATIONS
IN QUANTUM SYSTEMS

I.M. KHALATNIKOV

The Academy of Sciences of the USSR, Moscow
The Landau Institute for Theoretical Physics

Kinetic theory deals with macroscopic systems involving a large number of particles. At Boltzmann's time the kinetic theory described the properties of systems consisting of particles subject to the laws of classical mechanics. The XX-th century quantum physics has developed a new field - quantum theory of solids. This theory studies the properties of macroscopic systems of particles, their motion being described by the laws of quantum mechanics. In the language of quantum mechanics, the states of such macroscopic quantum systems are defined by their energy levels. A weakly excited state of such a system (beyond the ground state), as can be proved, may be described as a superposition of some elementary excitations, characterized by a certain dependence of their energy on the momentum. This dependence $\varepsilon(p)$ is called the energy spectrum of the system. The well known Debye crystal theory is based on these concepts, in which the role of elementary excitations is taken by phonons characterized by a phonon spectrum

$$\varepsilon = c\,p \qquad (1)$$

Elementary excitations have a behaviour similar to that of particles. The application of the laws of statistical physics to these quasiparticles allows to calculate all the thermodynamical values of the system, as it has been done, for instance, in the Debye theory. However, the fact, that the kinetic theory may be applied to quasiparticles, and that in this way the kinetic properties of quantum systems can be studied, has been understood much later. Apparently, this has been realized only in the thirties, and in the forties it has been widely developed. Namely, in those

years, after the discovery of helium superfluidity, there
appeared a new field of research - the physics of quantum
liquids. And it is in the explanation of kinetic properties of quantum liquids that the triumph of the elementary
excitation concept took place.

I would now like to treat in detail the application
of the Boltzmann kinetic equation to the theory of quantum
liquids. As it is well known, there are quantum liquids of
Fermi and Bose types, their names being given according to
the type of statistics, to which the elementary excitations
are subject to. L. Landau first put forward the theory for
the Bose-liquids, which are characterized by a superfluidity
property (1941). Only 15 years later it was Landau again
who proposed the theory of Fermi-liquids, in which the
superfluidity property is absent. It is interesting to note,
that the Boltzmann kinetic equation happened to be the
starting point for the construction of the Fermi-liquid
theory. The application of this equation to Fermi excitations first led to some difficulty - Landau could not manage
to satisfy the conservation laws of momentum. Let me begin
with this paradoxical situation. The kinetic equation for
elementary excitations can be written in the same way as
at Boltzmann's time. The distribution function of elementary excitations satisfies the equation

$$\frac{\partial n}{\partial t} + \frac{\partial n}{\partial \vec{r}} \frac{\partial H}{\partial \vec{p}} - \frac{\partial n}{\partial \vec{p}} \frac{\partial H}{\partial \vec{r}} = I(n) \tag{2}$$

where I(n) is the collision integral of quasiparticles,
H is their Hamiltonian, that coincides with the energy in
the absence of superfluidity. In the presence of superfluid motion with the velocity $\vec{v}_s$ the Hamiltonian H is

equal to

$$H = \varepsilon + \vec{p}\vec{v}_s \qquad (3)$$

where $\varepsilon(p)$ is energy in the frame, in which the superfluid component is at rest. Let us see how the conservation laws of momentum and energy are satisfied. For this purpose equation (2) should first be multiplied with the momentum component p_i, and then integrated over all p-space. The momentum is conserved in each simple process of collision, therefore, the integral of the right hand side of (2) is equal to zero. Thus, we get

$$\frac{\partial}{\partial t} \int n p_i d\tau + \int p_i \frac{\partial H}{\partial p_k} \frac{\partial n}{\partial r_k} d\tau - \int p_i \frac{\partial n}{\partial p_k} \frac{\partial H}{\partial r_k} d\tau = 0 \qquad (4)$$

or after a simple transformation

$$\frac{\partial}{\partial t} \overline{n p_i} + \frac{\partial}{\partial r_k} \overline{n p_i \frac{\partial H}{\partial p_k}} + n \overline{\frac{\partial H}{\partial r_i}} = 0 \quad . \qquad (5)$$

Here the line denotes the integration over p-space. In order that the momentum of the whole system be conserved, it is necessary that the time derivatives of the total momentum $\vec{P} = \overline{n\vec{p}}$ should be equal to a divergence of a certain tensor

$$\frac{\partial}{\partial t} \overline{n p_i} + \frac{\partial \Pi_{ik}}{\partial r_k} = 0 \quad . \qquad (6)$$

Equation (5), however, has not such a property, and the term

$$\overline{n \frac{\partial H}{\partial r_i}} \quad ,$$

generally speaking, cannot be given by

$$\frac{\partial \Pi_{ik}}{\partial r_k} .$$

Consider first the case of the Fermi-liquid. The term

$$\overline{n \frac{\partial H}{\partial r_i}}$$

may be transformed into the form

$$\overline{n \frac{\partial \varepsilon}{\partial r_i}} = \frac{\partial}{\partial r_i} \overline{n\varepsilon} - \overline{\varepsilon \frac{\partial n}{\partial r_i}} . \qquad (7)$$

Then the energy of an elementary excitation should be defined in a correct way. Define the energy of an elementary excitation as a functional derivative of the total energy of the system with respect to the distribution function

$$\varepsilon = \frac{\delta E}{\delta n} \quad \text{or} \quad \delta E = \int \varepsilon \delta n \, d\tau . \qquad (8)$$

Thus we assume that the excitation energy in the general case is a functional of the distribution function $\varepsilon[n]$. It should be stressed that the energy ε could also depend on the liquid density directly, so that (8) could involve the term $\frac{\partial E}{\partial \rho} \delta \rho$ as well. However, (as it will be seen in what follows), we are forced to express the energy ε by the function n only, in order that the momentum conservation law be satisfied. It should also be noted, that the application of the model of non-interacting Fermi-excitations, to He^3 in particular, leads to a contradiction with experiment, therefore, we necessarily have to take into account interactions of Fermi-excitations from the very beginning, and thus the definition (8) is the only possible one. Taking (8) into account, we may rewrite (7) as follows

$$\frac{\partial}{\partial r_i}(\overline{n\epsilon} - E) \quad . \qquad (7')$$

(Note that due to the interaction excitations, $\overline{n\epsilon} \neq E$). With the aid of (7') the expressions for Π_{ik} can now be obtained

$$\Pi_{ik} = \overline{np_i \frac{\partial H}{\partial p_k}} + \delta_{ik}(\overline{n\epsilon} - E) \qquad (9)$$

and, therefore, the momentum conservation law is fulfilled.

Quite another solution should be used for the problem of the momentum conservation law for the Bose-liquid. Substituting the Hamiltonian (3) into (5) we obtain the equation of motion

$$\frac{\partial}{\partial t}\overline{np_i} + \frac{\partial}{\partial r_k}\overline{np_i(\frac{\partial \epsilon}{\partial p_k} + v_{s_k})} + \overline{n(\frac{\partial \epsilon}{\partial r_i}} + \frac{\partial}{\partial r_i}\vec{p}\vec{v}_s) = 0 \quad (10)$$

defining the rate of the total momentum of the relative motion of normal and superfluid parts (p is the excitation momentum in the frame, in which the superfluid part is at rest).

Now write the equation of motion for the total momentum

$$\vec{j} = \overline{\vec{p}n} + \rho\vec{v}_s \quad . \qquad (11)$$

Since the total momentum should be conserved, its time derivative is equal to a divergence of the symmetric tensor of the momentum flux Π_{ik}

$$\frac{\partial j_i}{\partial t} + \frac{\partial \Pi_{ik}}{\partial r_k} = 0 \ . \tag{12}$$

The momentum flux tensor Π_{ik} in the rest frame may be expressed through its value π_{ik} in the frame moving with the velocity $\vec{v}_s$

$$\Pi_{ik} = \pi_{ik} + v_{s_k}\overline{np_i} + v_{s_i}\overline{np_k} + \rho v_{s_i} v_{s_k} \ . \tag{13}$$

Subtract equation (10) from equation (12) and use the continuity equation

$$\frac{\partial \rho}{\partial t} + \frac{\partial}{\partial r_k}(\overline{np_k} + \rho v_{s_k}) = 0 \ . \tag{14}$$

As a result we get

$$\rho \frac{\partial v_{s_i}}{\partial t} + v_{s_k}\frac{\partial v_{s_i}}{\partial r_k} + \frac{\partial \pi_{ik}}{\partial r_k} - n\frac{\partial \varepsilon}{\partial r_i} - \frac{\partial}{\partial r_k}\overline{np_i \frac{\partial \varepsilon}{\partial p_k}} = 0 \tag{15}$$

Now the condition of potentiality of the superfluid motion rot $\vec{v}_s = 0$ should be used. From this condition it follows, that the sum of the last four terms in (15) must be equal to the product of the density ρ with the gradient of a certain function. The derivative $\frac{\partial \varepsilon}{\partial r_i}$ is likely to be rewritten as $\frac{\partial \varepsilon}{\partial \rho}\frac{\partial \rho}{\partial r_i}$. Furthermore, the tensor π_{ik} at T = 0 in the absence of excitations should be equal to $p_o \delta_{ik}$ (p_o is the pressure at T = 0). From these requirements the tensor π_{ik} follows in a unique way:

$$\pi_{ik} = \overline{np_i \frac{\partial \varepsilon}{\partial p_k}} + \delta_{ik}(p_o + n\frac{\partial \varepsilon}{\partial \rho}\rho) \tag{16}$$

and according to thermodynamical identities we have

$$\frac{\partial p_o}{\partial r_i} = \rho \frac{\partial \mu_o}{\partial r_i} \qquad (17)$$

(μ_o is the value of the chemical potential at $T = 0$). Substituting (16) into equation (15) and, taking (17) into account, we get

$$\frac{\partial \vec{v}_s}{\partial t} + \vec{\nabla}(\mu_o + \overline{n \frac{\partial \varepsilon}{\partial \rho}} + \frac{v_s^2}{2}) = 0 \quad . \qquad (18)$$

Thus we have obtained the equation of motion of a superfluid component. Hence, in the case of a superfluid liquid the kinetic equation (2) involves two external fields ρ and $\vec{v}_s$, which are defined by two equations - the continuity equation (14) and the equation of motion (18). Let us finally write out this self-consistent system.

$$\frac{\partial n}{\partial t} + \frac{\partial n}{\partial \vec{r}} \frac{\partial H}{\partial \vec{p}} - \frac{\partial n}{\partial \vec{p}} \frac{\partial H}{\partial \vec{r}} = I(n) \quad , \qquad H = \varepsilon + \vec{p}\,\vec{v}_s$$

$$\frac{\partial \rho}{\partial t} + \text{div}(\int n\vec{p}\,d\tau + \rho \vec{v}_s) = 0 \qquad (19)$$

$$\frac{\partial \vec{v}_s}{\partial t} + \vec{\nabla}(\mu_o + \int n \frac{\partial \varepsilon}{\partial \rho} d\tau + \frac{v_s^2}{2}) = 0 \quad .$$

Summarize the foregoing. The application of the kinetic equation to elementary excitations in quantum systems causes some difficulty with the conservation of the momentum of the system. The only possibility to overcome this difficulty for Fermi-systems without superfluidity consists in the assumption that the energy of the system is only a functional of the distribution function. In Bose-systems with superfluidity, a new function, the velocity of the superfluid component, plays the role of an external field in a kinetic equation. This external field should satisfy the equation, that conserves the total momentum of the system.

It can be seen from the foregoing, that the Boltzmann kinetic equation describes not only the kinetic properties of macroscopic systems, but the ground for new theories as well.

In conclusion let us shortly investigate the energy conservation law. It is easy to see, that taking into account everything said above, the energy conservation law would be fulfilled automatically. For this purpose let us calculate the time derivative of the total energy of the system and express all the time derivatives with the help of equations (19) by coordinate derivatives. Then, collecting all the terms under the sign div, we obtain the expression for the energy flux Q. The total energy E consists of the kinetic energy

$$E_k = \rho \frac{v_s^2}{2} + \vec{v}_s \overline{n\vec{p}} \quad , \tag{20}$$

the internal energy

$$\delta E_i = \int \varepsilon \, \delta n \, d\tau \tag{21}$$

and the zero energy E_0 (T = 0), defined by the identity

$$dE_0 = \mu_0 \, d\rho \quad . \tag{22}$$

The time derivative of the kinetic energy according to (19) is equal to

$$\frac{\partial E_k}{\partial t} = -(\rho v_{s_i} + \overline{np_i})\frac{\partial}{\partial r_i}(\mu_o + \overline{n\frac{\partial \varepsilon}{\partial \rho}} + \frac{v_s^2}{2}) - \frac{v_s^2}{2}\frac{\partial}{\partial r_i}(\rho v_{s_i} + \overline{np_i})$$

$$- v_{s_i}\frac{\partial}{\partial r_k}\overline{np_i(\frac{\partial \varepsilon}{\partial p_k} + v_{s_k})} - v_{s_i}(\overline{n\frac{\partial \varepsilon}{\partial r_i}} + \overline{np_k\frac{\partial v_{s_k}}{\partial r_i}}) . \quad (23)$$

To obtain the derivative of the internal energy we multiply the left and right hand sides of the kinetic equation by ε and integrate over p-space. The integral of the right hand side is zero, since in collisions the energy is conserved and we get

$$\overline{\varepsilon \frac{\partial n}{\partial t}} + \overline{\varepsilon \frac{\partial n}{\partial r_k}\frac{\partial H}{\partial p_k}} - \overline{\varepsilon \frac{\partial n}{\partial p_k}\frac{\partial H}{\partial r_k}} = 0 . \quad (24)$$

Using (22), (23), and (24), we obtain the time derivative out of the total energy

$$\frac{\partial E}{\partial t} = - \mathrm{div}\{ (\overline{n\vec{p}} + \rho\vec{v}_s)(\mu_o + \overline{n\frac{\partial \varepsilon}{\partial \rho}} + \frac{v_s^2}{2}) + \overline{nH\frac{\partial}{\partial \vec{p}} H}\} . \quad (25)$$

Thus, the total energy E is conserved and its flux is defined by the vector under the sign div in the right hand side of (25).

In the case of Fermi-systems the validity of the energy conservation law may be proved much easier. Multiplication of the kinetic equation with ε and integration over momenta yields directly the following result

$$\frac{\partial E}{\partial t} + \mathrm{div} \int n\varepsilon \frac{\partial \varepsilon}{\partial \vec{p}} d\tau = 0 . \quad (26)$$

Finally it should be noted that in case of solutions of Bose- and-Fermi-liquids we have a situation, where both the external field $\vec{v}_s$, and the functional dependence of energy on the density of excitations should be necessarily introduced for the description of the superfluid part. The corresponding theory is a peculiar hybrid of the theories considered above for Fermi and Bose systems.

References

I.M. Khalatnikov. Theory of Superfluidity, Nauka, Moscow, 1971.

ERGODIC THEORY

Ya.G. SINAI

Landau Institute for Theoretical Physics
Academy of Sciences
USSR

ABSTRACT

Apparently there are at least two reasons why ergodic theory has the honour to be discussed at this solemn Conference. The first reason follows from the fact that Ludwig Boltzmann is one of the founders of ergodic theory. The term "ergodicity" was introduced by Ludwig Boltzmann in his paper [1] in 1887 (see also his "Lectures on Gas Theory"). In fact it is connected with another remarkable discovery by Boltzmann of how dynamical laws of motions can lead to statistics. As a result, ergodic theory is a branch of mathematics which investigates statistical properties of dynamical systems. The second reason is that the concepts of ergodicity and mixing of Ludwig Boltzmann and J. Gibbs have become very fruitful and given many new notions, ideas and results of a general character. The mathematics clarified the essence of these concepts and gave deeper penetration into many problems. This led to some progress in the solution of these problems.

Part I. Statistical Properties of Dynamical Systems

> "Das wird nächstens schon besser gehen,
> wenn Ihr lernt, alles zu reduzieren
> und gehörig zu klassifizieren"
> Goethe, "Faust"

We begin with a discussion of what we understand by statistical properties of dynamical systems. Any system of ordinary differential equations of motion

$$\frac{dx_1}{dt} = f_1(x_1,\ldots,x_n)$$
$$\ldots\ldots\ldots\ldots\ldots\ldots \quad (1)$$
$$\frac{dx_n}{dt} = f_n(x_1,\ldots,x_n)$$

is naturally connected with the one-parameter group of transformations $\{S_t\}$ of the phase space X where the transformation S_t shifts each point x_o t units of time along its trajectory. In ergodic theory one considers actions of one-parameter groups of transformations without assuming that such groups are generated by a system of differential equations (1). Such generalization is very useful because it leads to very deep connections of ergodic theory with the theory of probabilities, functional analysis, number theory, etc. One-parameter groups, generated by systems (1) or more generally by vector fields on manifolds are called often classical dynamical systems.

Any point x_o defines uniquely its trajectory (Laplace's determinism). Statistics appears when one considers not the evolution of separate points, but the evolution of other objects naturally connected with the group $\{S_t\}$.

The first class of objects is the function space X^* of "good" functions $f(x)$ on the phase space X. The group $\{S_t\}$ generates the group $\{S_t^*\}$ acting on X^* by the formula $(S_t^* f)(x) = f(S_t x)$.

The second class of objects is the space of probability distributions X^{**} on X. The group $\{S_t\}$ generates the group $\{S_t^{**}\}$ acting on X^{**} by formula

$$(S_t^{**} \lambda)(A) = \lambda(S_t A) , \quad A \subset X \qquad (2)$$

for any $\lambda \in X^{**}$.

Statistical properties of dynamics (i.e. of the group $\{S_t\}$) are formulated as properties of corresponding groups $\{S_t^*\}$, $\{S_t^{**}\}$. We propose below six such properties. In some sense each succeeding property is stronger than the previous ones.

I.1. Existence of an Invariant Probability Distribution

A distribution $\lambda \in X^{**}$ is called invariant or stationary for the group $\{S_t\}$, if

$$S_t^{**} \lambda = \lambda , \quad -\infty < t < \infty .$$

If such an invariant distribution exists, the ergodic theorem of Birkhoff can be applied to permit us to average over time: the limits

$$\bar{f}_+(x_0) = \lim_{T \to \infty} \frac{1}{T} \int_0^T f(S_t x_0) dt , \quad \bar{f}_-(x_0) = \lim_{T \to \infty} \frac{1}{T} \int_0^T f(S_{-t} x_0) dt ,$$

$$\bar{f}(x_o) = \lim_{T\to\infty} \frac{1}{2T} \int_{-T}^{T} f(S_t \, x_o) \, dt$$

exist and coincide with probability 1 (for λ) for each integrable f.

The general theory of Bogoliubov-Krylov gives the existence of at least one invariant distribution for any compact phase space X and continuous group $\{S_t\}$. In the Hamiltonian case or, more generally, in the case of dynamical systems connected with variational principles, there is a natural Liouville invariant distribution on manifolds of constant energy (microcanonical distribution). For further discussion only the case of compact manifolds of constant energy is essential.

The problem of the existence of an invariant distribution is highly non-trivial in the case of dynamical systems with infinitely many degrees of freedom, for example, systems generated by non-linear hyperbolic equations. To demonstrate the point let us consider oscillations of a one-dimensional non-linear string defined by the Hamiltonian

$$H = \int_0^1 (\rho(x) u_t^2 + k(x) u_x^2 + \varepsilon u^4) \, dx \tag{3}$$

with fixed boundary conditions. For $\varepsilon = 0$ the equations of motion are linear and the energy of any separate harmonic is conserved. Thus, if I_k, ϕ_k are "action-angle" variables corresponding to k-th harmonics, then the general form of the invariant distribution in the absence of degeneracy is

$$d\lambda = \rho(I_o, I_1, \ldots) \prod_k dI_k \, d\phi_k \, .$$

We know of no general theorems which give the existence of at least one invariant measure for this case when $\varepsilon \neq 0$. Moreover it is not known what ρ can be imbedded in a continuous family of densities ρ_ε where ρ_ε gives an invariant distribution for corresponding ε.

If one takes natural finite-dimensional approximations of the whole system and the microcanonical distribution on the manifold H = const then it is easy to show that the microcanonical distribution has no meaningful limit. At the same time the famous experiments by Fermi-Pasta-Ulam [2] on computers indicate that an invariant distribution must exist.

I.2. Ergodicity and Mixing

It is well known that the dynamical system $\{S_t\}$ with the invariant distribution λ is ergodic or the invariant distribution λ is ergodic for the dynamical system $\{S_t\}$ if for any integrable function f

$$\bar{f}(x_0) = \int f \, d\lambda$$

with probability 1 (for λ), i.e. time averages are equal space averages almost everywhere. As was mentioned above, the concept of ergodicity has been introduced by Ludwig Boltzmann, but in a different form.

The first examples of ergodic dynamical systems appeared in number theory and the theory of almost-periodic functions. For example, the group of shifts of an n-dimensional torus $S_t(x_1 \ldots x_n) = (x_1 + \omega_1 t \pmod 1), \ldots, x_n + \omega_n t \pmod 1)$ is ergodic if and only if the ω_i are rationally independent. This group is often called quasi-periodic motion. The ergodicity of this and similar dynamical systems is of great significance for the proof of many important theorems, e.g., the theorem of H. Weyl on the equidistribution of fractional parts of polynomials. These dynamical systems have a unique invariant measure (strict ergodicity) and Birkhoff's ergodic theorem is valid everywhere (for continuous f).

The ergodicity of quasi-periodic motion on the torus is essential for the solution of Lagrange's problem about mean motions. This problem has appeared in Lagrange's astronomical investigations. One considers a curve $z(t) = z_1 e^{i\omega_1 t} \ldots + z_n e^{i\omega_n t}$, $-\infty < t < \infty$ on the complex plane C^1. If this curve does not pass through 0 then one can construct the continuous function arg z(t). Lagrange's problem is to find $\lim \frac{\arg z(t)}{t}$. The existence of this limit follows from a slightly stronger form of the ergodic theorem. The value of the limit can also be found by using ergodic considerations. A beautiful discussion of all these problems can be found in the book by S. Sternberg, "Celestial Mechanics", p. I (see also [3], [4], [5]).

There will be several other examples below which show the importance of quasi-periodic motions on tori.

Mixing is stronger than ergodicity. A dynamical system $\{S_t\}$ is mixing if for any two bounded functions f, g, the time correlations

$$\langle S_t^* f, g \rangle - \langle f, 1 \rangle \langle g, 1 \rangle = \int f(S_{-t} x) g(x) d\lambda - \int f d\lambda \int g d\lambda$$

go to zero as $t \to \pm\infty$. There are well-known descriptions of mixing in books by Gibbs [6] and Arnold and Avez [7].

Assume $f > 0$. Then one can construct a "non-equilibrium distribution" μ_o, for which $d\mu_o/d\lambda = f$. For the distribution in time t $\mu_t = S_t^{***}\mu_o$

$$\frac{d\mu_t}{d\lambda} = f(S_t x) \quad .$$

It follows from mixing that any "non-equilibrium" distribution which is absolutely continuous with respect to λ tends to the "equilibrium distribution" λ. Now let us describe one example which shows what can be hidden under the brevity of mathematical language. Assume one is given a system of an infinite number of classical particles with the pairwise interaction with finite range on the real line R^1, appearing from thermodynamical limit transition. The equilibrium distribution is the limit of the microcanonical distribution. This limit always exists in the one-dimensional case (see, e.g. [8]). It depends on three parameters: $\lambda = \lambda(\rho,h,v)$ where ρ is the density, h is the energy per particle, and v is the mean velocity. It is possible to construct a natural dynamics in the phase space of such an infinite system which leaves invariant any distribution $\lambda(\rho,h,v)$ (see [8], [9]).

Any non-equilibrium distribution which is absolutely continuous with respect to $\lambda(\rho,h,v)$ is given by density $f = d\mu_o/d\lambda$. Any function f can be represented as a limit of functions depending on coordinates and momenta of a finite number of particles. It is clear that such a "non-equilibrium" distribution has nothing in common with thermodynamics. From the point of view of ergodic theory the investigation of this system is apparently a very difficult problem (see part III of this lecture).

From the point of view of physics we do not possess any example of a dynamical system where mixing can be proved by simple considerations. However there are many mathematically simple examples of such systems. Now we shall give an example which is very popular in information theory. Assume X to be a space of all infinite texts $\omega = \{\ldots\omega_{-n}\ldots\omega_0\ldots\omega_n\ldots\}$ where any ω_i belongs to some finite alphabet A. The shift $T : T\omega = \omega'$, $\omega'_i = \omega_{i+1}$ is a natural transformation of X. The distribution λ invariant under T generates the stationary random process of probability theory.

If the separate ω_i are statistically independent (Bernoulli distribution) then T is mixing. The same is true if λ is a general Markov distribution or sufficiently close to it. These examples, although they look abstract, are not far from quite concrete dynamical systems with elastic collisions from the point of view of ergodic theory (see part II).

I.3. Entropy of Dynamical Systems (Kolmogorov's Entropy)

The entropy in statistical mechanics appears usually as a coefficient in asymptotic expressions for the logarithm of a probability or the logarithm of a number of possible configurations with prescribed properties. The entropy of a dynamical system appears as a coefficient in asymptotic expressions for the logarithm of probability of an interval of a trajectory when $t \to \infty$. Therefore this entropy can be called a dynamical entropy.

The **concept** of dynamical entropy has been introduced by A.N. Kolmogorov [10], [11] in 1958. Here we give its definition. Assume the phase space X to be partitioned in cells $C_1,\ldots C_r$ and $\tau > 0$ is fixed. Then one can introduce probabilities $p(i_0,\ldots,i_n)$ for point x_0

$$S_{k\tau} x_0 \in C_{i_k}, \quad k = 0,1,\ldots,n.$$

The number $H_n(\xi|\{S_t\}) = -\sum p(i_0,\ldots,i_n) \ln p(i_0,\ldots,i_n)$ shows how many types of trajectories appear in the dynamical system. For any partition $\xi = (C_1,\ldots,C_r)$ there exists the limit

$$\frac{H_n(\xi|\{S_t\})}{n\tau} = h(\xi|\{S_t\}) \quad \text{(Macmillan)}.$$

The upper bound of $h(\xi|\{S_t\})$ over all finite ξ does not depend on τ and is called the entropy of the dynamical system $\{S_t\}$. It is denoted usually as $h(\{S_t\})$. If $h(\{S_t\}) < \infty$ then one can take max instead of sup (Krieger).

If the system $\{S_t\}$ is ergodic then $\ln p(i_0,\ldots,i_n) \sim n\tau h(\xi|\{S_t\}))$ in probability, or $p(i_0\ldots i_n) \approx \exp(-n\tau h \cdot (\xi|\{S_t\})$. Consequently h is big when the number of possible types of trajectories is also big. This shows that positive entropy is connected with instability.

The quasi-periodic motion on the torus has entropy 0. The Bernoulli shift has entropy $-\sum p_i \ln p_i$ where p_i is the probability of a separate letter in the alphabet A. Other examples will be described in the second part of this lecture.

Let U be a small cell of the phase space X. If $\{S_t\}$ is mixing then $S_t U$ eventually becomes uniformly distributed over the whole phase space: Concrete examples show that the necessary time for diam $(S_t U) \sim 1$ has order $-\ln \text{diam } U/(h(\{S_t\}))$. We must mention here the papers by the late Soviet physicist N.S. Krylov [12] who introduced concepts which were close to entropy and employed them for the estimation of the time of relaxation.

In a short period after Kolmogorov's paper, there appeared an extensive entropy theory of dynamical systems. Recently very important results in this field have been obtained by the American mathematician D. Ornstein [13] but we do not have the time to discuss them in detail here.

I.4. K-Systems

In several cases it is easier to prove from the beginning some stronger property than mixing because this property can be expressed in local terms in some sense. The dynamical systems which have this stronger property are called K-systems. The corresponding definition was introduced by A.N. Kolmogorov [10] in his first paper on entropy.

In the case of discrete time, K-systems can be defined as systems for which $h(\xi|\{S_t\}) > 0$ for any non-trivial ξ. One can express this in another way by demanding that, for any finite partition $\xi = (C_1,\ldots,C_r)$, the conditional probability for $x_o \in C_{i_o}$ given all inclusions $x_t \in C_{i_t}$, $t < -T$ tends asymptotically to $\lambda(C_{i_o})$ as $T \to \infty$. In other words the distribution for a position of a point at time $t = 0$ asymptotically does not depend on its movement in the distant past. This shows the Markovian character of K-systems.

K-systems are ergodic, mixing, and have positive entropy. The corresponding group of unitary operators has countably-Lebesgue spectrum.

I.5. Strong Mixing and the Central Limit Theorem

Following M. Rosenblatt [28] let us say that a system $\{S_t\}$ satisfies the strong mixing condition if for a sufficiently large set of partitions ξ the motion for $t < 0$ is asymptotically independent from the whole motion in $t > T$, $T \to \infty$. We do not explain the words "a sufficiently large set of partitions ξ" because it is quite clear in every particular case. Let us explain statistical independence of the future motion and the past motion. For any fixed partition ξ each subset

$$A = \{x_o : S_{-t_p} x_o \in C_{i_p}, S_{-t_{p-1}} x_o \in C_{i_{p-1}}, \ldots,$$

$$S_{-t_1} x_o \in C_{i_1}\}$$

for $t_p > t_{p-1} > \ldots > t_1 \geq 0$ is connected with the motion in the past $t < 0$. One can introduce in the same way the subsets connected with the motion for $t > T$.

In some problems it is necessary to estimate the difference

$$\frac{1}{T} \int_0^T f(S_t x) dt - \int f d\lambda$$

which tends to zero if the system is ergodic. For systems with strong mixing this difference has order $T^{-1/2}$ and after multiplication by $T^{1/2}$ has asymptotically a normal (Gaussian) distribution. The strong mixing condition is a stronger statistical property than the K-property.

I.6. Exponential Decay of Correlations

In the case of dynamical systems which are mixing, time correlations $<S_t f,g> - <f,1><g,1>$ tend to zero. The importance of exponential decay of correlations has been stressed in the lecture of Lebowitz [18].

The spectral theory of dynamical systems gives some upper bounds for the decrease of these correlations. Many examples show these bounds are exact for "bad" non-smooth functions. For good functions there must exist some general form of asymptotics. In particular the exponential decay of correlations occurs if one can construct a so-called Markov partition for the given dynamical system (see [19], [20], [21]). We shall describe Markov partitions in a slightly different way than in [19], [20].

Let $\xi = (C_1, \ldots, C_r)$ be a partition of phase space in cells and let the time t take the values $t = n\tau$, $-\infty < n < \infty$. One can construct conditional probabilities

$$\lambda(C_{i_0} | S_{-\tau} C_{i_{-1}} \cap \ldots \cap S_{-n\tau} C_{i_{-n}}) = \frac{\lambda(C_{i_0} \cap S_{-\tau} C_{i_{-1}} \cap \ldots \cap S_{-n\tau} C_{i_{-n}})}{\lambda(S_{-\tau} C_{i_{-1}} \cap \ldots \cap S_{-n\tau} C_{i_{-n}})}.$$

In the pure Markov case these probabilities depend on i_0, i_{-1} only. More often these conditional probabilities tend uniformly with exponential velocity to their limits when $n \to \infty$. We call partitions with this property "Markov partitions".

In several cases it is possible to deduce exponential decay of correlations from the existence of Markov partitions.

I.7. Concluding Remarks

There are some cases where it is difficult to prove the positiveness of the dynamical entropy. The topological entropy of the dynamical system [23] introduced by Alder, Konheim and McAndrew [23] is less informative but easier for calculation. We give its definition, which is also due to Kolmogorov (see [24], and also [19] where it appears independently). Let the phase space X be a complete metric space X with the metric d. For any $T > 0$ we shall define the metric d_T by expression

$$d_T(x,y) = \max_{0 \leq t \leq T} d(S_t x, S_t y) \;;$$

X with the metric d_T is also a compact metric space. Thus for any $\varepsilon > 0$ one can take the $N_\varepsilon(T)$ equal to the maximal number of points x_k for which $d_T(x_{k_1}, x_{k_2}) \geq \varepsilon$. Now

$$h_{top}(\{S_t\}) = \varlimsup_{\varepsilon \to 0} \varlimsup_{T \to \infty} \frac{\ln N_\varepsilon(T)}{T} \;.$$

Properties of the Topological Entropy

1. If $h_{top}(\{S_t\}) > 0$, then there exists an invariant distribution with positive dynamical entropy (Dinabourg, Goodman).

2. If $\nu(t)$ is the number of periodic trajectories with period not more than t, then for some classes of dynamical systems

$$\lim_{t \to \infty} \frac{\ln \nu(t)}{t} = h_{top}(\{S_t\})$$

(Bowen, Margulis).

3. If the dynamical system has homoclinic or heteroclinic points then $h_{top}(\{S_t\}) > 0$ (Poincaré, Smale, Shilnikov).

4. Topological entropy has some applications in the problem of three bodies (Alekseev).

In some cases it is easier to estimate h_{top} than h, but the inequality $h_{top} > 0$ is less informative than the inequality $h > 0$.

The general question of when a given 0-1 sequence can be considered to be a random sequence was considered recently by A.N. Kolmogorov [25] and his students [26]. The approach was based on the structure of algorithm which can produce the given sequence. This approach can be applied in ergodic theory. The corresponding characteristics are connected with the entropy.

Part II. Statistical Properties of Classical Dynamical Systems

> Time present and time past
> Are both perhaps present in time future,
> And time future contained in time past.
> (T.S. Eliot)

One of the main problems of ergodic theory is to investigate the statistical properties of classical dynamical systems. Many achievements have been obtained in this field after Kolmogorov's famous lecture at the Amsterdam Mathematical Congress [27]. Kolmogorov initiated two important directions in ergodic theory. One of them is connected with the consideration of Hamiltonian systems close to integrable ones; the other is the entropy theory of dynamical systems.

II.1. Kolmogorov-Arnold-Moser's Theory (KAM-Theory)[29-33]

In KAM-theory one considers a Hamiltonian system with a finite number of degrees of freedom obtained by a small perturbation of an integrable Hamiltonian system. In such a system the Hamiltonian can be represented in the following form:

$$H(p,q) = H_0(p) + \mu H_1(p,q) + \mu^2 H_2(p,q) + \ldots$$

μ a small parameter. According to Poincaré, the investigation of this case is the main problem of dynamics.

When $\mu = 0$, the system is integrable, and the phase space decomposes on invariant tori with quasi-periodic motion on each of them except for some degenerate submanifolds. S. Smale [34] considered recently the topological structure of such decompositions.

The main result of KAM-theory asserts that for sufficiently small μ the majority of the invariant tori are still present for the perturbed system but are displaced slightly in the phase space only. Thus for small μ such systems are non-ergodic. On the invariant tori there is ergodicity without mixing. Entropy is equal to 0.

KAM-theory has many applications. Here are some of them.

1. Free motion of a particle constrained to move on a convex surface close to an ellipsoid.

The uniform motion on geodesic lines on an ellipsoid is an integrable Hamiltonian system (Jacobi). It is proved that for small perturbations of the metric there exists a subset of positive measure in the unit tangent bundle consisting of invariant tori with quasi-periodic motion.

2. The neighborhood of an elliptic fixed point for a smooth area preserving transformation of plane.

Let $(p,q) \to (f_1(p,q), f_2(p,q))$ be a transformation of the plane for which $f_1(0,0) = f_2(0,0) = 0$,

$$\frac{\partial f_1}{\partial p} \frac{\partial f_2}{\partial q} - \frac{\partial f_1}{\partial q} \frac{\partial f_2}{\partial p} \equiv 1 \quad .$$

The point $(0,0)$ is called elliptic if the matrix of the linear part of the transformation has complex proper values. It is proved under very general conditions that such a transformation is stable. That means that for sufficiently small neighborhoods there exist invariant curves encircling the point $(0,0)$. Moreover, the set of such curves has positive area.

3. The motion of rigid body with a fixed point.

The stability of rapid rotation of a rigid non-symmetric gyroscope with an arbitrary fixed point can be proved from KAM-theory.

4. The external conservation of an adiabatic invariant in one-dimensional Hamiltonian systems.

For a one-dimensional Hamiltonian system with the Hamiltonian $H(p,q;\lambda)$ ($\lambda = \mu t$ is slow time), $H(p,q;\lambda+2\pi) = H(p,q;\lambda)$ it is proved that the action I is the external adiabatic invariant. As a consequence the magnetic moment is an adiabatic invariant for the motion of charged particles in axially-symmetric magnetic fields.

5. The existence of quasi-periodic motions in the problem of three bodies and the plane problem of n bodies.

6. The reducibility of linear systems of differential equations with quasi-periodic coefficients.

KAM-theory is based upon a new method of the perturbation theory. This method has much in common with the Newton method for the solution of non-linear equations $F(x) = 0$ (see [30], [32], [35]). Usually Newton's method converges very rapidly. This makes it possible to overcome the difficulties connected with small denominators.

Invariant subsets in KAM-theory are not domains. Between invariant tori there are trajectories of quite different nature. In the "instability zones" of Birkhoff, there are homoclinic and heteroclinic trajectories. Thus the topological entropy of these systems is positive. Therefore invariant distributions with positive Kolmogorov entropy exist for such systems (see above). However, it is unknown whether these distributions are absolutely continuous.

II.2. Statistical Properties of Unstable Dynamical Systems (Anosov systems).

Many authors have pointed out that instability leads to statistics (G. Birkhoff, E. Hopf, M. Born, N.S. Krylov and others). Sometimes one hears that instability is necessary for ergodicity. The corresponding mathematical theory was created during the last ten or twelve years. The concepts of entropy and K-system play significant roles in this theory.

The formal definition of unstable dynamical systems is due to D.V. Anosov [36] who called them Y-systems[+]. We

---

[+] As we shall explain below, any point has separatrices in the case of Anosov systems. In informal discussions these separatrices are called ("ус "- moustaches) (✖). That is the origin of the name "Y-system".

shall call them Anosov's systems. The topological theory of Anosov systems is connected with general structural stability theory of dynamical systems as developed by S. Smale and others [37]. We shall deal here only with ergodic properties of these systems.

If the system $\{S_t\}$ is unstable then two nearby points x, y should diverge when $t \to \infty$ or $t \to -\infty$. If a smooth invariant distribution exists it is not possible for any point x_0 that all points y in a small neighborhood of x_0 diverge when $t \to \infty$. As a consequence there should exist points y whose semi-trajectories tend to the semi-trajectory of the point x_0 when $t \to \infty$ or $t \to -\infty$. This property can be taken as a definition of Anosov systems. Namely, a classical dynamical system $\{S_t\}$ with phase space X, dim X = n, is called an Anosov system if for every point $x_0 \in X$ there exist two submanifolds $\Gamma^{(s)}(x_0)$, $\Gamma^{(u)}(x_0)$, for which

$$\Gamma^{(s)}(x_0) = \{y: \text{dist}(S_t x_0, S_t y) \leq Ce^{-\alpha t} d(x_0, y), t > 0\},$$

$$\Gamma^{(u)}(x_0) = \{y: \text{dist}(S_t x_0, S_t y) \leq Ce^{\alpha t} d(x_0, y), t < 0\},$$

C, λ be positive constants. Moreover, dim $\Gamma^{(s)}(x_0)$, dim $\Gamma^{(u)}(x_0)$ do not depend on x_0, dim $\Gamma^{(s)}(x_0)$ + dim $\Gamma^{(u)}(x_0)$ = n - 1 and $\Gamma^{(s)}(x_0)$, $\Gamma^{(u)}(x_0)$ are continuous in x_0 in a natural way. The existence of such families of submanifolds $\{\Gamma^{(s)}(x_0)\}$, $\{\Gamma^{(u)}(x_0)\}$ determine all statistical properties of Anosov systems.

The typical example of an Anosov system is free motion (geodesic flow) on a compact surface of negative curvature. There have been many profound investigations of this dynamical system. Hedlund [38] and Hopf [39], proved ergodicity of this system. (Hopf also proved ergodicity for the multi-dimensional manifolds of constant negative curvature and mixing in the two-dimensional case.) Gelfand and Fomin [40] found the spectrum of this dynamical system for surfaces of constant negative curvature by using Lie group representation theory.

We now have stronger results for the whole class of Anosov systems.

1. Anosov systems are ergodic (Anosov).

2. With one natural exception all Anosov systems are K-systems. Therefore they are mixing, have positive entropy, countably-Lebesgue spectrum (Sinai).

3. In the case of discrete time there exist Markov partitions. Thus the exponential decay of correlations and the central limit theorem are valid in this case (Bowen, Sinai, Gordin).

I think that 3. is valid for continuous time also, but the construction of Markov partitions is still not done in this case except the three-dimensional case (Ratner)[+].

Thus Anosov systems have the nicest possible statistical properties. However, they are only model systems. In more concrete situations there can occur more complicated types of behaviour.

In conclusion of this paragraph I want to illustrate the Markov character of dynamics in Anosov systems in the two-dimensional case. There exist the partitions of the phase space X into cells $C_1, \ldots, C_n$ for which all intersections $C_{i_0} \cap TC_{i_1} \cap \ldots \cap T^n C_{i_n}$ have very regular form. Namely, if the C_j are parallelograms in some sense then any intersection $C_{i_0} \cap TC_{i_1} \cap \ldots \cap T^n C_{i_n}$ is also a parallelogram

---

+) Added in proof: I was informed recently that R. Bowen constructed Markov partitions for the case of continuous time.

for which one pair of sides is parallel to corresponding sides of C_{i_0} and another pair of sides lie inside the corresponding sides of C_{i_0} (see Fig. 1)

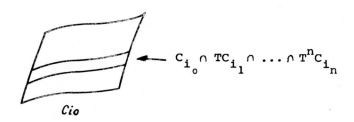

$$C_{i_0} \cap TC_{i_1} \cap \ldots \cap T^n C_{i_n}$$

Fig. 1

This permits us to introduce a stochastic Markov operator P generated by one-sided conditional probabilities $\lambda(C_{i_0} | TC_{i_1} \cap \ldots \cap T^n C_{i_n} \ldots)$. These conditional probabilities depend on all indices $i_0, i_1, \ldots$ but can be approximated uniformly with exponential velocity by functions depending on a finite number of indices. Because of this fact one can apply some concepts of statistical mechanics to Anosov systems (Ruelle [43], Sinai [21]).

II.3. Classical Dynamical Systems with Elastic Collisions

Let us call Born's model (see [44]) the dynamical system generated by a motion of a point with mass $m > 0$ among immobile spherical molecules inside a box with plane walls (see Fig. 2). The moving point reflects elastically from the molecules and the walls. The so-called "wind-tree" models treated recently by G. Gallavotti [45] have a similar character.

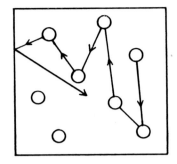

Fig. 2

Born's model is unstable because of the dispersing character of reflections from the round molecules[+], but this instability is not so uniform as in the case of Anosov systems. For example, there can be trajectories without any reflections from the molecules. Nevertheless the main property of Anosov systems is valid here with probability 1 (see [46]). To be more exact let the phase space X consist of linear elements $x = (q,v)$, where q are coordinates, v is velocity, $||v|| = 1$. Then for almost every point $x_0 = (q_0, v_0)$ (see Fig. 3) there exists a curve $\gamma^{(s)}$ consisting

Fig. 3

[+] N.S. Krylov [12] was the first to notice the analogy between geodesic flows on surfaces of negative curvature and systems with elastic collisions.

of linear elements orthogonal to a convex curve $\tilde{\gamma}^{(s)}$ and such that

$$\text{dist}(S_t y, S_t x_o) \leq Ce^{-\lambda t} \text{dist}(y,x) \tag{4}$$

for all $t \geq t(x_o)$. Here C, λ are constants which do not depend on x_o. Thus in Born's model there exist families of exponentially contracting semi-trajectories. The curves are orthogonal curves of such families. In non-Euclidean geometry one calls such curves horocycles. We use this term in our situation also.

It is possible to construct similar curves $\gamma^{(u)}$ changing t to $-t$. For $\gamma^{(u)}$ we shall have instead of (4):

$$\text{dist}(S_t y, S_t x_o) \leq Ce^{\lambda t} \text{dist}(y,x_o) \quad , \quad 0 < t_1(x_o) \leq -t \; . \tag{5}$$

Fig. 4

Let us deduce ergodicity from the existence of stable and unstable horocycles. Our considerations here are by necessity non-rigorous. The rigorous proof needs at least two difficult mathematical theorems of a technical nature. We employ one idea of E. Hopf [39]. For any continuous function f, the time averages

$$\bar{f}_+(x_0) = \lim_{T \to \infty} \frac{1}{T} \int_0^T f(S_t x_0) dt, \quad \bar{f}_-(x_0) = \lim_{T \to \infty} \frac{1}{T} \int_0^T f(S_{-t} x_0) dt$$

exist and coincide with probability 1. It follows from the definition of $\gamma^{(s)}$, $\gamma^{(u)}$ that $\bar{f}_+(x_0) = \bar{f}_+(y)$, $y \in \gamma^{(s)}$; $\bar{f}_-(x_0) = \bar{f}_-(y)$, $y \in \gamma^{(u)}$. Also $\bar{f}_+(x_0) = \bar{f}_+(y)$, $\bar{f}_-(x_0) = \bar{f}_-(y)$, if $y = S_t x_0$. Let us take two sufficiently close points x_1, x_2 and assume that it is possible to find a path from x_1 to x_2 consisting of the curve $\gamma^{(s)}(x_1)$, a trajectory interval $\gamma^{(t)}$ and the curve $\gamma^{(u)}(x_2)$. It is

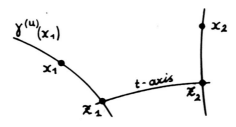

Fig. 5

possible to choose $\gamma^{(t)}$ so well (by one of technical mathematical theorems) that $\bar{f}_+(z_1) = \bar{f}_-(z_1)$, $\bar{f}_+(z_2) = \bar{f}_-(z_2)$. Then

$$\underbrace{\bar{f}_-(x_1) = \bar{f}_+(x_1)}_{\text{with prob.1}} = \underbrace{\bar{f}_+(z_1)}_{\text{Hopf}} = \underbrace{\bar{f}_+(z_2)}_{\text{technical theorem}} = \underbrace{\bar{f}_-(z_2) = \bar{f}_-(x_2)}_{\text{Hopf}}.$$

Thus $\bar{f}_-(x)$ is constant almost everywhere on a small neighborhood of the point x_1. From this ergodicity follows easily.

It is possible to prove that Born's model has stronger statistical properties. In particular, it is a K-system (see [46]). However almost nothing is known about the strong mixing condition and exponential decay of correlations. Recently L.A. Bunimovitsch [47] obtained some results about the central limit theorem for Born's model. He also constructed partitions which can be called Markovian. Unfortunately, all these partitions were infinite. Apparently there are no finite natural Markov partitions in this case.

The author is now preparing a big paper devoted to ergodic properties of general systems with elastic collisions. One should stress that the **form** of walls is important for our methods. We cannot apply them to systems with elastic collisions in volumes of general form. Also the generalization of the case of an infinite number of particles seems to be very difficult.

II.4. Several Open Problems

Anosov systems and integrable systems belong to opposite poles from the point of view of statistical properties. Unfortunately we have no example where it is possible to prove the presence of both types of behavior. The first candidate is Fermi-Pasta-Ulam's system of coupled non-linear oscillators. Physical considerations and computer experiments (Chirikov, Ford, Sagdeev, Zaslavski, etc.) show that in general one has a part of the phase space consisting of invariant manifolds with quasi-periodic motions on them. This part has a form of "islands" (so-called "stability islands") dispersed over the whole phase space. Between these islands there lies an ergodic layer with strong statistical properties. Unfortunately we have no model where this picture can be rigorously proved.

One of the most striking problems concerns the application of ergodic theory to turbulence. Two recent papers by A.M. Obuchov should be mentioned [52], [53]. He introduced an interesting class of dynamical systems with a finite number of degrees of freedom which can be considered as approximations to the motion of an ideal fluid. Namely, he investigated the systems of the following form:

$$\frac{dx_i}{dt} = \sum a^i_{k_1 k_2} x_{k_1} x_{k_2} \quad , \quad i = 1,\ldots,n \quad .$$

It is easy to formulate conditions under which this system has a positively-definite first integral

$$H = \sum c_{k_1 k_2} x_{k_1} x_{k_2}$$

and an invariant distribution with density

$$\rho = \sum r_{k_1 k_2} x_{k_1} x_{k_2} \quad .$$

For small values of n, n = 3,4, it is possible to reduce these systems to integrable ones. For larger n almost nothing is known. As Professor Obuchov told me, Dr. Orszag constructed examples of such systems where all fixed points are unstable. The topological entropy of these systems is unknown.

Recently O.A. Ladyzenskaya [54] constructed a dynamical system with compact phase space corresponding to the Navier-Stokes equation. In the paper by D. Ruelle and F. Takens the authors considered systems of ordinary differential equations which have some properties modelling turbulence. [65]

Part III. On Interrelations Between Ergodic Theory and Kinetic Theory

> "Physical law should have mathematical beauty."
> P.A.M. Dirac

Kinetic theory considers statistical properties of dynamical systems with a large number of degrees of freedom. It seems reasonable to construct from the very beginning the corresponding dynamical systems with an infinite number of degrees of freedom. The systems which appear in this way have some specific properties and differ from dynamical systems with an infinite number of degrees of freedom generated by non-linear partial differential equations. In particular all degrees of freedom are equivalent (symmetric in some sense) in these systems.

Already the first steps are difficult. It is easy to describe a corresponding phase space X^∞ for an infinite system of classical particles in R^k with a short range potential. However the non-trivial problem consists in the description of a class of natural probability distributions on X^∞ (from thermodynamical point of view). One should distinguish between two-phase and one-phase situations. The one-phase case is simpler and only will be discussed below.

In one-phase case the limits of microcanonical,

canonical and grand canonical distributions coincide and
do not depend from the sequence of boundary conditions
chosen in the process of thermodynamical limit transition
[55], [56]. Thus we obtain a three-parameter family of
equilibrium distributions $\lambda(\rho,h,v)$, where ρ is the density,
h is the energy per particle and v is the mean velocity.
It is interesting that the set of parameters coincide with
the set of natural first integrals of motion.

Using the thermodynamical limit transition one can
construct a large class of natural distributions on X^∞ corresponding to short-range potentials (not necessary pair
potentials) and parameters (the domain of convergence of
the virial expansions). The resulting distributions are
invariant under the group of space translations and have
good mixing properties with respect to this group. We shall
call them Gibbsian distributions (only in this lecture).

Now let us consider a fixed short-range pair interaction. The problem is to construct the dynamics in the
phase space. In the case under consideration the ordinary
theorems of existence and uniqueness of solutions for the
systems of differential equations are not valid. When the
velocities of particles grow sufficiently quickly at infinity there can occur infinite collapses during infinitesimally short time. The difficulty was overcome by O.
Lanford [9] and myself [8] in the one-dimensional case.
It was shown in these papers that it is possible to find
a set of probability one for any equilibrium distribution
$\lambda(\rho,h,v)$ where the dynamics has the following form: at
time t = 0 one can decompose all particles into finite
groups (thus the number of groups is infinite) and integrate the equations of motion on the interval $0 \leq t \leq 1$

for every group separately in such a way that particles from different groups will not interact. At time t = 1 one can construct another decomposition of particles into finite groups with the same property and so on. Thus for every finite interval $0 \leq t \leq T$ the system of differential equations of motion decomposes into an infinite set of finite separate systems. It is natural to assume that the dynamics has similar structure in all one-phase systems. Unfortunately it is not proved until now despite attempts by several mathematicians.

The cited results of Lanford and me are non-sufficient in one-dimensional case also. It follows from the preceeding that there is no natural group $\{S_t\}$ in X^∞ because it is not possible to construct a trajectory passing through every point of the phase space X^∞. However one can try to construct the group $\{S_t^{**}\}$ acting on a subspace of "good" distributions.

Namely let μ_0 be a distribution in X^∞. Let us assume that it is possible to find a set $A_{\mu_0} = A \subset X^\infty$, $\mu_0(A) = 1$ on which there exists a group of measurable transformations $\{S_t\}$ of A onto itself such that the equations of motion are valid along every trajectory of this group. Then one can define $S_t^{**} \mu = \mu_t$ by formula

$$\mu_t(C) = \mu(S_{-t}(C \cap A)), \quad C \subset X^\infty .$$

Conjecture 1. The group $\{S_t^{**}\}$ acts on the set of Gibbsian distributions.

This conjecture is not proved at all. One should have in mind that $S_t^{**} \mu_0$ is not in general Gibbsian if μ_0 is Gibbsian. Therefore the class of Gibbsian distributions is not sufficient for the construction of dynamics.

Another problem is connected with the set of fixed points of the group $\{S_t^{**}\}$. In the one-dimensional case it is known that all equilibrium distributions $\lambda(\rho,h,v)$ are fixed points for this group. For the ideal gas of an infinite number of particles in R^k or for the gas of one-dimensional hard rods there are many other fixed points of this group. The reason lies in the absence of non-trivial collisions in these cases. The ergodic properties of these systems have been considered by Volkovyssky and Sinai [58], de Pazzis [61], Sinai [59], kinetic properties by Lebowitz, Percus and Sykes [63].

Conjecture 2. In the general case the set of fixed points for the group $\{S_t^{**}\}$ coincides with the set of equilibrium distributions.

Some partial results connected with this conjecture have been obtained quite recently.[+]

Conjecture 3. In the space of "good" distributions with fixed values of ρ,h,v the equilibrium distribution $\lambda(\rho, h,v)$ is a stable fixed point for the group $\{S_t^{**}\}$.

The proof of conjecture 3 can lead to mathematically rigorous approach to irreversibility.

+) Added in proof: It was shown recently that equilibrium distributions are the only invariant distributions in the space of all Markov distributions with finite memory (Gurevitsh, Sinai, Suchov).

Conclusion

"So kann es dem Mathematiker geschehen, daß er, fortwährend beschäftigt mit seinen Formeln und geblendet durch ihre innere Vollkommenheit, die Wechselbeziehungen derselben zu einander für das eigentlich Existierende nimmt und von der realen Welt sich abwendet."
                                         Ludwig Boltzmann

Every mathematician who tries to deal with problems on the boundary of mathematics and theoretical physics should keep this in mind.

Let me express my sincere gratitude to the Organizational Committee for the honour to present this lecture. I am very grateful to Professor D. Ruelle who kindly agreed to submit the text to this meeting and to Professors O.E. Lanford and D. Ruelle for their help in preparing this text for publication.

References

[1] L. Boltzmann, J. Math. 100, 201 (1877).

[2] E. Fermi, I.R. Pasta, S.M. Ulam, "Studies of Non-Linear Problems", Los Alamos Scientific Laboratory, Report LA-1940 (1955).

[3] P. Hartman, "Mean motions and almost periodic functions", TAMS Vol. 49, 1939, 64-81.

[4] Jessen, Tornehave "Mean motions and almost periodic functions", Acta Math. Vol. 77, 1945, 137-279.

[5] H. Weyl, "Mean motion", I, II, Amer. J. of Math. Vol. 60 (1938), Vol. 61 (1939), 143-148.

[6] J. Gibbs, Elementary Principles in Statistical Mechanics, Yale Univ., 1902.

[7] V.I. Arnold and A. Avez, Problèmes ergodiques de la Mécanique Classique, Gauthier-Villars, Paris, 1967.

[8] Ya.G. Sinai, Teor. i mat. fizika, 11, (2), 1972.

[9] O. Lanford, Comm. Math. Phys. 9, 176 (1968), 11, 257 (1969).

[10] A.N. Kolmogorov, Doklady A.N. USSR, 119, 861-864, 1958.

[11] A.N. Kolmogorov, Doklady A.N. USSR, 124, 754-755, 1959.

[12] N.S. Krylov, Raboty po obosnovaniu statisticheskoi fiziki, publ. by "Acad. Nauk SSSR", 1950.

[13] D. Ornstein, An Application of Ergodic Theory to Probability Theory. Preprint, 1972.

[14] I.A. Ibragimov, Yu.V. Linnik, Statsionarnye i statsionarno svyazannye sluchainye velichiny i protsessy, publ. by "Nauka", 1968.

[15] V.A. Rokhlin, Ya.G. Sinai, Doklady A.N. USSR, 141, (5) 1038-1041, 1961.

[16] Ya.G. Sinai, Doklady A.N. SSSR, 133, (6), 1303-1306, 1960.

[17] M.E. Ratner, Doklady A.N. USSR, **186**, (3), 1969.
[18] I.L. Lebowitz, Hamiltonian Flows and Rigorous Results in Non-equilibrium Statistical Mechanics. IUPAP Conference on Stat. Mech., March 1971, preprint.
[19] R. Bowen, Am.J.Math., Vol. 92 (1970), 725-747.
[20] Ya.G. Sinai, Funkts. anal. i ego pril. Vol. 2 (1968) 61-82.
[21] Ya.G. Sinai, Uspekhi M.N. **27**, (4), 1972.
[22] A.M. Kogan, Prikladnaya matematika, mekhanika, **29**, (1), 122-133 (1965).
[23] R. Adler, A. Konheim, M. McAndrew, TAMS, Vol. 14, 309-319 (1965).
[24] E.I. Dinaturg, Izv. A.N. SSSR, **35** (2), 323-366, 1971.
[25] A.N. Kolmogorov, Problemy peredachi informatsii **1**, (1), 1968.
[26] A.Z. Zvonkin, A.L. Levin, Uspekhi M.N., **25** (6), 1970.
[27] A.N. Kolmogorov, Amsterdam Congress on Math., 1954, 315-333.
[28] M. Rosenblatt, Proc. Nat. Acad. Sci. USA, Vol.37, (3) 1961.
[29] A.N. Kolmogorov, Doklady A.N. **98** (4), 527-530, 1954.
[30] V.I. Arnold, Uspekhi M.N. **18** (6), 91-196, 1963.
[31] V.I. Arnold, Uspekhi M.N. **18** (5), 13-40, 1963.
[32] J. Moser, Ann. Scuola Norm. Sup. Pisa, 1966.
[33] N.N. Bogoliubov, Yu.A. Mitropolskii, L.M. Samoilenko, Metod uskorennoi skhodimosti v nelineinoi mekhanike, publ. by "Naukova Dumka", Kiev, 1969.
[34] S. Smale, Invent. Math. **10**, 4, 1970, 305-331; 11, 1, 1970, 45-64.
[35] S. Sternberg, Celestial Mechanics, Parts I, II, Benjamin, 1969.

[36] D.V. Anosov, Trudy M.I.A.N. SSSR im. Steklova, v.90(1967).
[37] S. Smale, Bull. Amer. Math. Soc. Vol. 73, 1967, 747-817.
[38] G. Hedlund, Bull. Amer. Math. Soc. $\underline{45}$, 1939, 241-246.
[39] E. Hopf, Ergodentheorie (Springer, Berlin), 1937.
[40] I.M. Gelfand, S.V. Fomin, Uspekhi M.N. $\underline{17}$ (1), 118-137, 1952.
[41] Ya.G. Sinai, Izv. A.N. SSSR $\underline{30}$ (1), 15-68, 1966.
[42] D.V. Anosov, Ya.G. Sinai, Uspekhi M.N. $\underline{22}$ (5), 103-167 (1967).
[43] D. Ruelle, Statistical Mechanics on a Compact Set with Action Satisfying Expansiveness and Specification, Preprint, I.H.E.S. 1972.
[44] M. Born, Physikalische Blätter, 11 (9), 49-54, 1955.
[45] G. Gallavotti, J. Math. Phys. $\underline{185}$, 308 (1969).
[46] Ya.G. Sinai, Uspekhi M.N. $\underline{25}$ (2), 141-193, 1970.
[47] L.A. Bunimovich, Doklady A.N. SSSR $\underline{204}$, 1972.
[48] G.M. Zaslavskii, Statisticheskaya neobratimost v nelineinikh sistemakh, publ. by "Nauka", 1970.
[49] G.M. Zaslavskii, B.V. Chirikov, Uspekhi F.N. $\underline{105}$ (1), 3-37, 1971.
[50] G.M. Zaslavskii, R.Z. Sagdeev, J. E.T.F. $\underline{25}$, 718, 1967.
[51] J. Ford, J. Math. Phys. $\underline{2}$, 387, 1961.
[52] A.M. Obukhov, Fluid Dynamics, Trans. Vol. 5, part II, 193-199, 1970.
[53] A.M. Obukhov, Doklady A.N. SSSR $\underline{184}$, (2), 309-312, 1969.
[54] O.A. Ladyzhenskaya, Kraevye zadachi mat. fiziki i smezhnye voprosy teorii funktsii, Vol. 6, publ. by "Nauka", Leningrad, 1972.

[55] P. Mazur, Proc. IUPAP, Meeting Copenhagen, 1966.
[56] R.A. Minlos, A.M. Khalfina, Izv. A.N. SSSR, ser. mat. 43 (3), 1971.
[57] D. Ruelle, Comm. Math. Phys. 9, 267-278 (1968).
[58] K.L. Volkovysskii, Ya.G. Sinai, Funkts. anal. i pril. 5 (4), 19-21, 1971.
[59] Ya.G. Sinai, Funkts. anal. i pril. 6 (1), 41-50, 1972.
[60] N.N. Bogoliubov, Problemy dinamicheskoi teorii v statisticheskoi fizike, publ. by "Gostekhizdat", 1946.
[61] De Pazzis, Ergodic Properties of a Semi-infinite Hard Rods System, Preprint, 1971.
[62] C. Foias, G. Prodi, Rendiconti del Seminario Mat. della Univ. di Padova XXXIX, 1967, 1-34.
[63] Lebowitz, Percus, Sykes, Phys. Rev. 171, 1, 1968, 224-235.
[64] D. Ruelle, Statistical Mechanics, Rigorous Results, Benjamin, 1969.
[65] D. Ruelle, F. Takens, Comm. Math. Phys. 20, 167-192 (1972).

ERGODIC THEORY

D. RUELLE

Institut des Hautes Études Scientifiques
Bures-Sur-Yvette

ABSTRACT

This talk is devoted to some remarks on the problem of the time evolution of systems containing a large number of particles. Do we understand approach to equilibrium? Sensitive dependence of solutions of differential equations on initial conditions. Time evolution of infinite systems. Evolution equations for dissipative systems.

0. Introduction

A striking feature of statistical mechanics is that very often, one is able to give a satisfactory mathematical formulation of the important problems of the theory, but these mathematical problems then appear formidably difficult to solve. Think for instance of the problem of ergodicity, or the problem of phase transitions. By contrast, the problem of describing the interactions between elementary particles, which is the main problem of elementary particle physics, can obviously not at this point be given a purely mathematical formulation. Statistical mechanics, thus, is remarkable in presenting a number of deep conceptual problems of a mathematical nature. The occurence of such questions, in fact, dates back to the first days of statistical mechanics, with Boltzmann and then Gibbs.

The central theme of Boltzmann's thinking was the time evolution of systems constituted of a large number of particles. The Boltzmann equation is a great step towards understanding this time evolution, but not a final one because its derivation involves uncontrolled approximations and raises serious conceptual problems.

The central theme of Gibbs' thinking was the description of equilibrium states by the well-known ensembles. The justification of the ensembles by ergodicity or something similar is still far from being achieved. But this is really time evolution again. The fact that the ensembles describe equilibrium is not really doubted and the mathematical problems connected with the use of ensembles (thermodynamic limit, equivalence of ensembles, etc...) either have been solved or are fairly well understood.

In a sense thus, one can say that Gibbs' approach has been more successful than Boltzmann's approach. But just because of this, Boltzmann's ideas are of more current interest, because the problems that confronted him are still those which confront us now. Let me give an example, which I find amusing. There has been quite a bit of interest recently (see D. Ebin and J. Marsden [2]) in a paper by V. Arnol'd [1] describing the time evolution of an inviscid fluid in terms of geodesics on an infinite dimensional group manifold. Reading Klein's book [3] on Ehrenfest, I discovered that this very idea constituted the subject given by Boltzmann to Ehrenfest for his Doctor's thesis.

Sinai and Lanford have already at this symposium talked about modern mathematical developments in the study of the time evolution of large systems. I shall devote the rest of this talk to a certain number of remarks on the same subject. I shall in fact come back to several points mentioned by Sinai and Lanford.

1. Approach to Equilibrium

One of the basic facts of life that statistical mechanics should explain, is why isolated macroscopic systems tend to equilibrium even though the microscopic equations of motion are reversible. In view of a theorem by Poincaré the time evolution from any configuration of particle positions and velocities will bring us back to a configuration close to the original one after a sufficiently long time. However, it is argued, this Poincaré recurrence time is so long that a Poincaré recurrence is never observed for a macroscopic system. The argument is certainly right, but constitutes no proof as long as the recurrence time cannot be computed. Suppose now that our system is ergodic, as was shown to be the case by Sinai for a system of billiard balls. More precisely, suppose that the Liouville measure on the mass shell is ergodic. The distribution of some observable with respect to time averages can then be computed from an ensemble. We are thus reduced to a problem of equilibrium statistical mechanics, namely to study the fluctuations of an observable with respect to an ensemble. Great progress has been made in this sort of questions recently: we know that (unless a phase transition forces them to fluctuate) macroscopic observables fluctuate very little: very, very rarely are they not very near their average value. As a consequence we may say that we largely understand <u>why</u> an ergodic system must tend to equilibrium. What we don't understand is exactly <u>how</u> it tends to equilibrium.

2. Sensitive Dependence of Solutions of Differential Equations on Initial Conditions

Consider a differential equation

$$\frac{d}{dt} x(t) = X(x(t)) \qquad (\not=)$$

for a vector x, with initial condition

$$x(0) = x_0 \; .$$

It happens sometimes that when x_0 is known with precision, $x(t)$ can be estimated with precision even for large t. If furthermore $x(t)$ describes the trajectory of a planet, or a satellite, this is considered a great success of Science. Mathematicians have however put recently more and more emphasis on the fact that a very sensitive dependence of x on initial conditions is not exceptional. It is also realized more and more that this phenomenon may be important for physics. This sensitive dependence consists in that for two solutions $x(t)$ and $x'(t)$ of $(\not=)$ such that $x'(0)$ is close to $x(0)$ the difference $|x'(t) - x(t)|$ in general grows exponentially with t, at least until $x'(t)$ is sufficiently far from $x(t)$ (see Sinai's report at this symposium).

It is believed that sensitive dependence on initial condition occurs for the time evolution of large systems of particles of the sort considered in statistical mechanics. I would like to mention some fairly trivial consequences of this.

(a) It is thought that approximations made on the interparticle interactions in statistical mechanics don't alter much the equilibrium behaviour (this can actually be proved in some cases). On the other hand these approximations will in general modify considerably the microscopic time evolution. In particular the regodicity or non-ergodicity may be deeply affected.

(b) Unavoidable disturbances of a large system of particles (due to insufficient thermal isolation, or to the influence of solar flares, etc..) will probably be amplified by its time evolution. This may have the effect of "washing out" some non-ergodicity of the system. The meaning of the remark is this: while one would be very happy to prove ergodicity because it would justify the use of Gibbs' microcanonical ensemble, real systems perhaps are not ergodic but behave nevertheless in much the same way and are well described by Gibbs' ensembles.

(c) One would very much like to know what corresponds in quantum mechanics to the sensitive dependence on initial conditions in classical mechanics. Perhaps the question does not make sense for "small" systems, but does for "large" ones.

3. Time Evolution of Infinite Systems

The time evolution of infinite classical one-dimensional systems has been studied by Lanford and Sinai (see their contributions to this symposium). Recently O. de Pazzis [5] has been able to treat certain two-dimensional systems of hard squares in the same spirit. Let me add some comments, which may soon be outdated by the rapid evolution of this subject.

(a) Some aspects of the quantum case seem here to be easier, and not more difficult than the classical case. This is because time evolution is given by the unitary operator e^{iHt}, which is very much like the $e^{-\beta H}$ in the definition of the equilibrium state. This fact leads to useful analyticity properties of Green's functions (KMS condition). In any case there is a satisfactory theory of time evolution for quantum lattice systems [6] and for dilute continuous gases [7]. In particular, infinite volume Green's functions are well defined for a dilute quantum gas, while the corresponding time-dependent correlation functions for a classical gas have only been shown to exist for one-dimensional systems.

(b) One expects that the infinite volume limits of the
states given by the Gibbs ensembles (infinite volume
equilibrium states) are invariant under time evolution. This has been proved to be the case whenever
the existence of time evolution could be established.
An apparently very difficult problem is to decide if
such an infinite volume equilibrium state is <u>ergodic</u>,
i.e. whether or not one can write $\mu = \frac{1}{2}(\mu_1 + \mu_2)$ where
μ_1 and μ_2 are again invariant under time evolution. In
fact one can formulate for μ properties such as <u>mixing</u>
with respect to time evolution, which are stronger
than ergodicity and correspond for infinite systems
to the sensitive dependence on initial conditions discussed above. These properties may hold for quantum
as well as classical systems. As long as one considers
systems with a finite number of degrees of freedom
however, the quantum systems are very different from
the classical ones, and don't seem to have interesting
ergodic properties.

4. Evolution Equations for Dissipative Systems

As far as one can check, the time evolution of infinite systems leaves the entropy constant (see in particular Lanford [4]). This is natural because the state of
the system is described completely by all its correlation
functions and has a <u>reversible</u> time evolution. It is thus
not sufficient to make a system infinite to make it irreversible: one also has to keep loosing information about
the system. How to loose information by coarse-graining
is clear in principle but it seems hard to do it neatly

in practice, avoiding atrocities like an oscillating entropy. On the other hand there are plenty of systems in nature - so-called dissipative systems - where a macroscopic description is possible in terms of local density, etc.., and where there is a steady entropy production. It is a very unfortunate fact that a fundamental derivation of the evolution equations for dissipative systems, for instance in hydrodynamics, has not yet been possible from statistical mechanics.

Let me conclude with a remark on turbulence. Takens and myself [8] have recently proposed a scheme by which the observed solutions of the hydrodynamical equations and sufficiently high Reynolds number depend sensitively on the initial conditions. We think that these solutions represent turbulent flow. Notice that this explanation of turbulence does not involve any breakdown of the macroscopic hydrodynamical equations (the Navier-Stokes equations for instance). In particular it does not require at all that we consider the microscopic time evolution which underlies the macroscopic evolution. However, since the solutions of the macroscopic equations which we consider have a very sensitive dependence on initial condition, any given solution will be strongly altered - after a sufficient time - by small perturbations such as molecular fluctuations. This shows that in the domain of turbulence as in statistical mechanics and for much the same reasons, a statistical treatment is necessary.

References

1. V. Arnold, Sur la géométrie différentielle des groupes de Lie de dimension infinie et ses applications à l'hydrodynamique des fluides parfaits.
   Ann. Inst. Fourier. Grenoble 16, 319-361 (1966).
2. D.G. Ebin and J. Marsden, Groups of diffeomorphisms and the motion of an incompressible fluid.
   Ann. of Math. 92, 102-163 (1970).
3. M. Klein, Paul Ehrenfest. North-Holland, Amsterdam, 1971.
4. O.E. Lanford, The classical mechanics of one-dimensional systems of infinitely many particles. I An existence theorem. Commun. math. Phys. 9, 176-191 (1968).
   II Kinetic theory. Commun. math. Phys. 11, 257-292 (1969).
5. O. de Pazzis. Preprint.
6. D.W. Robinson, Statistical mechanics of quantum spin systems. I. Commun. math. Phys. 6, 151-160 (1967).
   II. Commun. math. Phys. 7, 337-348 (1968).
7. D. Ruelle, Analyticity of Green's functions of dilute quantum gases. J. math. Phys. 12, 901-903 (1971).
   Definition of Green's functions for dilute Fermi gases. Helv. Phys. Acta. To appear.
8. D. Ruelle and F. Takens, On the nature of turbulence. Commun. math. Phys. 20, 167-192 (1971).

ERGODIC THEORY AND APPROACH TO EQUILIBRIUM
FOR FINITE AND INFINITE SYSTEMS

OSCAR E. LANFORD III[*]

Department of Mathematics, University of California
Berkeley, California 94720

and

Institut des Hautes Études Scientifiques
91-Bures-sur-Yvette, France

---

[*] Alfred P. Sloan Foundation Fellow. Preparation of these notes was also supported in part by NSF grant GP-15735.

In this lecture, I will have something to say about:
a) Classical ergodic theory for finite classical systems
b) Time development and approach to equilibrium for infinite classical and quantum systems.

The first topic is a very large one, and I will give only a very sketchy survey. For more details, see Sinai's contribution to this Symposium or the book of Arnold and Avez [3]. The second topic is unfortunately much less well understood, and much of what I have to say will of necessity be speculative.

I begin by recalling some (idealized) history. To justify the use of the microcanonical ensemble in statistical mechanics, Boltzmann gave something like the following argument: A measurement on a classical system gives the value of some function on the phase space for the system. Measurements, however, are always made over intervals of time which are very long compared to the time it takes things to happen on the microscopic level, so what is really measured is the average value of the function over a long period of time, which he idealized as the average over all time. He then made the bold conjecture that the energy surface is composed of a single trajectory (ergodic hypothesis). If this is so, then the time average of any function is constant (i.e., independent of the starting point) and equal to the average of that function over the energy surface. In this way, one obtains a justification for the microcanonical ensemble which involves no probabilistic hypotheses at all.

The ergodic hypothesis according to which the entire energy surface consists of a single trajectory was soon realized to be implausible and eventually shown to be false.

Some attempts were made to rescue the argument using the "quasi-ergodic hypothesis", i.e., assuming that a single trajectory is dense on the energy surface. These attempts were not very successful, and the matter rested until the fundamental work of von Neumann and Birkhoff in the early 1930's. What Birkhoff proved was essentially the existence of time averages: If we let $d\mu$ denote the normalized microcanonical measure on the energy surface, and if we denote the point at time t on the solution curve starting at x at time zero by x_t or $T^t x$, then for any integrable function f:

a) Except for a negligible* set of initial points x

$$\lim_{T \to \infty} \frac{1}{2T} \int_{-T}^{T} dt \, f(x_t) = \bar{f}(x)$$

exists.

b) $\bar{f}$ is integrable and

$$\int \bar{f}(x) \, d\mu(x) = \int f(x) \, d\mu(x)$$

c) $\bar{f}(x_t) = \bar{f}(x)$, i.e. $\bar{f}$ is a "constant of the motion".

Now the quasi-ergodic hypothesis implies that the only continuous "constants of motion" are constant on the energy surface. Suppose we make the stronger hypothesis that every measurable invariant function is constant on the energy surface (except perhaps on some set of points of measure zero.) Applying this hypothesis to the function $\bar{f}$, we conclude that $\bar{f}$ must be a constant, and, by b), the constant must be equal to $\int f \, d\mu$. Thus we get

---

* By a negligible set of points we mean a set of μ-measure zero.

$$\lim_{T\to\infty} \frac{1}{2T} \int_{-T}^{T} dt\ f(x_t) = \int f(y)\ d\mu(y)$$

(where the left-hand side depends formally on x and the right-hand side does not, and where the equality may fail to hold in some set of measure zero.) Thus, we again have a justification of microcanonical averaging. Note, however, that the purely mechanical character of the original argument has been lost; the equality of time and ensemble averages holds only on the complement of a set of initial configurations of measure zero. To apply the result to physics, we need to make the weak probabilistic hypothesis that a property which holds on the complement of a set of measure zero always holds in practice.

Birkhoff's theorem, then, reduced the problem of justifying microcanonical averaging to proving that the only invariant measurable functions on the energy surface are essentially constant. A simple technical argument shows that the only invariant functions are constants if and only if the only invariant measurable subsets of the energy surface are either sets of measure zero or complements of sets of measure zero. If this condition holds, we say the time development T^t (more properly, the pair $(T^t, d\mu)$) is ergodic (or metrically transitive.) The mathematical question now becomes: Are realistic isolated Hamiltonian systems ergodic?

Before discussing this question, we introduce a stronger property of time development, the mixing property, which is closely connected with approach to equilibrium. Intuitively, we want to say that the time development is mixing if any non-negligible set F gets spread uniformly over the energy surface at large times. More precisely, this should mean the following: Given any other non-negligible set E, the fraction $\mu(E \cap T^t F)/\mu(E)$ of E occupied by $T^t F$ should for large t be approximately equal to the fraction of the whole energy surface occupied by $T^t F$, which is $\mu(T^t F) = \mu(F)$. Thus, we say the time-development is <u>mixing</u> if

$$\lim_{t \to \infty} \frac{\mu(E \cap T^t F)}{\mu(E)} = \mu(F)$$

for all E, F. It is easy to see that the mixing property implies ergodicity. The notion of mixing was invented by Gibbs, who thought of the time-development on the energy surface as analogous to incompressible fluid flow and mixing as describing what happens when two miscible liquids are stirred together.

Mixing implies a strong instability property of the equations of motion. Pick an initial point x_0 and a small neighborhood U of x_0. Since U is spread uniformly over the energy surface by the time development, we can for sufficiently large t find points x in U such that $T^t x$ is as far from $T^t x_0$ as is compatible with staying on the energy surface. In other words, an imprecise measurement of x_0 gives no useful information at all at large times; this crude argument shows already that mixing implies a loss of information with time.

To make this idea more precise, consider a non-equilibrium statistical state of the form $\rho(x)d\mu(x)$, where $\rho(x)$ is a positive function with integral 1. (We will refer to such a function ρ as a <u>density</u>; it is essential that the densities we consider are ordinary functions, not more general distributions containing δ-functions or other singularities.) If we start at time zero in this statistical state and let the system evolve to time t, we get a statistical state with density $\rho(T^{-t}x)$. It is easy to show that, if T^t is mixing, if ρ is any density, and if f is any bounded integrable function, then

$$\lim_{t \to \infty} \int f(x) \, \rho(T^{-t}x) \, d\mu(x) = \int f(x) \, d\mu(x) \quad .$$

In other words, the statistical state $\rho(T^{-t}x)d\mu$ converges as $t \to \infty$ (or as $t \to -\infty$!) to the equilibrium statistical state $d\mu$ in the sense that mean values of all bounded measurable functions converge to the equilibrium mean values. A mixing system, therefore, exhibits approach to equilibrium, but only in the sense that mean values converge, not in the (physically more realistic) sense that individual points on the energy surface look more and more like "typical" points for the equilibrium measure.

Since we have a kind of approach to equilibrium, we must also have irreversibility. The irreversibility occurs in the following way: For any finite t, the initial density $\rho(x)$ can in principle be recovered from the density at time t $\rho(T^{-t}x)$. However, all trace of the initial state is lost in taking the limit $t \to \infty$. Thus, the irreversibility occurs in the passage to the limit. Similarly, one can introduce the entropy of a statistical state with density ρ by

$$s = - \int \rho(x) \, \log \rho(x) \, d\mu(x) \quad .$$

The entropy does not change with time, but nevertheless the entropy of the limiting state is strictly larger than

that of the initial state. What is happening is something
like this: The entropy of a statistical state measures in
a certain sense the amount of order in that state. The
total amount of order present remains constant but, as time
goes on, it gets shifted into more and more subtle kinds
of correlations which are less and less accessible to
measurement. In the limit, the correlations become so
subtle that they are not there at all. Thus, we get a
mathematically precise version of coarse-graining (if not
quite the physically correct one.)

If the system is ergodic but not mixing, we need
not have approach to equilibrium in the above sense. Nevertheless, all that can happen is that the state "oscillates
about equilibrium", and the oscillation can be eliminated
by time averaging. To be precise, we have for any initial
density ρ

$$\lim_{T\to\infty} \frac{1}{2T} \int_{-T}^{T} dt \, [\int f(x) \, \rho(T^{-t}x) \, d\mu(x)] = \int f(x) \, d\mu(x)$$

for all bounded f.

We come now to the question of what is known about
when classical dynamical systems give ergodic time developments. The investigation of this question has turned out
to be very difficult, but there has been some genuine progress in the past twenty years. There is a widespread
conventional wisdom about ergodicity which holds, first,
that reasonable systems are ergodic unless they have smooth
constants of motion in addition to the energy, and, second,
that a non-ergodic system can be made ergodic by making
an arbitrarily small perturbation on the Hamiltonian. This
conventional wisdom has proved to be false. Consider, for

example, an assembly of some number (at least two) of non-interacting anharmonic[†] oscillators. This system is not ergodic since the energies of the individual oscillators are constants of the motion. The conventional wisdom suggests that, by adding an arbitrarily small amount of a sufficiently general coupling, we can make the system ergodic. Kolmogorov and Arnold, however, have proved that this is not the case. Specifically, they prove the following: Let H_0 be the Hamiltonian for the non-interacting system of oscillators and let H_1 be any analytic function on the phase space. Then for all sufficiently small ε the system with Hamiltonian $H_0 + \varepsilon H_1$ is not ergodic. Moreover, the non-ergodicity does not come about because of the presence of continuous constants of the motion. It seems that each energy surface (of dimension $2n-1$, n the number of oscillators) contains a very large number of invariant n-dimensional surfaces on which the motion looks like that of a system of n non-interacting oscillators. These n-dimensional surfaces fill a non-zero fraction of the energy surface but are nowhere dense, i.e., do not fill completely any open set. They are surrounded by an open set on which the motion is apparently much more irregular. The detailed picture of what happens in this open set is strikingly complicated, even for two oscillators.

In a more positive direction, Sinai has shown that a system of finitely many perfectly elastic billiard balls is mixing, and, in fact, has even better ergodic-theoretic properties. See Sinai's lecture for a more complete discussion of both these results and an extensive list of

---

[†] Anharmonicity is needed because the argument requires that the frequence of oscillation change with amplitude.

references.

For finite quantum systems, there seems to be no hope of obtaining any kind of satisfactory mathematical result on the approach to equilibrium. This is so because the quantum Hamiltonian for a system with a reasonable interaction and contained in a finite box has discrete spectrum, so the expected value of any measurement is an almost periodic function of time. In practice, this fact is irrelevant since the level spacing for a macroscopic system will surely be extremely fine; nevertheless, in order to get sharp theorems it is necessary to pass to infinite systems. We therefore turn now to the investigation of the new features which arise for infinite systems, beginning with infinite classical systems.

The first problem which must be faced is that of obtaining an existence and uniqueness theorem for solutions of the differential equations governing the time development. This has turned out to be fairly difficult, and a satisfactory solution has been given only for one-dimensional systems with bounded finite-range two-body forces. There are also two trivial cases which can be handled, the non-interacting gas (in any number of dimensions) and the one-dimensional hard-rod system (where particles simply exchange velocities when they collide.) We will not try to deal here with these problems, but will act as if they are solved; what we have to say at least applies to one-dimensional systems.

The next problem is to find the infinite-system analogue of the microcanonical ensemble. This is again not so straightforward; the phase space is infinite-dimensional, and the "energy surface" should correspond to the

set of configurations with a given finite value of the
energy per particle and thus with an infinite total energy.
We can, however, proceed as follows: Pick values for the
density and for the energy per particle. Construct a
sequence of bigger and bigger finite systems with the
chosen average density; for each of these, construct the
microcanonical measure with the chosen average energy per
particle[†]. Then "pass to the thermodynamic limit" to get
a probability measure on the phase space for the infinite
system. We will refer to measures obtained in this way as
<u>equilibrium measures</u> for the infinite system and will
again denote them systematically by $d\mu$. The above limit
need not be unique if phase transitions occur. However,
all classical systems for which the time evolution problem
can be solved have no phase transitions, so we will ignore
this complication for the moment.

      Equilibrium measures are, as expected, invariant
under the time evolution, so we can investigate the ergodic
properties of the time evolution with respect to them. In
particular, we would like to interpret what it means physically for the system to be ergodic or mixing. On the basis
of what we have seen for finite systems, our first guess
would be that, if the system is mixing, we should have
average convergence to equilibrium. This guess is only half
correct, for the following reason: Just as for finite
systems, mixing implies that any statistical state ρ <u>ob-
tained by multiplying the equilibrium state</u> μ <u>by a density</u>
converges back to the equilibrium state as $t \to \infty$. For
finite systems it is certainly not unreasonable to consider

[†] The technicalities actually require the use of a small
interval of energies per particle, rather than a single
value.

only statistical states of this form. For infinite systems, however, most interesting initial states are not obtainable in this way and are in fact concentrated on sets which have measure zero with respect to the equilibrium measure. Suppose we consider, for example, an initial statistical state in which the particles are uncorrelated and uniformly distributed in space but have a non-Maxwellian velocity distribution. Such a state, which is exactly the sort one wants to study in investigating approach to equilibrium, can never be obtained by multiplying an equilibrium state by a density. (Intuitively, this is so because such a density, if it existed, would have to contain a product of infinitely many factors of the ratio of the initial single-particle distribution to the Maxwell distribution. One easily sees that such a product must diverge.) There is in fact a very simple general theorem which says that a spatially homogeneous non-equilibrium statistical state can never be obtained by multiplying an equilibrium state representing a pure phase by a density.

We are thus led to ask which states are obtained by multiplying the equilibrium state by a density. The answer seems to be roughly the following: Such states describe a situation perturbed significantly from equilibrium only in a bounded region of space. This characterization has to be interpreted with a certain amount of care; what is really meant is that the "weight" assigned to each configuration differs from the equilibrium weight by something depending only on the part of the configuration in a bounded region of space. If the equilibrium state has long range correlations built into it, as for example if it represents a crystal, then such a local perturbation can change the global character of the state. The characterization in its simple heuristic form should, however, be correct for equilibrium states representing gases.

Thus we see that to say that the time-evolution is mixing with respect to an equilibrium state means that localized disturbances damp out with time, i.e., that we have <u>return to equilibrium</u> but not <u>approach to equilibrium</u>. What, then, is the appropriate mathematical formulation of the notion of approach to equilibrium? In other words, what sort of initial statistical states can be expected to the equilibrium state as $t \to \infty$? Since the investigation of this question is in a very primitive state, no very definite answer can be given. Nevertheless, it seems fairly likely that, to have any hope of proving approach to equilibrium, one should start with a state which

a) does not differ too radically locally from the equilibrium state, i.e., whose correlation functions contain no δ-functions.
b) is "clustering" in the sense that measurements in widely separated regions of space are approximately statistically independent.
c) has some mild spatial homogeneity properties

We will see shortly the role played by these conditions in the "approach to equilibrium" of a non-interacting classical gas.

It will probably come as no surprise that the problems of approach and return to equilibrium for realistic systems are extremely difficult and practically untouched. Nevertheless, there are two model systems for which satisfactory return to equilibrium results can be proved. These are the non-interacting gas (treated by Sinai and Volkovyssky [8]) and the one-dimensional hard-rod system (de Pazzis [4], Sinai [7]). In both these cases, the time development is mixing (in fact, is a K system) for all equilibrium measures. These results, however, are completely trivial physically; the localized disturbances simply stream off to infinity where they are no longer visible.

For the non-interacting gas, one can prove a correspondingly trivial approach to equilibrium result. Start with an equilibrium state described by correlation functions

$$\rho_n(q_1,v_1;\ldots;q_n,v_n) \quad .$$

The state at time t is then described by the correlation functions:

$$\rho_n(q_1-v_1t,v_1;\ldots;q_n-v_nt,v_n) \quad .$$

If we assume
a) The ρ_n's have no δ-functions in velocity
and
b) The ρ_n's are clustering, so

$$\rho_n(q_1,v_1;\ldots;q_n,v_n) \approx \rho_1(q_1,v_1)\rho_1(q_2,v_2)\ldots\rho_1(q_n,v_n)$$

when the q_i's are far apart
then for large t the time-dependent correlation functions become approximately

$$\rho_1(q_1-v_1t,v_1)\rho_1(q_2-v_2t;v_2)\ldots\rho_1(q_n-v_nt;v_n) \quad .$$

For very large t, averaging $\rho_1(q_0-vt,v)$ over a small volume in velocity space around v_0 is almost the same as averaging $\rho_1(q,v_0)$ over a large volume in position space around q_0-v_0t. If, now,

c) $\rho_1(q,v)$ has enough spatial homogeneity so this average becomes independent of q_0

then the time-dependent correlation functions, averaged over a small region in velocity space, converge as $t \to \infty$ to

$$\bar{\rho}_1(v_1)\bar{\rho}_1(v_2)\ldots\bar{\rho}_1(v_n) \quad ,$$

where $\bar{\rho}_1(v)$ denotes the average of $\rho_1(q,v)$ over q. This is about as well as we can hope to do; we obtain limiting correlation functions which are time-independent, spatially homogeneous, and represent a system of non-interacting particles. Since there are no collisions, there is no mechanism for making the velocity distribution Maxwellian.

We turn now to quantum systems. Since we will use the C^*-algebra formulation of quantum mechanics of infinite systems, we begin with a brief introduction to this formalism. For more details, see Ruelle [6]. The underlying principle is that the quantum mechanics of an infinite system should have states described by unit vectors or density matrices on some Hilbert space H and observables corresponding to self-adjoint operators on H; in contrast to the situation for finite systems, we do not know how to write down concretely the Hilbert space H and identify particular operators on H with familiar kinds of measurements. To get around this difficulty we adopt the point of view that the infinite systems we want to consider are made up out of finite systems which we understand and that the measurements we want to consider are either measurements on finite subsystems or limits of these. A typical example is provided by the Heisenberg model on an infinite lattice; the finite subsystems are Heisenberg models on finite subsets of the lattice. The general principles of quantum mechanics suggest that, if Λ is any finite subsystem, we can write the Hilbert space H as:

$$H = H_\Lambda \otimes H'_\Lambda \quad ,$$

where H_Λ is the well-understood Hilbert space for the finite system and H'_Λ is the Hilbert space of the infinite system with Λ removed. (The space H'_Λ is just as mysterious as H, but we ignore this fact). The tensor product "$\otimes$" is to be understood as a product in the same sense as the space of functions of two variables is a product of two spaces of functions of one variable. Similarly, if A is a self-adjoint operator on H_Λ corresponding to a particular measurement on the finite system Λ, then the operator $A \otimes 1$ on H corresponds to the same measurement on the subsystem Λ of the original infinite system. With this in mind, we let A_Λ denote the algebra of all bounded operators on H of the form $A \otimes 1$, where A is no longer assumed to be self-adjoint; A_Λ is interpreted as the algebra generated by all measurements which can be carried out in Λ. Now let A denote the norm closure of the union over all finite subsystems Λ of A_Λ; A is to be interpreted as the algebra generated by all measurements on finite subsystems and is called the <u>algebra of quasi-local observables</u>. The point is now the following: Although A was constructed above assuming the existence of a Hilbert space theory with reasonable properties, it can also be constructed abstractly, just using the properties of finite systems. Moreover, the abstract A obtained in this way is the same as (more precisely, is isomorphic to) the A constructed from any Hilbert space theory with the above properties. Thus, although the complete structure of the possible Hilbert space theory is obscure, the algebra A is at least well-defined and is therefore a natural starting-point for the theory of infinite quantum systems.

Now what about states? We said before that states should correspond to vectors or, more generally, density matrices on H. Even if we have a state which is given by a vector on H, it may look like a density matrix when restricted to a finite subsystem. Thus, the states of the infinite system are specified by giving, for each finite subsystem Λ, a density matrix ρ_Λ on H_Λ, with an obvious consistency condition on ρ_{Λ_1} and ρ_{Λ_2} if Λ_1 is a subsystem of Λ_2. A state in this sense defines a positive linear functional ρ on A by $\rho(A) = \text{tr}(\rho_\Lambda A)$ for any $A \in A_\Lambda$. Conversely, given a positive linear functional ρ on A, there is a general abstract construction (the Gelfand-Segal construction) which gives a Hilbert space H_ρ and a representation of A on H_ρ so that ρ is given by a vector in H_ρ. Thus, the Hilbert space theory can readily be constructed once the C^*-algebra and the state are known.

To get the time evolution, we suppose we have for each finite subsystem Λ a Hamiltonian H_Λ. We can therefore make the associated Heisenberg picture time evolution, which carries the observable A to

$$\alpha_\Lambda^t(A) = e^{iH_\Lambda t} A e^{-iH_\Lambda t} .$$

If we let Λ become infinitely large, we do not expect H_Λ to have a limit in any reasonable sense, but there is at least some chance that $\alpha_\Lambda^t(A)$ converges for all $A \in \mathcal{A}$. If it does, we get a Heisenberg picture time-development for the infinite system

$$A \to \alpha^t(A) ,$$

where α^t is a one-parameter group of automorphisms of $\mathcal{A}$. It should be remarked that such a description of time evolution by a one-parameter group of automorphisms of the algebra of quasi-local observables is probably not general enough to apply to interacting continuous systems, but it can be proved to apply to such lattice systems as the Heisenberg model.

Equilibrium states for the infinite system can be constructed as limits of finite-volume equilibrium states, just as in the classical situation, i.e., we should have

$$\mu(A) = \lim_{\Lambda \to \infty} \frac{\text{tr}(A\, e^{-\beta H_\Lambda})}{\text{tr}(e^{-\beta H_\Lambda})} .$$

Alternatively, the infinite volume equilibrium states may be described as the states satisfying the Kubo-Martin-Schwinger boundary condition with respect to the time-evolution automorphisms. (Technically, one doesn't quite know that these two descriptions of equilibrium states are equivalent.) There may be more than one equilibrium state for a given temperature, corresponding to the occurence of phase transitions.

Now what should approach and return to equilibrium mean for quantum systems? The analogue of the classical situation is clear; we want to determine when, starting from some non-equilibrium state ρ, we have

$$\lim_{t \to \infty} \rho(\alpha^t(A)) = \mu(A)$$

or, allowing for time averaging, when

$$\lim_{T \to \infty} \frac{1}{2T} \int_{-T}^{T} dt \, \rho(\alpha^t(A)) = \mu(A)$$

for all $A \in \mathcal{A}$. If we are to distinguish between approach and return to equilibrium, we need to decide which states are locally perturbed from equilibrium. The precise definition is debatable, but we take one which has at least the advantage of simplicity: We regard a state ρ as a local perturbation on an equilibrium state μ if it is a uniform limit of states of the form

$$A \to \frac{\mu(B^*AB)}{\mu(B^*B)}$$

or, equivalently, if it is given by a density matrix on the Hilbert space constructed from the equilibrium state μ. For such states, there is at least a mean ergodic theory, due in this context to Radin [5]:

Theorem: Let μ be an equilibrium state, ρ a density matrix state on the Hilbert space for μ. Then:

a) $\hat{\mu}(A) = \lim_{T \to \infty} \frac{1}{2T} \int_{-T}^{T} \rho(\alpha^t(A)) dt$ exists for all $A \in \mathcal{A}$.

b) $\hat{\mu}$ is a density matrix state on the Hilbert space for μ and is invariant under α^t.

We would like, of course, for $\hat{\mu}$ to be equal to μ. This cannot, however, be expected to be true in general, as the following argument shows: Let μ_1, μ_2 be two equilibrium states corresponding to distinct phases which can coexist at the same temperature, and let $\mu = \frac{1}{2}\mu_1 + \frac{1}{2}\mu_2$. Then μ represents a situation in which one or the other of the two phases is present, each with probability $\frac{1}{2}$. By making local measurements, we can tell which phase is present and hence we should be able to produce pure phase 1 by such measurements. Thus, we expect μ_1 to be a density matrix state on the Hilbert space for $(\mu_1+\mu_2)/2$ [*]. But μ_1 is constant in time; hence, cannot average back to $(\mu_1+\mu_2)/2$.

To eliminate this loophole, we want to consider only equilibrium states which represent pure phases. Moreover, by a similar argument, if the equilibrium state represents a crystal we want to fix the position and orientation of the crystal. It can be argued that this means we want to consider only those equilibrium states which are extremal points of the set of all equilibrium states.

Now the question becomes: Given an extremal equilibrium state μ and a density matrix state ρ on the Hilbert space of μ, is the time-average of ρ equal to μ? It follows from a result of Araki and Miyata ([2]; see also [1]) that this is true provided that

$$\lim_{t \to \infty} [\alpha^t(A), B] = 0 \qquad (\dagger)$$

---

[*] Whether or not one believes the heuristic argument, it is simple to prove that μ_1 is in fact a density matrix state on this Hilbert space.

for all A, B in $\mathcal{A}$, i.e., provided that two localized measurements made at very different times tend to commute. Thus, if (†) holds, any state obtained by making a local perturbation on a pure phase averages back to the pure phase, so investigating return to equilibrium is reduced to investigating (†). It is possible to exhibit reasonable models in which (†) holds and only slightly less reasonable ones in which it doesn't. What the situation is for realistic systems (e.g., the nearest neighbor isotropic Heisenberg model) is completely obscure, even on a heuristic level.

References

[1] H. Araki, Commun. Math. Phys. $\underline{14}$ (1969) p. 120.
[2] H. Araki and H. Miyata, Publ. Res. Inst. Math. Sci. Kyoto Univ. Ser. A $\underline{4}$, (1968), p. 373.
[3] V.J. Arnold and A. Avez, <u>Problèmes Ergodiques de la Mechanique Classique</u>, Paris: Gauthier-Villars (1967).
[4] O. de Pazzis, Commun. Math. Phys. $\underline{22}$, (1971) p. 121.
[5] C. Radin, Commun. Math. Phys. $\underline{21}$, (1971) p. 291.
[6] D. Ruelle, <u>Statistical Mechanics: Rigorous Results</u>, New York: W.A. Benjamin (1969).
[7] Ya.G. Sinai, Funct. Anal. and Appl. $\underline{6}$, (1971) p. 41.
[8] Ya.G. Sinai and K. Volkovyssky, Funct. Anal. and Appl. $\underline{5}$, (1971) p.19.

ERINNERUNGEN AN BOLTZMANNS VORLESUNGEN

K. PRZIBRAM

Boltzmann hielt seine Vorlesungen im baufälligen Hörsaal des alten physikalischen Instituts in der Türkenstraße 3. Er pflegte eine jede mit einem sonoren "Nun!" zu beginnen, in jener hohen Stimmlage, die so garnicht zu seiner wuchtigen Statur paßte.

Die Vorlesungen waren von wunderbarer Klarheit. So hoch auch Boltzmanns genialer Geist über dem Thema schweben mochte, so vermied er es doch immer, über die Köpfe seiner Hörer hinweg zu reden.

Wie sehr er bemüht war, seinen Vortrag dem Fassungsvermögen der Hörer anzupassen, mag folgendes Beispiel zeigen. Um in der Elastizitätstheorie den Begriff der Flächenkraft recht nahe zu bringen, sagt er: "Denken Sie sich an diese Fläche enorm viele Häkchen befestigt," und begann geduldig auf die Basisfläche eines auf die Tafel skizzierten Prismas wirklich sehr viele kleine Haken zu zeichnen, die natürlich zur Anbringung von Gewichten dienen sollten.

In der Mechanik erklärte er uns, daß ein Mensch von einem der kleineren Asteroiden aus eigener Kraft in den Weltraum hinausspringen könnte, und führte zur Belebung des Gesagten eine energische Bewegung aus, als setze er selbst zu so einem Sprung an.

Als es ihm in einer Vorlesung über die Maxwellsche Theorie, auf die er bekanntlich das Faust-Zitat "War es ein Gott, der diese Zeichen schrieb?" angewendet hat, gelungen war, etwas zu seiner eigenen Zufriedenheit recht klar herauszuarbeiten, meinte er: "So! Das muß doch jetzt jedes elektromagnetische Wickelkind verstehen!"

In der Hitze des Vortrages gebrauchte Boltzmann bisweilen kuriose Wortverbindungen, so wenn er ein zu vernachlässigendes Glied einer Gleichung als "riesig klein" bezeichnete. Doch derlei belebte den Vortrag nur noch mehr.

Ich hörte auch Boltzmanns Vorlesungen über Naturphilosophie, die er als Nachfolger seines wissenschaftlichen Gegners Ernst Mach in einem großen, stets überfüllten Hörsaal hielt.

Es wird wohl in einer dieser philosophischen Vorlesungen gewesen sein und nicht in einer physikalischen - Einsteins allgemeine Relativitätstheorie lag ja damals noch in weiter Ferne -, daß Boltzmann auf mehrdimensionale und gekrümmte Räume zu sprechen kam.

Darauf bezog sich ein Distychon, ich weiß nicht mehr von wem gedichtet, in einer Franz Exner gewidmeten Kneipzeitung:

"Tritt der gewöhnliche Mensch auf den Wurm, so wird er sich krümmen;

Ludwig Boltzmann tritt auf: siehe, es krümmt sich der Raum!"

Das war ein Scherz, aber man kann kaum treffender die gewaltige Wirkung schildern, die von Boltzmanns überragender Persönlichkeit ausging und sich auch in seinen Vorlesungen offenbarte.